AF608506

Simple Models of Many-Fermion Systems

Joachim Alexander Maruhn ·
Paul-Gerhard Reinhard · Eric Suraud

Simple Models of Many-Fermion Systems

Prof. Dr. Joachim Alexander Maruhn
Universität Frankfurt
FB 13 Physik
Inst. Theoretische
Physik/Astrophysik
60054 Frankfurt
Germany
maruhn@th.physik.uni-frankfurt.de

Prof. Dr. Paul-Gerhard Reinhard
Universität Erlangen
Institut für Theoretische Physik II
Staudtstr. 7
91058 Erlangen
Germany
reinhard@theorie2.physik.uni-erlange

Prof. Dr. Eric Suraud
Université Toulouse III
IRSAMC
Lab. Physique Quantique
118 route de Narbonne
31062 Toulouse
France
suraud@irsamc2-ups-tlse.fr

Additional material to this book can be downloaded from http://extra.springer.com.
Password: 978-3-642-03838-9

ISBN 978-3-642-03838-9 e-ISBN 978-3-642-03839-6
DOI 10.1007/978-3-642-03839-6
Springer Heidelberg Dordrecht London New York

Library of Congress Control Number: 2009940109

Cover design: eStudio Calamar S.L.

Printed on acid-free paper

Springer is part of Springer Science+Business Media (www.springer.com)

Preface

The term "finite Fermi systems" usually refers to systems where the fermionic nature of the constituents is of dominating importance but the finite spatial extent also cannot be ignored. Historically the prominent examples were atoms, molecules, and nuclei. These should be seen in contrast to solid-state systems, where an infinite extent is usually a good approximation. Recently, new and different types of finite Fermi systems have become important, most noticeably metallic clusters, quantum dots, fermion traps, and compact stars.

The theoretical description of finite Fermi systems has a long tradition and developed over decades from most simple models to highly elaborate methods of many-body theory. In fact, finite Fermi systems are the most demanding ground for theory as one often does not have any symmetry to simplify classification and as a possibly large but always finite particle number requires to take into account all particles. In spite of the practical complexity, most methods rely on simple and basic schemes which can be well understood in simple test cases.

We therefore felt it a timely undertaking to offer a comprehensive view of the underlying theoretical ideas and techniques used for the description of such systems across physical disciplines. The book demonstrates how theoretical can be successively refined from the Fermi gas via external potential and mean-field models to various techniques for dealing with residual interactions, while following the universality of such concepts like shells and magic numbers across the application fields.

We assume a familiarity with electrodynamic and quantum theory as presented in the usual introductory theory courses. Many-body techniques are for the most part developed in the book itself, although some prior acquaintance might be useful. They are, however, kept at a relatively low level throughout the book, staying within the confines of elementary second quantization without any resort to field theoretic techniques.

The accompanying software allows applying some of the models in simplified scenarios that are still sophisticated enough to contain all the essential qualitative ingredients, on which more refined models may be built even by the creative reader.

The authors would like to acknowledge the numerous fruitful discussions with several colleagues on teaching topics related to this book. One of us (E. Suraud) would make a special mention to P.M. Dinh and P. Quentin who were especially close in these teaching topics. Two of us (P.-G. Reinhard and E. Suraud) would also like to thank the Humboldt foundation and the French ministry of research through the Gay-Lussac prize for generous support allowing close collaboration which helped work out this book.

The authors are also indebted to Mrs. A. Steidl for competent assistance in producing figures.

Frankfurt, Germany J.A. Maruhn
Erlangen, Germany P.-G. Reinhard
Toulouse, France E. Suraud
July 2009

Contents

Chapter 1
The Variety of Finite Fermion Systems and Their Basic Properties

Many-fermion systems are all around us. Any form of massive matter is a discrete assembly of atoms, each of them constituted of a dense nucleus surrounded by a diffuse electron cloud. Nuclei themselves are built from interacting neutrons and protons, which are ultimately small compounds of quarks. Electrons, neutrons, protons, and quarks are fermions. Atoms themselves may also be fermions, if their total spin, built from those of electrons, neutrons, and protons, is half-integer (in units of $\hbar$), otherwise they are bosons. There are a few other bosonic particles in nature, especially the particles mediating elementary interactions but, apart from exceptions, most assemblies of massive particles are built from fermions.

Theoretical methods describing matter thus primarily have to deal with interacting fermions. Because of the variety and complexity of the systems under study, these methods require elaborate approximation schemes. A key aspect is the finiteness of the system under study. In particular the extent of the system fixes its long-range behavior and thus the boundary conditions of the problem. This has consequences for the development of a theoretical description. In this book, we focus on the case of finite systems composed of Fermions. We will discuss generic methods to treat such systems and illustrate them in terms of simple models, which allow to track the basic mechanisms in detail without being blinded by the complexity of a realistic simulation. The modeling covers a variety of physical systems, atoms, some atomic compounds (e.g., molecules), nuclei, but also compact stars, quantum dots and fermion traps. We will not address the case of elementary particles and quarks, which for a proper treatment requires elaborate relativistic approaches. Most of our discussions are restricted to non-relativistic models.

Nuclei are built from neutrons and protons. They are self-bound and stable per se, with no external field. The situation is different in atoms where the binding of the electron cloud is provided by the external attractive Coulomb field of the nucleus. Electrons are also the basic fermionic degrees of freedom of molecules. These cover, in fact, a huge world of different compounds, varying largely in bonding type, composition, and size. A particular class therein are *clusters*, which are composed of N-fold repeated small building blocks, so to say finite pieces of bulk material. Among these of special importance are the *metal clusters*. These are distinguished by a cloud of valence electrons which belongs to the system as a whole (metal

J.A. Maruhn et al., *Simple Models of Many-Fermion Systems*,
DOI 10.1007/978-3-642-03839-6_1,

bonding) and which thus provides the most pronounced fermionic effects, as we will see later on.

Quantum dots are electronic systems tailored by a proper arrangement of semiconductors to yield a dedicated (and tunable) confining potential. There is a different class of atomic compounds where the whole atoms represent the basic degrees of freedom. One example are *^{3}He droplets* with ^{3}He atoms as elementary building blocks (remember that helium is an especially inert rare gas with an extremely small binding capability). Helium droplets are dense, self-bound objects. Another quite different example are assemblies of fermionic atoms confined by dedicated external fields in an electromagnetic trap. Whole atoms here are the basic degrees of freedom because the systems are very dilute and thus involve only very low energies.

In this book we shall cover many-fermion systems constituted of electrons (atoms, molecules, metal clusters, quantum dots), of neutrons and protons (nuclei), and of fermionic atoms (^{3}He droplets, atomic traps). These are all interacting many-fermion systems which can only be treated by approximate solutions of the N-body Schrödinger equation. Many-body theory has developed a rich toolbox of powerful and efficient methods for that. We will here focus on a few generic methods, demonstrate their basic mechanisms in simple, analytically solvable models, and establish connections to the above-mentioned systems. Even with that restriction of scope, both in terms of constituents and in terms of finiteness, there remains a vast field of possible topics and one needs a guiding principle to organize the discussions.

The most useful sorting scheme for our purposes is given by the complexity of the approximation. We will thus successively consider models in which the interaction is treated at increasingly detailed levels. Predominant is the picture of fermions which move independently in a common mean field. The most advanced approach here are the *self-consistent mean-field models*, which incorporate the interaction as well as possible. There are, of course, several simplifications to that elaborate treatment. On the other side are the even more advanced methods which take care of correlations, i.e., of effects which cannot be incorporated into a mean-field picture. The outline thus is as follows: We start in Chap. 2 from non-interacting particles, continue in Chap. 3 to a description inside a common external field, and in the subsequent chapters proceed to the various elaborate methods where the interaction is taken into account explicitly. Chapter 4 briefly presents a treatment in highly reduced spaces. Chapter 5 discusses the widely used self-consistent mean-field methods. Chapter 7 presents the family of quasispin models, which provide powerful tools for testing many-body theories. Chapter 8 continues with mean-field methods for collective dynamics. Approaches beyond the mean field will be briefly addressed in Chap. 9 (coherent correlations).

A clarifying word is in order here about the notion of "correlations." We will use it here in the sense of many-body quantum theory. We define correlations as everything which cannot be accounted for by an independent-particle picture, however, elaborate it may be. Let us consider a many-fermion problem. As we shall see in many places all over this book, it can often be described to first order by considering that the fermions evolve in a common potential well, stemming either from an external agent or from an average of interactions between themselves, or both. This covers what is meant by *independent-particle picture*. But even if this independent-particle picture often provides an even quantitatively correct description of the fermion system, it usually overlooks some subtle details as soon as one truly wants to account for interactions between constituents. There thus remains the so-called *residual interaction* effects reflecting the difference between the average and the exact treatment of interactions. These differences may sometimes lead to huge effects in some observables, even at a qualitative level. The term "correlations" thus gathers all the various mechanisms induced by the residual interactions, i.e., interaction effects not accounted for in the mean field.

1.1 Fermions in the Universe

1.1.1 The Four Elementary Interactions

There are four elementary interactions in the universe: the gravitational, electromagnetic, and the strong and weak nuclear interaction. These interactions act under different physical conditions and on different particles. Figure 1.1 provides a gross schematic picture of interactions and scales of objects on which they act. Historically the longest known interaction is gravitation, which acts on any massive particle everywhere. It is the weakest of all forces and becomes effective only when huge masses are involved. Apart from compact stars, which we will discuss in the context of the Fermi-gas model, gravitation will not be considered further here. Next in the scale of strength comes what is called the "weak interaction." It is involved in rare

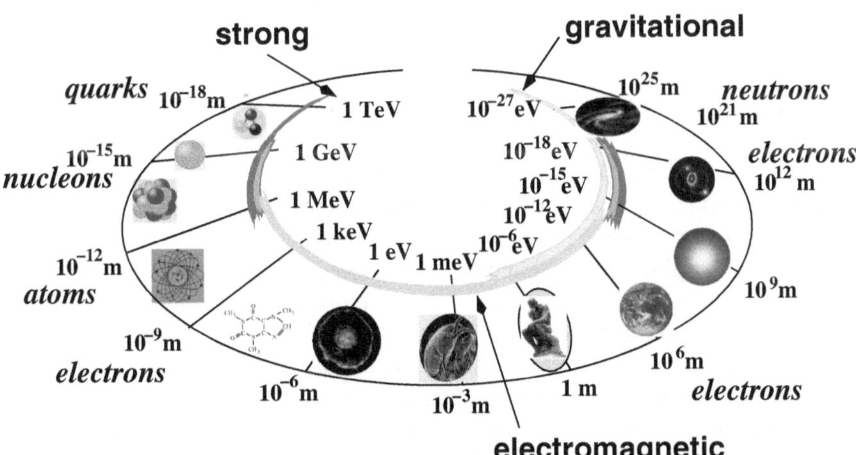

Fig. 1.1 Schematic representation of distance and energy scales in relation to corresponding typical physical systems and dominant associated interactions. The associated many-fermion systems are also indicated for completeness. Adapted from [25]

reactions at the nuclear or particle scale like, β-decay [66]. It will not be discussed further here. In the center of the discussion will stay the electromagnetic interactions, especially the Coulomb interaction, which act on any charged particle. The Coulomb interaction dominates in most systems ranging from atoms to macroscopic objects and also plays a role in atomic nuclei. It is thus at the core of nuclear, atomic, and molecular physics as well as chemistry and material science. The Coulomb interaction has infinite range, which makes its handling delicate. Effects of positive and negative charges compensate at large distances reducing often the cumbersome long-range effects. The strong interaction is responsible for the nuclear forces. It has short range and is confined inside atomic nuclei. In principle, the strong interaction applies to the quarks constituting neutrons and protons. In practice, one derives from that an effective nucleon–nucleon force to deal with protons and neutrons as "elementary" building blocks of the nucleus. Besides these four basic interactions, we often find "derived interactions" for an effective description of systems made of composite, but inert, constituents. A prominent example is given here by the atom–atom interaction between rare gases, see Sect. 1.1.7.

This usual sorting of interactions into gravity, weak, Coulombic, and strong has to be taken with a grain of salt. The actual impact of an interaction does also depend on the size and density of a system. A simple example is the Coulomb force whose effect depends critically on system size due to the long range. For example, small atoms can be treated very well by a self-consistent mean field with a few correlation corrections in perturbation theory. Perturbation theory with Coulomb forces, however, becomes invalid for large systems [43]. More dramatic is gravity which is usually ignorable, but becomes the leading agent if stellar masses are involved (see Sects. 1.1.8 and 2.4). We will distinguish here from a practical perspective: An interaction is considered strong if it does not allow a self-consistent mean-field approach as zeroth-order description. That happens to be the case, e.g., for the nuclear force and the interaction between ^{3}He atoms.

1.1.2 Basics of Nuclei

Nuclei constitute the cores of atoms and are themselves composite systems, constituted of protons and neutrons. The latter were identified only in the early 1930s. More recently, in the 1960s, it was realized that both protons and neutrons are themselves composite objects constituted of quarks. Quarks are fermions but need such high energies to be studied/identified that they require a fully relativistic quantum treatment, well beyond the scope of this book. Thus here we consider nuclei in terms of neutrons and protons as effectively being the elementary constituents of a non-relativistic nuclear many-body problem.

The nuclear interaction binding nucleons together is attractive at medium range (about $1\,\mathrm{fm} = 10^{-15}\,\mathrm{m}$) and strongly repulsive at shorter distance (due to the Pauli principle between the quarks constituting the nucleons). The attractive part is sufficiently strong to provide a net binding of several nucleons to one compact nucleus. The strong repulsion produces huge correlations between the nucleons which renders the detailed structure of the bound state very involved requiring the most advanced many-body theories for a nearly appropriate description [90, 78]. For the description of ground-state and low-energy processes, the short-range correlations can be eliminated. The interacting nucleon problem can thus be reformulated in terms of a softer "effective" nuclear interaction in the nuclear medium. This explains the long mean-free path of nucleons inside the nucleus (typically of the order of the nuclear radius) and makes the nuclear many-body problem much simpler than originally envisioned. Indeed, the long mean-free path provides a key justification for the mean-field picture of nuclei in which nucleons move quasi freely inside the global field created by all nucleon–nucleon interactions. In a similar way, the success of the shell-model picture in which an ad hoc external field mocking up the mean field is used, also finds its origin in this quasi independence of nucleons inside the nucleus.

Nuclei are self-bound objects not requiring any external confining field. The average nuclear potential which is felt by each nucleon is thus totally self-consistent. The strong repulsive core of the nuclear interaction defines a closest packing which applies to nuclei of any size. This means that the average distance between the nucleons, and correspondingly the central density, is about the same in all nuclei. This is called a saturating system. It allows to view nuclei in a simplest picture as finite drops of a (hardly compressible) nuclear fluid somewhat like water. Saturation has an immediate consequence for the scaling of the nuclear extent with nucleon number A. Figure 1.2 shows the systematics of nuclear radii R as determined in

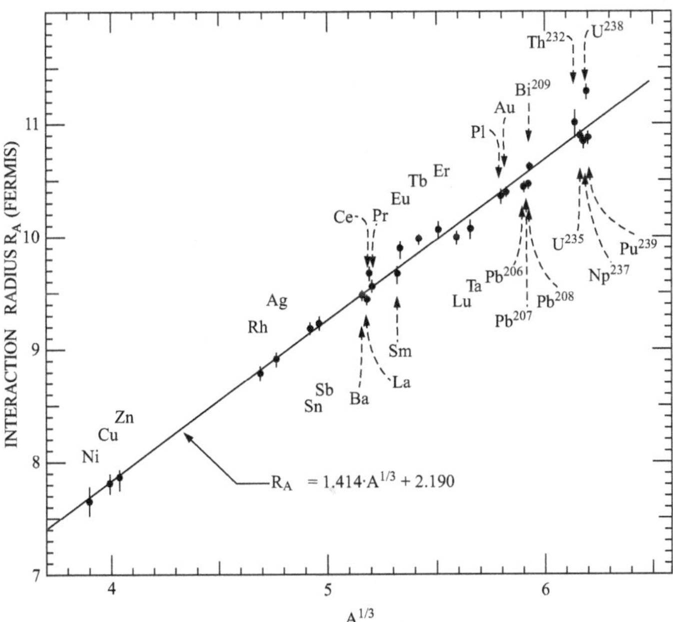

Fig. 1.2 Nuclear radii as a function of total nucleon number A. The abscissa scale is $A^{1/3}$ which allows to clearly see the linear dependence of radii on this quantity

electron scattering experiments. They exhibit a remarkable scaling law with nucleon number A: $R \simeq r_0 A^{1/3}$ where $r_0 \simeq 1.14\,\text{fm}$ is a constant almost identical for all not too small nuclei. This scaling property is related to the saturation property. It means that each nucleon occupies the same elementary "volume" $V_0 = 4\pi r_0^3/3$. Adding a nucleon to a nucleus will lead to a typical increase of its volume by V_0, independent of possible internal rearrangements. This implies that the average density inside nuclei is about the same for all nuclei $\rho_0 = 3/(4\pi r_0^3) \approx 0.16\,\text{fm}^{-3}$. This value of the nuclear density is known as nuclear matter equilibrium density, namely the density of the ground state of a hypothetical infinite phase of proton–neutron symmetric nuclear matter (without Coulomb effects).

Saturation also has consequences for the spatial distribution of the density: it becomes basically constant inside the nucleus. Figure 1.3 shows detailed charge density distributions for a variety of nuclei. It is obvious that the central charge density for the larger nuclei approaches a nearly constant value in the nuclear interior, still allowing for a smooth transition zone at the surface. Smaller nuclei are dominated by the surface such that the "constant" interior region is very small.

Neutrons and protons have nearly the same mass and, once Coulomb effects are set aside, exhibit nearly identical binding properties between each other. This is reflected in several physical quantities, such as spectroscopic properties of nuclei in which one proton is replaced by one neutron or vice versa. This approximate symmetry is known as the isospin symmetry and widely used in nuclear and particle physics [92]. We shall use it at many places in this book and often treat nucleons generically, not specifying whether we consider neutrons or protons. The major difference lies in the fact that protons are charged and neutrons not. This has consequences for the stability of nuclei. The strong interaction alone would prefer nuclei with equal number of protons Z and neutrons N. Increasing Z enhances the Coulomb repulsion and eventually destabilizes the nucleus. On the other hand, increasing the neutron number N at frozen Z stabilizes the nucleus. This explains why heavier nuclei have a large neutron excess. However, one cannot enhance N

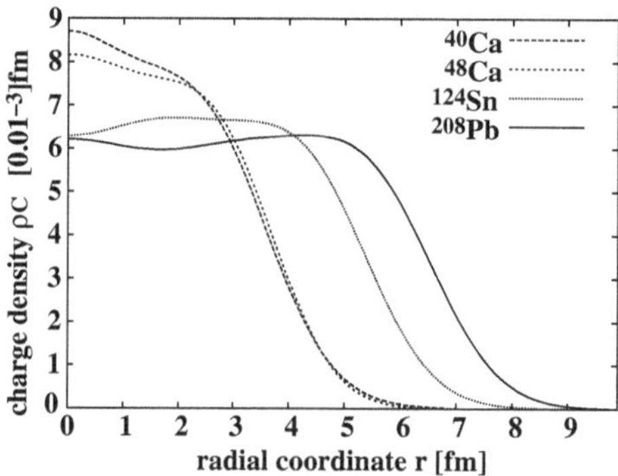

Fig. 1.3 Radial charge density distributions $\rho_C(r)$ for a variety of nuclei as indicated. The data are taken from electron scattering experiments [37]

arbitrarily. There thus exists an upper limit to observable Z because Coulomb pressure renders (super-)heavy nuclei unstable against fission.

Having identified both the nuclear density and the nuclear radius finally fixes the gross characteristics of the nuclear potential. The average depth of the potential can be estimated from infinite nuclear matter. It is fixed by the nuclear density and thus, up to details, about the same for all nuclei. The spatial extent is related to the nuclear radius, as the nucleon–nucleon interaction is short range. The emerging picture is thus a simple potential well with about fixed depth and extent $R \simeq r_0 A^{1/3}$. That shape is generic for all saturating Fermion systems as will be shown in Fig. 1.13. The shape strongly differs from the atomic central field. We shall come to that in the next section.

Nucleons move in a common mean field and thus the single-nucleon spectra have a shell structure. A simple but reasonably good approximation for the self-consistent potential is provided by the spherical harmonic oscillator augmented by a strong spin–orbit interaction as a crucial ingredient. Nucleons obey the Pauli principle and thus the shells are to be filled state by state. Particularly stable configurations appear if a spherical shell (with angular momentum degeneracy) is just being closed. This gives rise to the well-known magic numbers and associated magic nuclei such as ^{16}O ($Z = 8, N = 8$), ^{40}Ca ($Z = 20, N = 20$), ^{48}Ca ($Z = 20, N = 28$), or ^{208}Pb ($Z = 82, N = 126$).

Doubly magic nuclei (nuclei with both Z and N magic numbers) are spherical objects because shell closure leads to spherical distributions. Open-shell nuclei may acquire large deformations. Nuclear shapes thus constitute a genuine quantum effect which has focused numerous studies. One was even able to identify super-deformed nuclei with typical axis ratios of 2 to 1. The microscopic description of such deformed nuclei is quite involved but gross effects can be understood on the basis of deformed oscillator potentials which exhibit, for specific values of axes ratios, new shell closures and thus qualitatively explain the occurrence and the stability of strongly deformed nuclei. We will address this question in Chap. 3.

Another interesting issue is the study of nuclear dynamics, e.g., in giant resonances and fission. The latter phenomenon is well known and constitutes the basis of many applications. It corresponds to the breaking of a heavy nucleus in two smaller nuclei. The process gains energy but is inhibited by a (fission) barrier which has to be overcome. This can occur naturally by tunneling or through external excitation as, e.g., neutron capture. The process is rather slow, on a nuclear timescale, and goes through deformations with very large amplitude. It is a beautiful example of how nuclei can reach large deformations, in a way quite similar to a charged liquid droplet. The nuclear liquid drop model can indeed be successfully used for analyzing gross properties of fission. Quantum shell effects are, nevertheless, crucial for establishing fission barriers. We will discuss that in Chap. 3. Other important dynamical effects can be found in giant resonances. These (low amplitude) motions correspond to a collective displacement or deformation of the nuclear fluid. They are sorted according to the multipolarity of the mode. The most famous one is the Giant Dipole Resonance (GDR) in which neutrons oscillate against protons. Next come the Giant Monopole Resonance which is a radial density oscillation and the Giant

Quadrupole Resonance in which the nucleus shape oscillates from oblate to prolate.[1] The GDR has been particularly studied and possesses a well-studied analog in the Mie plasmon resonance of metal clusters (Sect. 1.1.5). We will address resonance modes in Chap. 8.

To summarize the specific features of the nuclear many-body problem: The originally singular nucleon–nucleon interaction is effectively softened inside the nuclei, which provides a solid justification for the picture of independent nucleons moving in a common self-consistent average potential. The nuclear potential has an almost constant depth and its extent is fixed by the nuclear radius, which scales with nuclear mass A as $R \simeq r_0 A^{1/3}$. This reflects the saturation property of nuclear matter. The nuclear shape may be strongly deformed depending on shell filling. Neutrons and protons have similar properties, except for the charge. The interplay between strong interaction and Coulomb force produces the observed richness of encountered situations.

1.1.3 Basics of the Electronic Cloud in Atoms

Atoms consist of a central, compact, charged nucleus (Sect. 1.1.2) surrounded by a diffuse, distant, neutralizing electron cloud. The nuclear radius of the order of several fm is negligible as compared to the extent of the electron cloud. In most atomic problems the nucleus can thus safely be reduced to a point charge Ze acting on electrons by its Coulomb field. The atomic number Z determines the number of electrons in the cloud and thus the chemical properties of the element. For heavier atoms, relativistic effects become important, especially the coupling between electronic spin and angular momentum (spin–orbit). Still spin–orbit effects are relevant mainly for quantitative details of bonding. We shall neglect these fine-structure aspects in this book.

The atomic problem then reduces to a non-relativistic many-electron problem in the central field of the nucleus. Electrons arrange themselves in a regular sequence of shells around the nucleus. Let us start from the over-simplistic picture of the pure nuclear Coulomb field. The single-electron energies are then sorted in shells n with energies $\varepsilon_n = Z^2 n^{-2}$ Ry and a high $2n^2$ degeneracy accounting for angular momentum $l = 0, 1, \ldots n-1$ and its z-component $m = -l, \ldots, +l$, and for spin $\sigma = -\frac{1}{2}, \frac{1}{2}$. This pure Coulomb spectrum is shown for $Z = 36$ in the left part of Fig. 1.4. Here and in the following we use the spectroscopic notation of angular momentum identifying $l = 0, 1, 2, 3, \ldots$ with $s, p, d, f, g, h, \ldots$ as indicated in the figure. The electronic configuration for given Z is obtained by successively filling the single-electron states. Closed shells correspond to particularly stable systems, the rare gases. The pure Coulomb field fails to predict the proper sequence of rare gases. This is demonstrated in Fig. 1.4 where the last Coulomb shell $n = 4$ cannot be

[1] In an axially deformed system, if the extent along the symmetry axis is longer than the perpendicular one (cigar-like shape), the deformation is called *prolate*, otherwise *oblate* (pancake-like).

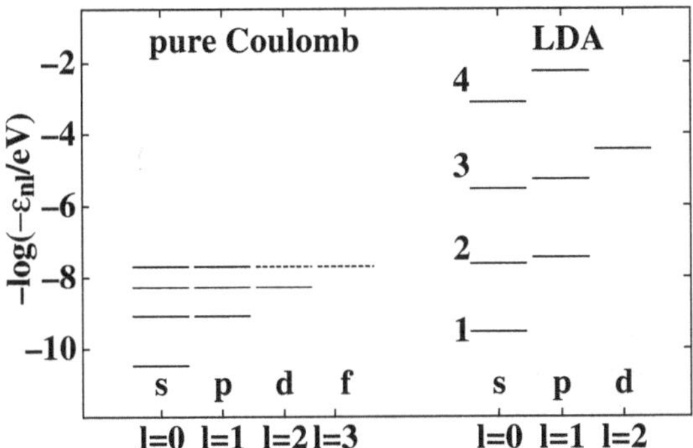

Fig. 1.4 Occupied electron levels in the Kr atom Kr ($Z = 36$). The single-electron energies are drawn on a logarithmic scale to cover the enormous span of energies between deepest bound 1s state and highest occupied orbital. The levels are disentangled with respect to orbital angular momentum l and the corresponding spectroscopic notation is indicated. The numbers beneath the groups of levels denote the principal quantum number n. The *left* panel shows the spectrum from the pure nuclear Coulomb potential $V = -Z^2e^2/r$. The *right* panel shows the more realistic spectrum of a self-consistent calculation (see Chap. 6) at the level of the local density approximation (LDA) (see Sect. 6.1)

completely filled with $Z = 36$ electrons, while it is known that $Z = 36$ corresponds to the rare gas Kr. The repulsive electron–electron interaction has to be taken into account. This can be done in a mean-field picture. The result of a density-functional calculation (see Sect. 6.1) is shown in the right part of Fig. 1.4. The effect on the single-electron energies as compared to the pure Coulomb field is dramatic. There is a strong up-shift of single-particle energies and the extremely high degeneracy is removed, leaving only the angular momentum and spin degeneracy of a central-field problem. This changes the filling pattern and a substantial energy gap (not shown) emerges above the last filled state in Kr rendering this element safely a rare gas.

To better understand the electronic structure of atoms we recall the "aufbau principle" [114], ignoring details like the very small deviation from sphericity of the mean field, thus always assuming a central field. We imagine that we build the atom electron by electron successively. The first electron will feel only the pure nuclear Coulomb field. The second electron will already experience both the nuclear attraction and the repulsion from the first electron. A balance is finally established between both competing effects leading to the formation of the most deeply lying electronic level, the so-called 1s shell, which has degeneracy 2 (two electrons in the shell with opposite spins). The third electron feels the Coulomb field of the attracting nucleus and the repulsion from the two already present 1s electrons. This amounts effectively to a screened nuclear charge $\simeq (Z-2)|e|$, with the 1s electrons remaining closely packed around the nucleus, and keeping the third one farther away. The case of the fourth electron will be somewhat similar to the case of the second one. It will form, together with the third electron, the so-called 2s electronic shell, again with zero angular momentum. The screening of the nuclear charge by the $1s$ shell makes the $2s$ shell much less bound than the $2s$ shell of the pure nuclear field (see Fig. 1.4). This effect is even enhanced by the Pauli principle, which tends to expel the $2s$ electrons from the region already occupied by the $1s$ electrons. Continuation of this aufbau principle will proceed similarly. Electrons do gather in shells, characterized by a so-called principal quantum number ($n = 1, 2, 3, \ldots$) and an orbital one ($s, p, d, f, \ldots$). Each shell may contain up to $2(2l + 1)$ electrons, the factor 2 allowing for pairs of opposite spins. Shell filling remains simple in light atoms (lighter than Ca ($Z = 20$)), but becomes more involved in heavy systems. It leads to the well-known sequence $1s|2s, 2p|3s, 3p|4s, 3d, 4p|\ldots$. Each major shell closure (indicated by a vertical bar) leads to the appearance of a peculiar class of especially bound atoms: the so-called rare gases known to be especially inert from the chemical point of view. The gain in electronic binding associated to shell closure explains the enhanced stability and corresponds to a particularly high value of the ionization potential (IP) indicating that it is very costly to remove one electron from a closed shell.

The filling of electrons in shells becomes more and more involved when considering larger atomic numbers, because polarization effects may lead to "irregular" behaviors: some shells may be left partially empty while less bound ones start to be filled. Hund's rules here provide, fortunately enough, robust guidelines for an estimate of the successive occupation of electronic levels [114]. The first rule states that electrons successively occupy the levels $1s$, $2s$, $2p$, $3s$, $3p$, $4s$, $3d$, $4p$, $5s$, $4d$, $5p$, $6s$, extending the above-mentioned sequence in light atoms. When d states are involved that monotonous shell filling is perturbed and tends to oscillate between d and s with increasing atomic number. The second rule gives some clues on the impact of degeneracy, which becomes larger and larger with increasing atomic number due to increasing angular momenta in electronic shells. But the problem already shows up in a small atom such as carbon. The occupied levels in carbon are $1s$, $2s$, and $2p$ with the corresponding electronic structure $1s^2 2s^2 2p^2$. The $2p$ state is 2×3-fold degenerate ($2p_x, 2p_y, 2p_z$ times spin), which for 2 electrons implies 15 equivalent choices of 2 occupied levels out of 6. The second Hund's rule helps solving that dilemma by stating that electrons occupy all degenerate sub-shells once *before* occupying any level twice. This means that, for example, in the case of C, the 2 $2p$ electrons will occupy two different orbitals, say $2p_x$ and $2p_y$, rather than twice $2p_x$. There remains to decide what is the relative orientation of spins in such sub-shells and this is fixed by the third Hund's rule which states that the ground-state configuration of an atom is the one with maximum net spin (namely aligned spins). This means that the spin-polarized structure emerges as more favorable as compared to a spin-saturated configuration. Hund's rules thus provide a guideline for predicting the electronic structure and the level filling in an atom, especially in light atoms, heavier atoms leading to many exceptions, foremost in relation to the polarization of the d shell. The case of noble metals such as Ag, Au, or Pt is quite typical in this respect. The discussion of such details is nevertheless somewhat specific and we shall not further discuss this point. It is sufficient to remember that in such noble metals the least-bound electronic shell (s shell) is significantly affected by the only slightly more bound d shell which, with its large degeneracy (a d shell may accept up to 10 electrons), may lead to strong polarization effects, thus making the overall picture quite complicated.

The filling sequence is also reflected in the systematics of atomic radii shown in Fig. 1.5. The last elements before shell closure (the halogens F, Cl, Br, J) have the smallest radii, while the first elements after closure (the alkalies Li, Na, K, Rb, Cs) have largest. It is, furthermore, noteworthy that atomic radii stay in the same range throughout, much different as compared to nuclei where radii scale with a power of system size (see Sect. 1.1.2).

Electrons in an atom thus constitute an interacting Fermion system in which the external nuclear Coulomb field plays the major role, responsible for the binding of the whole atom and strongly influencing the shell structure of atoms. The repulsive electron–electron interaction can be handled to first order in terms of a central mean field which then allows to deduce simple rules for the sequence of elements and their basic properties. Spin–orbit effects are essential for explaining the details of electronic structure, but will not be discussed much in this book.

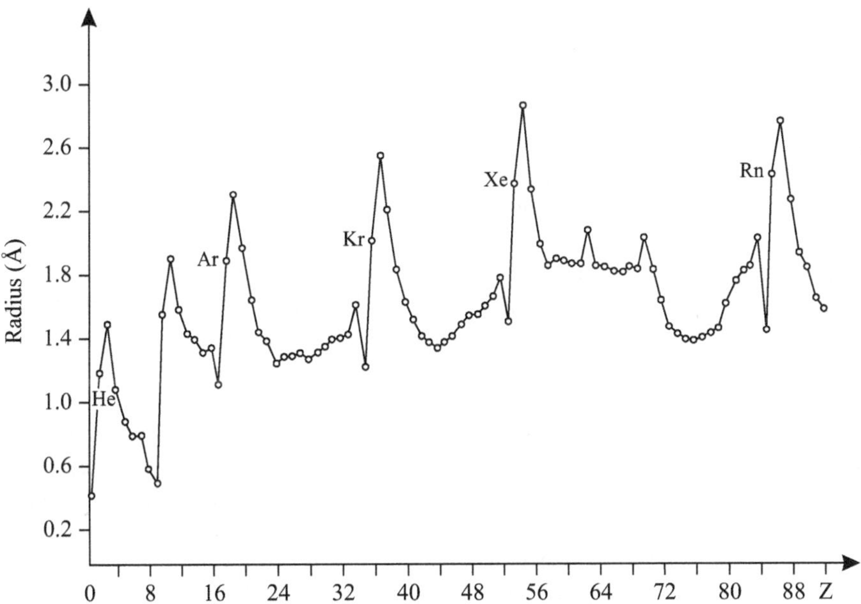

Fig. 1.5 Atomic radii (in Å) as a function of atomic number Z up to $Z = 92$. The overall evolution is a soft increase of radii as a function of Z. This evolution is strongly modulated by shell effects, whence the marked oscillatory pattern. Halogens have consistently small radii, while alkalies with their weakly bound s valence electron, have consistently large radii. The positions of the noble gases are indicated in the figure

1.1.4 Basics of Molecular Binding

As is well known, the successive filling of electronic shells provides the microscopic basis for the understanding of chemical properties of elements and also a simple understanding of the simplest dimer molecules binding together two atoms. To understand such features is one of the goals of the Mendeleev classification [95], which explains the regularity observed in the chemical reactivity of apparently different atoms. Because chemical binding is by nature a low-energy phenomenon, involving at most a few eV energy, these chemical similarities reflect the behavior of the tiny fraction of least-bound electrons in atoms, called valence electrons. The Mendeleev classification, based on valence electron properties, thus provides a gross map of which atoms may possibly bind with which other and how. Atoms with valence electrons belonging to shells with the same angular momenta ($s, p, d\ldots$ shells) behave similarly at the chemical level, which is not so surprising as the angular momentum fixes the shape of the electronic wave functions and thus the way electrons from two distinct atoms can interact with each other. The degree of filling also plays an important role by determining the capability of an atom to accept or to release valence electrons. Rare gases (He ($Z = 2$), Ne ($Z = 10$), Ar ($Z = 18$), Kr ($Z = 36$), Xe ($Z = 54$), and Rn ($Z = 86$)) with electronic shell closures are thus especially stable, chemically inert, and have a high ionization potential. In spite of this, rare gases can bind with each other. The bonding is faint and arises from the polarization of one atom's electronic cloud by the other one. It is known as rare gas or van der Waals bonding and is, e.g., responsible for the formation of helium droplets as discussed in this book. Nevertheless, apart from this example of helium we shall only little consider this type of system in the following.

Next to the rare gases there are two categories of especially active atoms with one electron more or one electron less than the neighboring rare gas. These atoms have a strong tendency either to grab or to release an electron, in order to attain a closed-shell electronic configuration. The alkaline atoms (Li ($Z = 3$), Na ($Z = 11$), K ($Z = 19$), Rb ($Z = 37$), Cs ($Z = 55$), Fr ($Z = 87$)) have one weakly bound valence electron belonging to a s shell on top of a rare gas "core." This valence electron has a small IP and prefers to leave its parent atom. Halogen atoms (F ($Z = 9$), Cl ($Z = 17$), Br ($Z = 35$), I ($Z = 53$), At ($Z = 85$)), in turn, exhibit a valence p-shell missing one electron for closure and tend to capture this electron from their surroundings. Halogens and alkalies thus have a strong tendency, when in each other's presence, to bind together by electron exchange, leading to the most robust chemical bond, the so-called ionic bond associating one negatively charged halogen to one positively charged alkaline. In ionic bonds electrons are exchanged between the two atomic partners and play little role as such. In this book, we shall not consider such systems in which there is no specific many-electron problem, at variance with other molecules or clusters.

Up to here, we have discussed 3 columns of elements (rare gas, alkalies, halogens), while the Mendeleev classification contains up to 18 columns. In fact, halogens and alkalies as "ends" of each row represent the extreme cases of tendencies which actually develop along each row of the classification. Starting from an alkaline atom and successively adding electrons (namely going in the direction of the next heavier halogen) will progressively attenuate the tendency of the atom to release electrons and increase its tendency to capture them from its surrounding, up to the extreme case of halogens. There is thus a continuous path, along a given row, between electron donors and electron acceptors. In the middle of the path lie atoms with an "equal" tendency to capture or to release electrons. These atoms play a central role in nature. Indeed the typical representatives of this class are carbon (C, $Z = 6$) and silicon (Si, $Z = 14$), which are building blocks of organic molecules (carbon) and electronic devices (silicon). These atoms bind together in a specific way known as covalent binding. In this case electrons neither stay on their parent atom (rare gas bond) nor are transferred to the partner atom (ionic bond) because there is no energetic reason to do so. Still, because the electrons are not too strongly bound to their parent atom (as in rare gases) they have a possibility to effectively "connect" to neighbor atoms. Electrons are finally shared between two atoms offering them the access to levels of comparable energy (whence the term "covalent") and the electron cloud binding the two atoms together gathers along the line joining the two atoms, again in a fashion to establish a maximum shell gap for these now molecular spectra.

If the valence electrons are sufficiently weakly bound to their parent atom the electronic wave functions can easily spread outside the parent atoms. The electronic wave functions may even become delocalized all around the two atoms and one then speaks of a metallic bond (as a precursor of the behavior of such systems in the bulk, discussed below). Alkaline atoms are the typical elements which establish metallic bonds between each other. They are called simple metals because the valence electron is well separated in energy from the more deeply bound (core) electrons. They thus provide a generic test case of metallic systems. Metallic binding is also observed in some other metals such as Cu, Ag, Au, or Pt, but binding is more complicated in such cases due to the presence of an energetically close d shell next to the valence s shell and causes strong polarization effects. We shall thus mostly refer, in the following, to the case of simple metals when speaking of metallic bond.

Bonding can even be directly visualized by means of scanning tunneling microscopy (STM) which provides detailed images of the electronic density. An example is provided in Fig. 1.6 for the case of O_2 molecules deposited on a TiO_2 surface, a prototypical transition metal-oxide model system in surface science. The figure exhibits a clear alternation of dark and light rows which can be attributed to the alignment of O (dark rows) and Ti (light rows) atoms. The continuous rows indicate electrons spreading between atoms and ensuring bonding. Figure 1.6 focuses on the diffusion of oxygen molecules on transition metal-oxide surfaces, an effect which plays a vital role for the understanding of catalysis and photo-catalysis on these materials. STM images are time resolved which allows to explicitly follow in time the diffusion process, as can be seen from the figure. These experiments show that when properly heated the system exhibits numerous O vacancies (here missing O atoms at the surface, open squares in the figure) on surfaces which trap electrons and act as adsorption sites for simple molecules such as CO and O_2. The time-resolved STM images then reveal that O_2 molecules reside on top of the Ti atoms

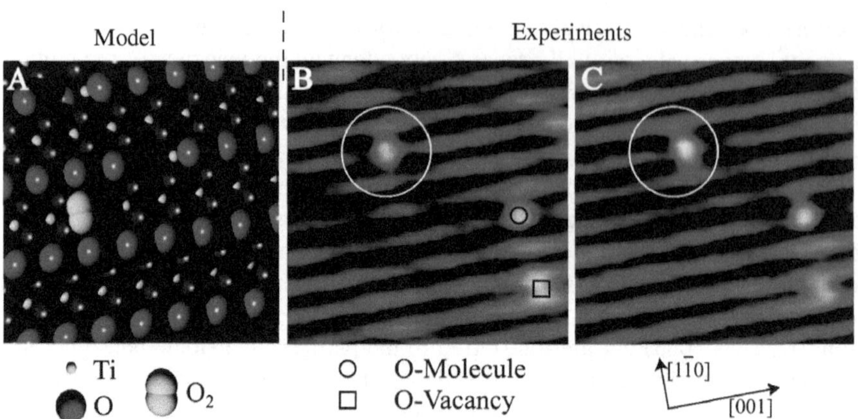

Fig. 1.6 Scanning tunneling microscopy (STM) time resolved images (*two right panels*) of the diffusion of O_2 molecules on a TiO_2 surface. The *left* panel provides a schematic ball model of the TiO_2 (110) surface. The two STM images are separated by about 0.45 s. Adapted from [112]

that constitute the troughs along the [001] direction in between the protruding O rows and diffuse (circles) only along the [001] direction.

The differences between the various types of bonding are, to a large extent, related to the energies of the valence electrons of the two atoms being bound. The more strongly bound the electrons, the less likely is a bonding between the two atoms. Note that this energy is directly connected to the degree of localization of the electrons, the deeper bound the electron the more localized its wave function, which brings a complementing viewpoint to this question and on the bonding classification. This also points out the fact that the classification into four bonding types (rare gas, ionic, covalent, metallic) is, to some extent, conventional. There are continuous paths between the various types of bonding, especially between rare gas, covalent, and metallic. In this book we are interested in finite fermion systems and thus concerned mostly with one type of bonding, namely metallic systems (except for the specific case of helium). It turns out that electrons in a metallic system indeed behave as many-fermion systems, requiring for their description the typical techniques for many-body problems. Metallic systems will thus constitute prime examples for finite electron systems and be used throughout this book. One of their simplest realizations is the so-called metal clusters discussed in Sect. 1.1.5. Simple approaches to other sorts of chemical bonding will also be discussed in Chap. 4.

1.1.5 Basics of Metal Clusters

Clusters are a special class of molecules. They are built from arbitrary repetition of the same building blocks, similar to bulk crystals, but finite. They constitute an intermediate state of matter between molecules and bulk matter. Clusters have been applied for centuries as dispersed metallic material in glass or as deposited pieces in photographic emulsions. Cluster physics was strongly developed as an independent field at the interface between physics and chemistry during the last few decades, once cluster production in dedicated sources had been made practical. One of the major features of clusters is the fact that their size can be varied almost at will, from a few through several hundreds of thousands of atoms up to the bulk material. They are usually constituted of one type of atom, sometimes of two for mixed systems, and possibly more in molecular clusters like water clusters $(H_2O)_N$. Clusters are classified according to the type of binding between their constituents. We have outlined in Sect. 1.1.4 that one can distinguish four basic types of chemical bonding.

Figure 1.7 shows examples of clusters for the four different binding types. The structure of the mixed system Na_5F_4 with ionic binding is involved, but the typical alternation of positively and negatively charged sites and the associated regular arrangement of ionic crystals is already visible. C_{60} is an example for a predominantly covalent cluster and shows the famous Fullerene pattern. The Ar cluster exemplifies a van der Waals system which requires extremely low temperatures to stay stable. Finally, Na_4 stands for a metallic cluster. Here we show produced the electron cloud explicitly to demonstrate how it extends smoothly all over the cluster.

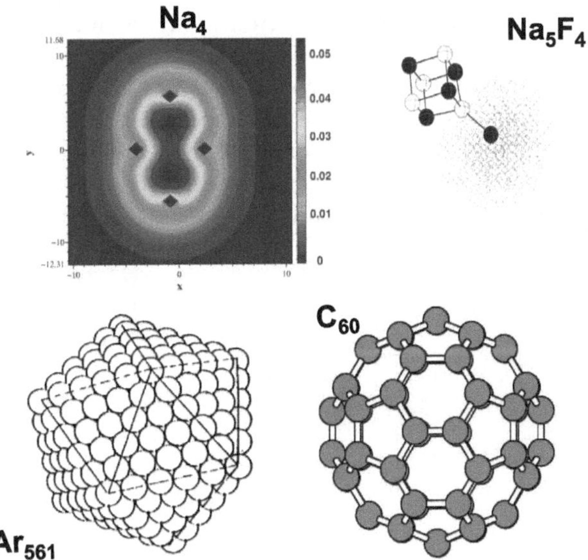

Fig. 1.7 Different geometrical views of four different clusters as indicated. *Upper left*: electronic density in the x–y-plane for Na_4; the positions of the four Na^+ are indicated by *rhombi*. *Lower left*: configuration of Ar atoms in Ar_{561}; the *dashed lines* indicate the emerging cuts through bulk fcc structure. *Upper right*: configuration of the mixed cluster Na_5F_4; the Na positions are indicated by *filled circles* and the F positions by *open circles*; the *shaded area* around the outer Na ion indicates a concentration of the electron cloud. *Lower right*: configuration of the C_{60} cluster; the *gray* balls indicate the C atoms and the bridges in between the leading covalent bonds

Among the four types, metallic clusters will play an important role in this book because the electrons move freely throughout the whole cluster, similar to nucleons inside a nucleus. This property is also responsible for similar scaling laws and trends observed in metal clusters. In a metal atom the valence electron is only weakly bound and has a large spatial extent. If several metal atoms are put together the electronic wave functions of these valence electrons will strongly overlap and recouple to explore the whole available space of the system together composing its binding properties. The valence electrons interact via the repulsive Coulomb interaction, but this repulsion is compensated by the attraction provided by the positively charged background of the remaining ionic cores (= atoms without valence electron). Altogether, the net classical Coulomb effect almost vanishes and only subtle quantum corrections, linked in particular to the exchange interaction (see Chap. 5), lead to a net binding of the system. Thus the valence electrons in such metallic systems effectively interact only weakly with each other and can be considered as a quasi-free electron system. The situation thus differs from the nuclear case to the extent that there exists an external confining field from the ionic cores. It also differs from the atomic case because the confining field is much softer and because the electron cloud determines the ionic configuration in a decisive manner.

The valence electrons move with momenta near the Fermi momentum, which corresponds to a spatial resolution of the order of the Wigner–Seitz radius r_s (typically 3–5 a_0). The fine details of the ionic background (typically at scales of $\approx 0.3 - 0.8$ a_0) are thus seen by the electrons only in an average manner. This motivates the *jellium approximation*, in which the ionic background is smeared out to a constant positive background charge. It is a standard approach in the theory of bulk metals [3], and the adaptation to a finite cluster is straightforward. From the bulk a finite element of constant positive charge is carved out. The roughest approach is a homogeneously charged sphere with sharp surface. More versatile, more realistic, and still easy to handle is a Woods–Saxon profile for the jellium density

$$\rho_{\rm jel}(\mathbf{r}) = \frac{3}{4\pi r_s^3}\left[1+\exp\left(\frac{|\mathbf{r}|-R(\vartheta,\phi)}{\sigma_{\rm jel}}\right)\right]^{-1} \quad , \tag{1.1}$$

$$R(\vartheta,\phi) = R_{\rm jel}\left(1+\sum_{lm}\alpha_{lm}Y_{lm}(\vartheta,\phi)\right) \quad ,$$

where the overall extent R_0 is to be adjusted such that the total charge is reproduced, i.e., $\int \mathrm{d}^3 r\, \rho_{\rm jel} = N_{\rm ion}$. The central density is determined by the bulk density $\rho_0 = 3/(4\pi r_s^3)$ of the given metal. To give an impression of the surface width: the transition from 90 to 10% bulk density is achieved within typically $4\sigma_{\rm jel}$.

Possible deformations are parametrized in $R(\vartheta,\phi)$ through the coefficients α_{lm}. The most important of these is α_{20} which produces axially symmetric deformation, where positive values lead to prolate (cigar-like) and negative values to oblate (pancake-like) shapes. The additional parameters lead to more complicated shapes, e.g., $\alpha_{2\pm2}$ allows three different radii in the three coordinate directions (see Sect. 3.2).

The similarity between metal clusters and nuclei has played an important role in the study of clusters. Modeling the cluster as a cloud of weakly interacting electrons inside an external confining potential immediately suggests that the electrons will gather in shells. This was indeed observed experimentally. Early measurements on small alkaline clusters have identified a series of "magic" electron numbers corresponding to enhanced stability and abundances. The observed series 2, 8, 20, 40, 58, 92... matches spherical harmonic oscillator shells up to 40 and can be explained by more realistic model potentials for the larger numbers. That will be discussed in Chap. 3.

One of the specificities of metal clusters is the fact that one can vary their size N deliberately, a parameter whose variations are rather limited both in atoms and in nuclei. Indeed the heaviest element has atomic number 114 and accounting for the various neutron/proton combinations in nuclei leads to about 3000 stable nuclei. Metal clusters, in contrast, can have sizes between about 2 and hundreds of thousands of atoms. For example, regularity (or irregularity) in abundance spectra (which is closely connected to stability) can thus be explored on a very large scale. Such studies have been performed and have exhibited several interesting features. First, one indeed observes strong electronic shell effects and this up to any size range. Second, for moderate sizes the amplitude of shell effects (in particular the shell separation) has a tendency to decrease for increasing system size. This is very similar to what has been observed in both atoms and nuclei. The surprise comes from the fact (impossible to explore in atoms or nuclei) that when further increasing the size one observes a revival of the amplitude of shell effects. There appears in fact a beat in the amplitude of the shell separation with increasing size [19], an effect predicted on a purely theoretical basis in the early 1970s [6, 20]. As a third point one should note that ions may also play a key role for the shell structure. Indeed while small to moderate sizes are dominated by electronic shell effects one can show that atomic shell effects play an increasing role with increasing size. The building blocks have a tendency to arrange themselves in a more or less regular way, somewhat like a precursor of the bulk solid and when a particularly compact and regular arrangement with flat surfaces is attained this leads to an enhanced stability. The occurrence of atomic shell effects is known in almost all materials, metals or non-metals. But while cluster geometry is observed whatever the material, electronic quantum effects are only possible in metals where the electronic mean-free path is typically larger than the cluster diameter.

An interesting property of clusters is their ionization potential (IP), which can be rather easily determined by irradiation with a tunable laser, and provides valuable information on electronic structure, especially in the case of simple metal clusters. Ionization potentials of Na and K metal clusters are shown as examples in Fig. 1.8. The curves exhibit clear jumps for well-defined values of the electron numbers, which correspond to electronic shell closures. At shell closure a gain in IP is observed and it drops sharply when adding one extra atom or electron (remember that Na and K are alkalies with one single valence electron per atom). These jumps

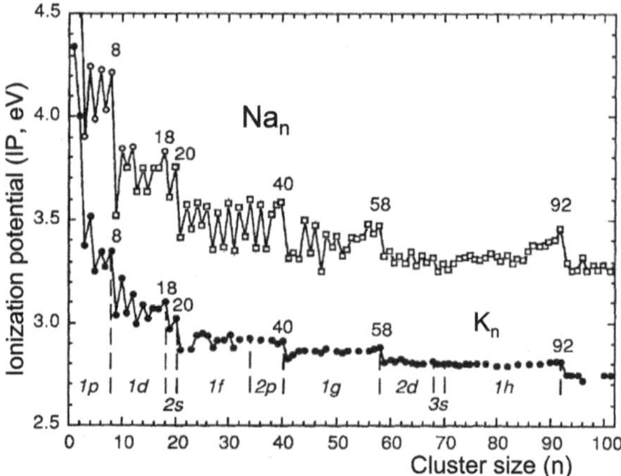

Fig. 1.8 Ionization potentials (IP) of Na and K clusters as a function of cluster size. The steps in the IP reflect electronic shell closures [117]

reflect the fact that the extra electron in the new open shell is much less bound that the ones belonging to closed shells. The amplitude of the effect decreases with increasing cluster size.

Ions also play a key role in metal clusters for determining cluster shapes. While in atoms the central nuclear field plays a key role and in nuclei the shapes result from pure nucleonic effect, the case of metal clusters lies somewhat in between in this respect. Indeed, a balance is established between electronic and ionic effects to reach the most stable structure for a given size. The basic mechanism is well known in simple molecules but becomes complicated in metal clusters. The number of possible ionic/electronic configurations (=isomers) increases very quickly with size, because the rather soft metallic binding allows rearrangements at almost no energetic cost. Thus a proper discussion of metal cluster shapes requires to take the ionic background into account. We are interested here mostly in the electrons (as being the active Fermions) and thus when necessary take the shape for granted, being obtained from more sophisticated calculations.

The metallic bond and associated long electronic mean-free path also leads to an important phenomenon in the optical response: the Mie plasmon resonance. When an electric field is applied to a metal cluster it tends to polarize the system by displacing the whole electron cloud with respect to the ionic background. When left free the cloud will then oscillate with respect to the ionic background with a well-defined frequency, which in simple metals lies in the range of visible light. The seminal work on this question goes back to the early twentieth century with the paper of Mie [71], in which the author proposed to study how small metal particles react to light and can be analyzed that way. Indeed the optical response characterizes both the constituent material and its size. It is nowadays a key tool to analyze cluster properties, not only size effects but also shapes. A remarkable feature of the Mie plasmon resonance is the collective character of the electronic motion, in which the electronic cloud oscillates coherently as one whole unit. Although dominating the entrance channel of an optical excitation, the lifetime of the resonance is limited by coupling to individual single-electron excitations and to possible ionic motion.

The notion "plasmon" for the Mie resonance is taken from plasma physics in which similar collective electronic oscillations are well known and called plasmons. The usage in metal clusters requires a word of precision. In an infinite plasma, electronic oscillations correspond to oscillations of the local electronic density throughout the volume. In finite systems the dominating oscillations result from a global displacement of the electron cloud, creating a charge enhancement or depletion on the surface. One thus distinguishes volume from surface plasmons. It should also be noted that the Mie plasmon resonance is pretty similar to the nuclear dipole giant resonance discussed above in which neutrons oscillate against protons. We shall come back to this point at several places in this book.

In summary, cluster physics covers several interesting mechanisms from a fundamental physical point of view. Clusters allow to deliberately vary the system size and so help to understand how matter builds up between atom and bulk and how bonding evolves with system size. Metal clusters, in particular, are perfect laboratories for a finite Fermion cloud of electrons with their pronounced shell effects and Mie plasmon resonance in the optical response.

1.1.6 Quantum Dots

Quantum dots are nanometer-sized dedicated structures in semiconductors in which a finite (possibly small) number of electrons can be trapped. They have something in common with atoms and are often called "artificial atoms". They have been studied in many physical devices such as transistors or diode lasers and it is even hoped that they might be used as elementary pieces for quantum computers. Before quickly describing them as finite fermion assemblies we briefly recall how they can be manufactured.

Bulk crystals exist as conductors or insulators. The difference is explained by the band structure. In solids, levels are very close to each other (at variance with finite systems) and gather in "continuous" bands rather than sets of discrete levels forming shells. The last occupied electron states build the valence band (these are usually the valence electrons of individual atoms forming the crystal). The first unoccupied electronic states form the conduction band. An *insulator* results if there exists a large energy gap between these two bands which inhibits electron transport. A conductor, on the other hand, is a system where there is no energetic distinction between occupied and unoccupied states. The resistivity of an insulator depends quantitatively on the size of the gap. For insulators with not too large a gap (typically below 2 eV), there appears some conductance through thermal agitation of electrons or appropriate doping. This is then called a *semiconductor*. The value of the gap varies from one semiconductor to the next. Let us now imagine that we combine different semiconductors with different gaps to what is called a semi-conducting heterostructure. One standard option is to build a "quantum well" from a layer of one material with low band gap embedded into surrounding layers of material with higher band gap. The electrons are more mobile in the low-gap region and feel a barrier toward the high-gap bounds. A sufficiently thin embedded layer then allows to create a 2D electron gas thus reducing dimensionality from 3D to 2D. By applying external electrical fields and/or special etching techniques, one can further reduce the dimension of the electron gas to 1D (quantum wire) and eventually 0D (quantum dot). All in all, it is nowadays possible to confine a (possibly very small) finite number of electrons in a virtually 0D potential well. The notion of "zero dimension" here does not strictly mean a point-like object. Quantum dots have typical extents in the nanometer range, but they can contain a very small number of electrons inside the confining well, whence the term "quantum dot." Figure 1.9 shows the STM image of a quantum dot with a schematic inset of the piling-up of constituting materials. The dot has electric contacts on its top and bottom.

Quantum shell effects play, of course, a key role in quantum dots and it was soon discovered that electrons indeed occupy discrete states. The number of electrons in the dot can also be controlled, and it is observed that some particular numbers of electrons are strongly favored, which is typical of shell closure effects. The many studies of these systems confirm that the confining potential is very close to a harmonic oscillator. There is additionally the electron–electron interaction to be taken into account as strong correlation patterns have also been observed. A quantum dot thus provides an interesting and versatile electron system. It has some similarity with atoms because of the dominant confining external field and the still non-negligible interactions between the electrons, but it goes far beyond that in flexibility. There is scalability, namely the capability to vary at will the number of electrons in the dot (somewhat like the variation of the number of atoms in clusters). There are also important effects connected to the impact of external magnetic fields which may strongly affect electronic energies. One should also mention that the shape of the

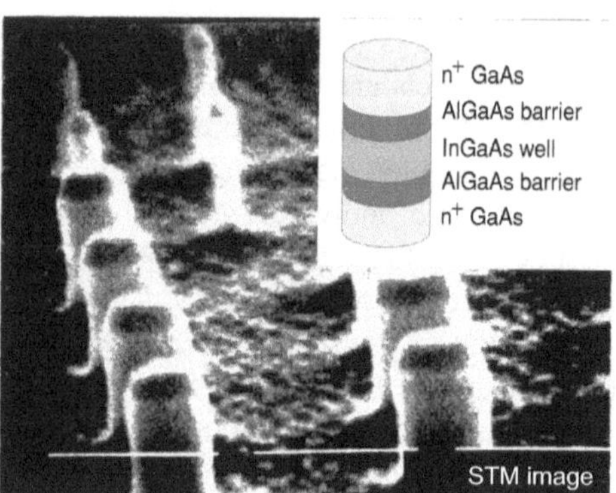

Fig. 1.9 Scanning electron micrograph image showing etched quantum dots. (The *white bars* have a length of 0.5 μm.) A schematic drawing of the double-barrier dot structure is shown as an inset. Adapted from [87]

dot is tunable (at least in two directions) by varying the values of the confining fields.

At first glance, quantum dots look like very complicated objects with all these different materials and interfaces. As already indicated above, however, the treatment can be reduced to the dynamics of a few interacting electrons in a harmonic well. One concentrates on the few mobile electrons in the semiconductor material. These move in a conduction band with a single-particle energy $\varepsilon(\mathbf{p})$ which is a function of the momentum $\mathbf{p}$. Only low $\mathbf{p}$ come into play such that we can expand $\varepsilon(\mathbf{p}) \simeq \varepsilon_0 + \mathbf{p}^2/2m^*$ which defines an effective mass $(m^*)^{-1} = \nabla_p^2 \varepsilon\big|_{\mathbf{p}=0}/3$. To a very good approximation, the active electrons thus move with the effective mass m^* of the given semi-conductor material [3]. This means that the semiconductor environment can be effectively taken into account for the electrons inside the quantum dot by attributing such an effective mass to them. Similarly, the Coulomb interaction is modified by the corresponding dielectric constant of the material, which can be accounted for by renormalizing the Coulomb coupling constant e^2 of electrons inside the quantum dot. For completeness and in order to recover relations between typical constants similar to the set of atomic units one also redefines the Bohr radius a_0 (see Appendix B.1) to an effective a_{B} and the Rydberg energy scale to an effective Ry*. These renormalized constants are often referred to as *dot units* (du) in the literature. A typical material is GaAs with the properties $m^*/m = 0.07$ and $e^{*2} = e^2/\varepsilon = 0.08$ which yields an effective Bohr radius of $a_{\mathrm{B}}^* = 9.8$ nm and effective hydrogen binding energy Ry*= 6 meV [87].

1.1.7 Basics of ^{3}He Droplets

Helium droplets are a particular class of clusters constituted of helium atoms. Helium, which is the lightest rare gas, is, like all rare gases, a rather inert material, among rare gases even the most inert one, with minimal interactions. Binding may still be possible though between two helium atoms, even if the bond length is tremendously large (several tens of times the hydrogen or carbon bond length), because of the very small binding energy. Helium was first studied in bulk phase and was the subject of many investigations due to its peculiar properties at zero temperature. Helium is the only material which does not crystallize at low temperatures under normal pressure, because He is a light atom and the ensuing larger quantum fluctuations inhibit localization. It shows superfluid behavior at sufficiently small temperatures of a few K. The dominant helium isotope is ^{4}He (99.99986%) which is a boson, composed of six fermions (two protons, two neutrons, two electrons). There also exists ^{3}He in small amounts which is a fermion being composed of five fermions (two protons, one neutron, two electrons). The total spin ^{3}He is half-integer as it should be for a fermion.

In recent decades helium has also been studied in finite pieces in the form of droplets. Most studies were performed on ^{4}He but a few also on ^{3}He. In the following we shall only discuss droplets of ^{3}He, as this is a fermionic object, alternative

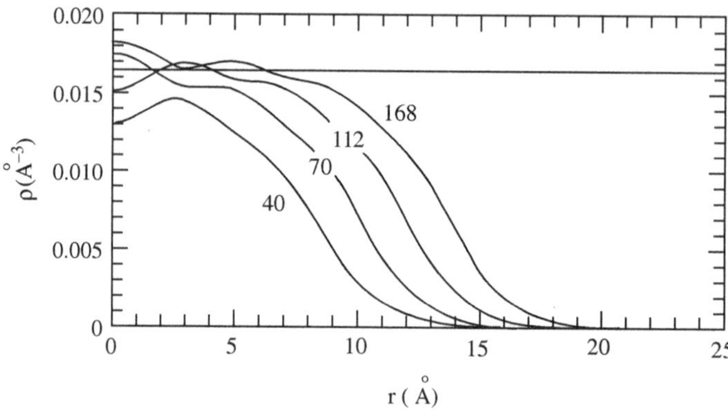

Fig. 1.10 Density profiles for ^{3}He clusters. It illustrates the evolution toward bulk properties. The large surface energy makes the convergence much slower than for nuclei or metal clusters

to nucleonic or electronic ones. The theoretical description of ^{3}He clusters can ignore the intrinsic degrees of freedom of the atoms and treat them as "elementary" particles with a given atom–atom interaction, which has many similarities to the nucleon–nucleon interaction. It also consists of a medium range attraction (van der Waals force) and a huge short-range repulsion (Pauli principle of He electrons). Thus we also encounter a saturating Fermi system with a well-defined bulk equilibrium density which is successively approached in the interior of large clusters.

Figure 1.10 shows a sequence of density profiles for ^{3}He droplets, estimated by a self-consistent mean-field model (unfortunately, there exist very few experimental results for ^{3}He droplets). The two largest systems nicely show the approach to bulk density (indicated by the horizontal line). The surface binding, however, is much weaker than for nuclei such that small systems tend to dissolve (compare with the nuclear density distributions in Fig. 1.3). The ^{3}He droplets constitute, in fact, the softest self-bound fermion systems. The experimental minimal observed size of ^{3}He droplets is still under debate. A typical order of magnitude is assumed to be around 30 atoms, in accordance with the theoretical results. More detailed observables have not yet been truly accessed. We shall thus discuss ^{3}He only occasionally as examples of fermion assemblies.

1.1.8 Degenerate Fermion Gas: Atom Traps, White Dwarfs, Neutron Stars

The fermion systems considered up to now cover different stages of interaction effects. Quantum dots to some extent offer adjustable interaction scenarios, but the most versatile tool in this respect are atomic clouds in traps, in which all physical parameters can be tuned in a wide range. The density can be adjusted by the filling conditions and external fields and the atom–atom interaction can be varied by cleverly exploiting the quickly changing cross sections of scattering resonances [14, 38]. A large interest lies in the design of an almost interaction-free system, close to the ideal of a fermion gas.

The interest in producing a model fermion gas is, in fact, not purely theoretical. One usually considers some forms of compact stars as practically non-interacting fermion systems. It is thus desirable to create some prototypes of fermion gases in the laboratory. Before considering atomic traps in more detail, let us briefly discuss the celestial case. One class of such stars close to fermion gases are White Dwarfs (WD) which constitute the final stage of evolution of low mass stars, like our sun. For these after the initial burning of hydrogen into helium, nucleosynthesis is bound to stop because of the small mass and the star starts to collapse. A stabilizing pressure against gravity is provided by the quantum pressure of the electron gas, namely the pressure due to the Fermi motion of a highly degenerate fermion gas. That Fermi pressure suffices to counterweight gravitation if the star mass is small enough (see Sect. 2.4). White dwarfs are mostly composed of hydrogen and a helium core with typical electron binding energies of a few eV. The star temperature is low, well below the eV range. Typical mass densities are of order $10^6 \mathrm{g\,cm}^{-3}$. This corresponds to a typical nucleon density of $\rho_{\mathrm{p,n}} \sim 0.5 \times 10^{30}\,\mathrm{cm}^{-3} = 0.5 \times 10^5\,\mathrm{a}_0^{-3}$ where $\mathrm{a}_0 \approx 0.53\,\AA$ is the Bohr radius (see Appendix B.1). Assuming equal fractions of proton and neutrons, the typical electron density is half of that, i.e., about $10^4\,\mathrm{a}_0^{-3}$. According to (2.19) this means that each electron on an average occupies a sphere with a radius (Wigner–Seitz radius) of about $r_s \sim 0.03\,\mathrm{a}_0$. This value is far below typical distances in matter under earth pressure ($r_s \sim 2 - 6\,\mathrm{a}_0$). In that extremely dense matter, the electrons are delocalized and form a Fermi gas with a Fermi energy of order $\varepsilon_{\mathrm{F}} \sim 50$ keV. The electron gas is thus practically at zero temperature, deep in what is called the degenerate regime where the Pauli principle determines the occupancies. The Coulomb energy also increases with density but more slowly than the kinetic energy. One can then safely consider that the electrons are non-interacting from a Coulomb perspective. Of course, the electron system is bound by the gravitational interaction together with all other constituents to form the star. The WD electrons thus constitute a degenerate fermion gas confined by an external potential.

Neutron stars are another sort of stars which are often discussed as an infinite phase of stabilizing fermions. They also constitute the latest stage of evolution of some stars, but this time much more massive ones. Again, the mother star has consumed all its nuclear fuel and runs into gravitational collapse. This latest phase of evolution leads to a neutralization of matter and the core of the finally exploding star gives birth to an extremely dense (about 10 times the density inside a terrestrial nucleus) star mostly composed of neutrons. With different energy scales, the sequence of arguments developed for the WD electrons can almost completely be transferred to the neutron star case, and the pressure stabilizing gravitational collapse is again mostly due to the fermionic nature of neutrons. This is thus again an almost infinite fermion gas confined by gravitational interaction. Nevertheless, there remains a major difference to WD: The neutron–neutron interaction has a huge repulsive core (as discussed in Sect. 1.1.2) which is never negligible. It is only the effective neutron–neutron interaction after eliminating the short-range correlations which becomes negligibly weak under neutron star conditions and which allows to consider neutron stars as a fermion gas similarly to WD.

Figure 1.11 illustrates the density and mass range for these two compact stars. There exists some variety in the predictions for the case of neutron stars [63] which, however, does not play a role at the huge scale of the present figure. The mass of neutron stars ranges up to 3–5 solar masses, while WD are limited to about 1.5 solar masses (Chandrasekhar mass). The density ranges are quite different with the WD still "dilute" enough to have electrons as major agents while neutron stars are dominated by nucleons. The figure also indicates the density of two other fermion gases, the electron gas in a metal (or metal cluster, respectively) which represents normal matter density under earth conditions and the typical density ranges in atomic traps which is extremely low.

Energy scales vary as much as the length scales shown in Fig. 1.11. Fermi energies are in the keV–MeV range for compact stars, in the eV range for the metallic electron gas, and shrink to the sub-Kelvin range for atomic traps ($1\,\mathrm{K} \sim 10^{-4}\,\mathrm{K}$). Constructing a degenerate gas of fermionic atoms hence requires very low temperatures. The physics of low temperatures has developed since the beginning of the twentieth century and received a substantial boost with the advent of traps. Bose–Einstein condensation of trapped atoms was attained in 1995. In that case bosonic atoms form a macroscopic state within a macroscopic fraction of the system, congregating in the lowest energy level of the system. Only a few years later, in 1999, the first (almost fully) degenerate gas of fermionic atoms in traps was experimentally observed. The production proceeds by confining a (non-degenerate) gas of fermionic atoms in a magnetic trap and progressively cooling it down (by laser cooling and finally evaporative cooling) in order to reach a temperature in the range of the Fermi

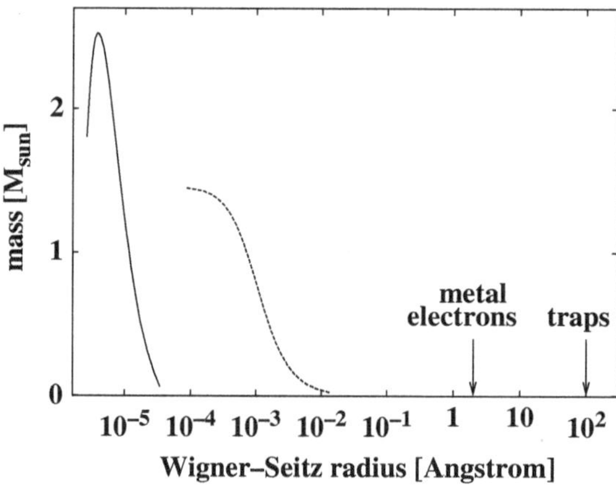

Fig. 1.11 Mass of two dense star types, neutron stars, and white dwarfs, versus inter-particle distance in terms of the Wigner–Seitz radius r_s. (The particle density is $\rho = 3/(4\pi r_s^3)$.) The Wigner–Seitz radius for two other typical Fermi gases, metal electrons, and atoms in a trap, is indicated for comparison

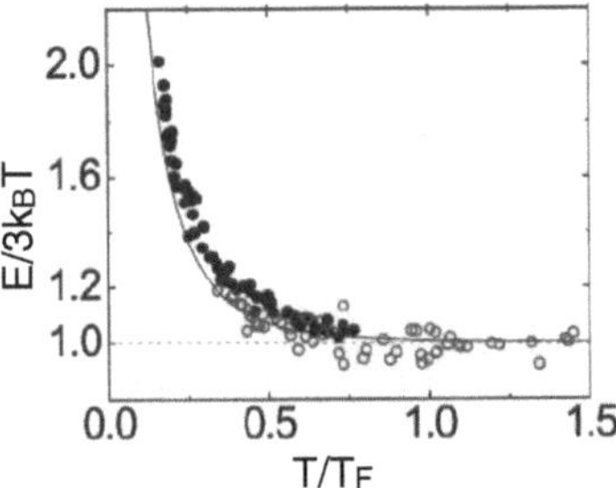

Fig. 1.12 Quantum degeneracy effects in trapped Fermi gases of ^{40}K atoms. The temperature is given in units of the Fermi temperature $T_F = \epsilon_F / k_B$. The average energy per particle, extracted from absorption images, is shown for two-spin mixtures (*filled* and *open circles*). In the quantum degenerate regime ($T < T_F$), the data agree well with the ideal Fermi-gas prediction (*solid line*). The *horizontal dashed line* corresponds to the result of a classical gas. Adapted from [65]

energy ϵ_F (Sect. 2.2). This is illustrated in Fig. 1.12, showing the average energy per atom as a function of the system temperature. The deviation from the classical estimate is obvious for sufficiently low temperatures, $T < T_F/2$, which corresponds to the expected transition regime where one reaches the degenerate fermion gas. The density of the system is tunable through the external confining potential and it is easy to reach a regime in which interactions are vanishingly small, all the more with overall neutral constituents.

Even more interesting is the fact that the interaction between constituents can be modified by an external magnetic field, which allows to tune the interaction from attractive to repulsive in a broad range of strengths. This delivers a laboratory for studying the influence of interactions in the formation of compound fermion systems. For zero interaction, we have the Fermi gas like in compact stars. Other regimes are equally interesting, e.g., the regime of pairing correlations (see Sect. 9.4).

As a final remark it should be noted that the tuning of the interaction in traps also allows to reach a quite particular regime for a fermion gas, the so-called unitary or resonant regime. By exploiting resonant scattering, one can design a situation where the scattering cross sections suddenly become very large, for then the details of the interaction do not matter anymore and its length scale characteristic can be taken as a "universal" infinite length. Such phenomena occur even in a very dilute system. This is then a very peculiar situation of a dilute (in which the range of the interatomic potential is small as compared to the interatomic distance) but strongly interacting (because the effective interaction length becomes much larger than the interatomic distance) system. All specific length scales associated to the interactions then disappear from the problem and the system is expected to exhibit a universal behavior. This unitary regime is also expected to occur in neutron stars which makes model studies in traps even more interesting. We shall discuss the unitary regime in the Fermi gas in Sect. 6.5.2.

1.2 Similar and Different

The presentations of Sect. 1.1 have unfolded the great diversity of fermion systems. There are, however, many profound similarities. It is the aim of this section to analyze these similarities and differences, as well as to put on a wider basis the theoretical approaches to be discussed in the following.

1.2.1 Overview of Many-Fermion Systems

Table 1.1 summarizes the fermion systems under consideration and introduced in Sect. 1.1. The column "Fermion" shows the quantum-mechanically active constituent. Note that some systems also have classically treated counterparts like the ionic cores in molecules or clusters. The column "size" provides typical constituent numbers and the last column indicates the possibility of a corresponding "bulk" material, i.e., the arbitrary scalability of a system. The ideal scalable systems are clusters which, indeed, have been produced at any size [59, 44]. That also includes droplets of ^{3}He which are special clusters, distinguished by the fact that the whole atoms are the active Fermions and that the bulk limit ends up in a Fermi liquid rather than in a solid. The bulk limit is somewhat ambiguous for molecules. They can grow huge, but in irregular manner. Everything which is composed of regularly repeated building blocks would belong to the category of clusters. Note that in practice clusters are often in a liquid state because they are produced at finite temperature. For then the bulk limit is also liquid-like. But remember that electrons, whatever the temperature at which the cluster has been produced (some hundreds of K at most), can always be considered to be at zero temperature because electronic energies lie in the eV range (1 eV $\sim 10^4$ K).

Table 1.1 Schematic overview over many-Fermion systems. Typical sizes N are indicated as well as constituents. The bulk limit is indicated in a key word and explained in the text

System	Fermion	Size(N)	Bulk
Self-bound fermion systems			
Nucleon	Quark	$N = 3$	Limited by color confinement
Nucleus	Nucleon	$N \lesssim 300$	Limited by Coulomb
Neutron star	Neutron	10^{57-58}	Practically bulk
White dwarf	Electron	10^{57-58}	Practically bulk
^{3}He droplet	^{3}He	$30 <\sim N$	Quantum liquid
Fermion systems bound by an external field			
Atom	Electron	$N \leq 112$	Limited by nuclei
Molecule	Electron		Irregular
Cluster	Electron	$3 \leq N \leq 10^{5-7}$	Corresponding solid
Artificially stabilized systems			
Quantum dots	Electron	Defined by construction	Limited by construction
Atoms in traps	Atom	Defined by construction	Limited by trap size

Nuclear matter is often considered as the bulk limit of nuclei. That, however, has to be taken with care. The steadily increasing Coulomb energy sets an upper limit of nuclear stability [18, 42]. Nuclear matter, although very useful for characterization, is a theoretical construction from which the Coulomb force has been removed to allow the bulk limit. Neutron matter exists at any size and is realized in neutron stars [39]. These are bound by gravitation which, being a long-range force, does not generate a universal equilibrium density (saturation). Atoms and nucleons exist only as finite systems. The atomic size is limited by the available nuclear charges and the nucleons are limited to three quarks by color confinement. Both artificially bound systems (Quantum dots, fermion traps) can, in principle, be extended to any size. The quantum dots tend to constant electron densities if the same carrier material is used for the different sizes while the trapped clouds can be squeezed and expanded deliberately by tuning the external fields.

In the following, we will not address the nucleon as an assembly of fermions (quarks), because that is touching the regime of particle physics and the relativistic domain. All other systems will show up here and there as examples (we shall take WD as examples for compact stars, leaving out the quite similar neutron stars). Most models are so generic that they can be applied for several of the above-listed cases.

1.2.2 Scales and Interactions

Table 1.2 gathers the typical length, momentum, and energy scales for the fermion systems of interest. The dominant interaction is also indicated for completeness. Scales show enormous variations, by 7 order of magnitude for length (r_s) and up to 20 for energy. One may thus wonder how it is possible to consider so different systems on a similar footing. Table 1.2 already contains the clue to resolve the apparent paradox. In most of these systems a characteristic length scale r_s exists, usually known as the Wigner–Seitz radius. It provides a scaling of the radius R of a finite system of N fermions as $R \sim r_s N^{1/3}$. This trend with nearly constant r_s is called saturation as it means that the system tends to have the same average density for all sizes. This is valid in particular for nuclei, metal clusters, and helium droplets. Interestingly these three systems scan the range of sizes/energies by providing extreme cases, small (fm) tightly bound (MeV) nuclei, and large (nm) loosely bound (sub meV) helium droplets, as well as a typical intermediate regime case with (0.1 to hundreds of nm) softly bound (eV) metal clusters representative of most electronic systems in materials. The Wigner–Seitz radius, furthermore, allows to introduce a typical momentum scale in terms of the Fermi momentum $\hbar k_{\mathrm{F}}$ which is the momentum of the least-bound fermion, and the Fermi energy ϵ_F which is the corresponding kinetic energy (see Sect. 2.2) and which thus sets a reference value for the energies in the system. From the Fermi momentum one can estimate the de Broglie wavelength in the ground state $\lambda_B \sim 2\pi/k_F \sim \pi r_s$, the large value confirming the essentially quantal nature of these systems. Indeed the de Broglie wavelength typically spreads over two interparticle distances which means that the typical wave function contains several tens of constituents within its quantal width.

Table 1.2 Orders of magnitudes for the gross properties of typical fermion systems, Wigner–Seitz radius r_s, Fermi momentum k_F, Fermi energy ε_F, and type of interaction. a_0 denotes the Bohr radius (see Appendix B.1). For systems which do not have a bulk limit, the overall radius R is given in place of r_s

System	Fermion	r_s	k_F	ε_F	Interaction
Self-bound fermion systems					
Neutron star	Neutron	1.1 fm	1.8 fm^{-1}	$\approx$ 60 MeV	Gravitation
White dwarf	Electron	0.01 a_0	2 10^2 a_0^{-1}	3 keV	Gravitation
Nucleus	Nucleons	1.2 fm	1.35 fm^{-1}	$\approx$35 MeV	Nuclear strong
^{3}He droplet	^{3}He atom	12 a_0	0.16 a_0^{-1}	2.7 K	Atom–atom
Fermion systems bound by an external field					
Atom	Electron	$R \approx 3$ a_0	0.4–0.8 a_0^{-1}	2–20 eV	Coulomb
Molecule	Electron	$R > 1$ a_0	0.4–0.8 a_0^{-1}	2–20 eV	Coulomb
Metal cluster	Electron	3–5 a_0	0.4–0.6 a_0^{-1}	2–5 eV	Coulomb
Artificially stabilized systems					
Quantum dot	Electron	$\approx$ 200 a_0	$\approx$ 0.01 a_0^{-1}	1.4 meV	Coulomb
Atomic trap	Atom	$\approx$ 200 a_0	$10^{-2} a_0^{-1}$	nK	Atom–atom

Most fermion systems under consideration here are described in first order as independent particles moving in a common mean field. At first glance this seems surprising, as almost all systems have interactions which are, in principle, singular at small distances. The Coulomb singularity is in practice moderate as it immediately allows a mean-field treatment (Hartree–Fock). Nonetheless, long-range correlations can grow large. They can effectively be included through density functional theory, see Sect. 6.1. Nucleons (in nuclei and neutron stars) and ^{3}He atoms have an interaction with a huge repulsive core which inhibits a direct mean-field treatment with the bare interaction. The strong short-range correlations have to be dealt with by elaborate many-body theories (Brückner–Hartree–Fock, hypernetted chain, etc. [90]). In the end, they can be eliminated effectively leaving a moderate in-medium interaction (see Sect. 1.1.2). As a consequence systems like nuclei, metal clusters, or helium droplets can be safely viewed as finite systems of moderately interacting fermions which evolve in a common average (mean-field) potential built from the piling-up of (effective) interactions between the constituents. This can be quantified in terms of the mean-free paths. In all cases, they turn out to be of the order of magnitude of the actual size of the system, so that one can adopt the view that the fermions evolve nearly independent from each other throughout the whole system, motivating a mean-field approach.

The simplest idea to understand how a finite system can be self-bound is to imagine that the constituent particles are bound inside a confining potential. Some cases are obvious, as for example atoms, in which electrons are bound in the field of the central nucleus if one neglects repulsive Coulomb effects between the electrons. In turn the case of nuclei is less obvious, as neutrons and protons bind together as a whole to form a nucleus without any "external agent". As we shall see below, though, the idea of a container is in fact quite physical, whatever the system, even if we shall have to discuss its origin in detail in some cases. We thus start the discussion

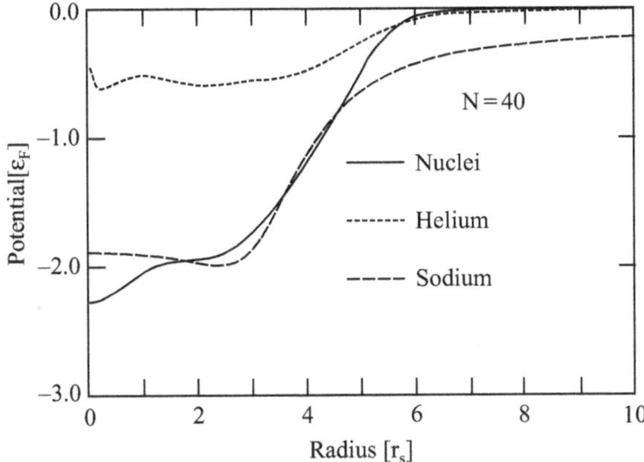

Fig. 1.13 The mean-field potentials for a Na cluster, a nucleus, and a helium droplet, for 40 particles (the results are only shown for the neutron part in the nuclear case). Natural units are used: lengths in units of r_s and potentials in units of ϵ_F. Adapted from [73]

on the basis of this simple picture of fermions inside a container potential, whatever its origin. The basic idea here is to average out the interactions to produce an average field. This is usually called a mean-field picture and it will constitute the foundation of numerous discussions in this book.

Rescaling distances by r_s and energies by ϵ_F, one can now compare such average potentials from apparently very different systems. A comparison is presented in Fig. 1.13 for the cluster Na_{40}, the nucleus ^{78}Sr (with 40 neutrons), and a helium droplet with 40 ^{3}He atoms. The results are plotted using "natural" units r_s and ϵ_F. The first point to be noted is that all three systems fit into one figure, i.e., have about the same scales when expressed in natural units. They moreover have the same spatial extent, which is directly connected to the "saturation scale" introduced by r_s. There are, of course, also some differences, especially in the depth of the potential wells and in the asymptotic behaviors. While the cluster and nucleus share a comparably deep potential, the helium droplet exhibits a much more shallow potential well reflecting the faintness of the interaction between two He atoms and the fact that $N = 40$ is at the lower bounds of binding for ^{3}He droplets. On the other hand, the helium droplet and the nucleus share the same exponential asymptotic behavior characteristic of a system dominated by a relatively short-range interaction (nuclear interaction for the nucleus, van der Waals for helium). On the contrary, the cluster exhibits a typical long-range Coulomb behavior. Still, up to these details, the comparison shows the overall similarity between the various systems.

1.2.3 Shells in Finite Systems

The occurrence of shells is a generic feature of finite fermion systems. Let us thus start with the general problem of a set of (effectively non-interacting) fermions in a potential. The potential sets definite boundary conditions to the wave functions so that only discrete energies are possible. This is the well-known quantization of energy levels in a potential. The sequence of energy levels (or spectrum) is in most cases non-trivial and requires elaborate numerical techniques to be worked out in detail. These energy levels often gather in bunches of degenerate energy values, called energy shells, the more so the higher the symmetry of the system. At very large energies or very large fermion numbers so many and so closely lying energy levels appear that one progressively switches from a discrete to a continuous picture. This is the path to band theory in bulk material involving a virtually infinite number of electrons and corresponds to continuum (unbound) states in finite systems. Apart from these extreme cases the energy spectra of finite systems thus usually exhibit pronounced shell structure. The details of the shell structure, especially the sequence of numbers corresponding to shell closures, is characteristic of the container potential and thus of the physical system under consideration. As mentioned above, symmetries here also play a key role.

Shell closure corresponds to the fact that a given number of particles completely fill all possible sub-levels in a shell of degenerate energies. This is associated to an energy gap to the next shell (the existence of such a gap actually defines the shells themselves). The appearance of an energy gap induces an extra stability at shell closure. The effect is well known in many systems. These are, for example, rare gases in atoms associated to electronic shell closures or magic nuclei associated to neutron or proton shell closures. But similar effects are also observed in metal clusters or quantum dots as this corresponds to a genuine quantum effect not specific to a particular system. Shell closures and associated shell effects will thus constitute a major issue in this book and will be discussed at many places both from a generic and specific point of view. We shall see in particular their role in the shape of the systems we shall study.

Shell effects are illustrated in Fig. 1.14 for four typical fermion systems discussed in this book: atoms, nuclei, metal clusters, and quantum dots. Except for nuclei, in which the fermions are neutrons and protons, the three other cases correspond to electronic shell closures. The associated "magic" numbers of course vary from one system to the next, due to different confining potentials. In the case of atoms (upper left panel of Fig. 1.14) shell closures can be directly read from the pronounced maxima of the ionization energies of neutral atoms for atomic numbers $Z = 2, 10, 18\ldots$ which correspond to noble gases He, Ne, Ar ... In this case the spectra are "structured" by the dominating spherical symmetry imposed by the strong (confining) Coulomb potential of the nucleus (which leads to large degeneracies). The case of nuclei is illustrated in the lower left panel of Fig. 1.14 in terms of nucleon separation energies (the analogue to ionization potentials in electronic systems). The steps in this observable reflect nuclear shell closures (2, 8, 20, 28 ...), in a way very similar to atoms. Remember that nuclei are self-bound so that the "confining" potential directly stems from the nucleon–nucleon interaction with no external factor. The upper right panel of Fig. 1.14 shows an example in simple (Na) metal clusters, this time considering as observable the abundance of species as a function of cluster size. Shell closure corresponds to a gain in energy/stability which directly translates into a larger abundance of the associated species/size. One indeed observes pronounced maxima for clusters with 2, 8, 20, 40, and 58 atoms.

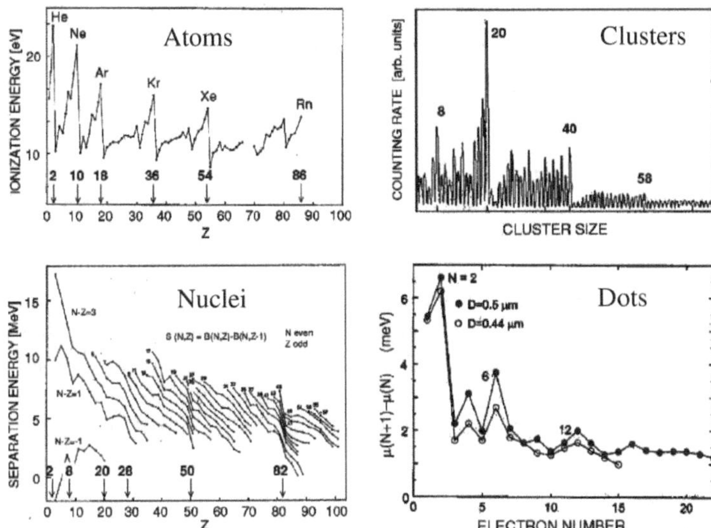

Fig. 1.14 Illustration of shell structure and magic numbers in finite fermion systems. *Upper left*, ionization potentials in atoms; *lower left*, separation energies in atomic nuclei (corresponding to ionization potentials); *upper right*, abundance spectra of metallic clusters (counting rate in arbitrary units, generic example of Na); *lower right*, differences in the chemical potential (corresponding to ionization potentials) of disk-shaped quantum dots. Adapted from [87]

The cluster case is mixed what concerns the confining potential, which "equally" reflects the (attractive) ionic component and the electronic contributions. The last case of quantum dots (lower right panel), in turn, is almost exclusively associated to a true external potential which takes a typical harmonic oscillator shape with the associated shell closure sequence 2, 6, 12..., appropriate for a 2D oscillator (remember the strongly deformed (disk-like) shape of the dot).

1.2.4 Beyond the Static Mean Field

As discussed in the previous Sect. 1.2.3 shell effects constitute a major issue in the physics of finite fermion systems. We will complement here the general analysis of the previous section by considering a few more observables in which again the quantization of fermion levels plays a major role, but which require more elaborate theoretical treatments beyond a mere static mean-field picture. We take as examples correlation effects in the electron emission from irradiated atoms, the optical response in complex molecules, and the level density parameter in nuclei.

1.2.4.1 Electron Correlations in Irradiated Helium

The helium atom constitutes a prototypical many-body problem with two densely packed, thus correlated electrons. It is the first rare gas, having exceptional stability and correlatively very faint interactions with its environment. It is thus an example of choice to perform studies of principle. The (very) small number of electrons is just large enough to account for possible correlation effects and still small enough to allow highly sophisticated calculations, even in the time-dependent domain. We consider here as illustration the case of an irradiation of an helium atom by a single photon. The photon energy is chosen large enough to allow for double ionization (i.e., emission of both electrons). Figure 1.15 shows the measured momentum distribution of one of the emitted electrons at fixed direction of emission of the other electron. Neglecting the (small) photon momentum (incoming photon provoking the double ionization), the vector momenta of the ion and both electrons have to be in one plane, for obvious kinematic reasons. The momentum distribution of Fig. 1.15 is plotted in this plane. The data are integrated over all orientations of the polarization axis with respect to this plane, and the x-axis has been chosen to be the direction of one of the electrons. The momentum distribution exhibits an interesting pattern reflecting deep correlation effects between the two emitted electrons. The first remarkable feature is the fact that the electron–electron repulsion leads to almost no intensity for both electrons in the same half plane (namely toward the positive x-axis). The second aspect reveals details of the wave functions. Indeed one expects that the two emitted electrons, which are in the continuum, have to be coupled in a very specific way which implies the appearance of a node in the square of the wave function at the point at which electron momenta are opposite. The effect can be directly spotted from the figure ("node" circle in the figure) . The latter aspect, in particular, proves the strong correlation effects observed in this electronic emission following a single-photon excitation.

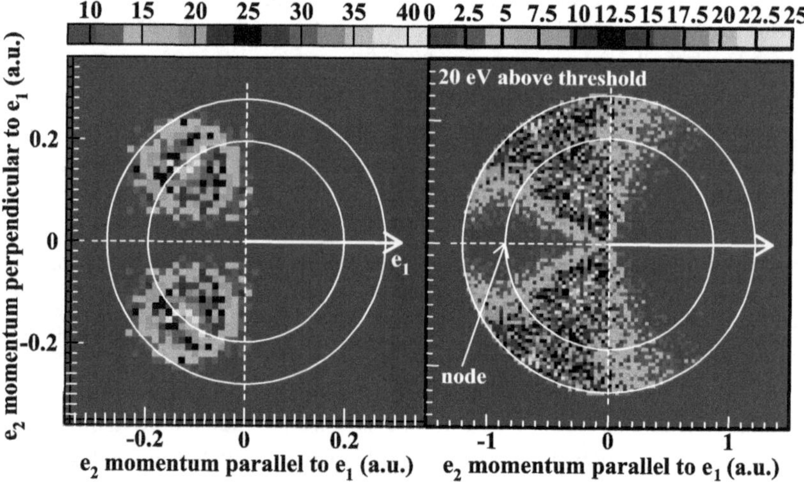

Fig. 1.15 Cross sections of the double ionization of He at 1 and 20 eV above threshold following one photon ionization by linearly polarized light [28]. The momentum distribution of electron 2 for fixed direction of electron 1 is plotted in the plane of the two electrons + ion. Data are integrated over all orientations of the polarization axis with respect to this plane. The outer circle corresponds to the maximum possible electron momentum, while the inner one to the case of equal energy sharing. From ("R. Dörner, 2009, Private communication")

1.2.4.2 Optical Response in Guanine

The optical response, i.e., the distribution of photo-absorption strength over photon energies, is a key tool for the investigation of structure and dynamics of electronic systems (atom/molecule/cluster). The observable is called optical response, as it often concerns frequencies in the visible part of the electromagnetic spectrum, i.e., when irradiated by light the system appears with certain colors. The optical response brings decisive information on the structure of the studied species. In metal clusters, for example, the dominant absorption stems from the Mie surface plasmon (Sect. 1.1.5), i.e., which corresponds to a collective oscillation of the electronic valence cloud against the ionic background, and consequently exhibits interesting systematic trends with size. In complex molecules it shows the various (individual) electronic contributions in a more subtle way. In all cases it can be viewed as the summed up response of individual electronic excitations (see Chap. 8 for details).

Figure 1.16 shows an example of optical response in the case of guanine, which is one of the main nucleobases found in the nucleic acids DNA and RNA. The figure exhibits an experimental spectrum which is compared to a theoretical calculation (with methods as explained in Chap. 8). The spectrum exhibits a series of peaks which result from the combination of electronic excitations (promotion of electrons from occupied to unoccupied levels). Two particular such levels are also shown in the figure, the HOMO (highest occupied molecular orbital) and the LUMO (lowest unoccupied molecular orbital) in terms of the corresponding electronic density (see also Fig. 1.19). Keep in mind that the resulting optical response is not a mere superposition of such elementary excitations. As, for example, in the case of an assembly of coupled strings, the system exhibits eigenfrequencies, indeed built out of the elementary excitations, but non-trivially re-coupled through interactions. The optical response in a complex molecule such as guanine is thus an involved landscape of strongly fragmented peaks which carry valuable information about the underlying molecular structure whose proper analysis is, however, a non-trivial task.

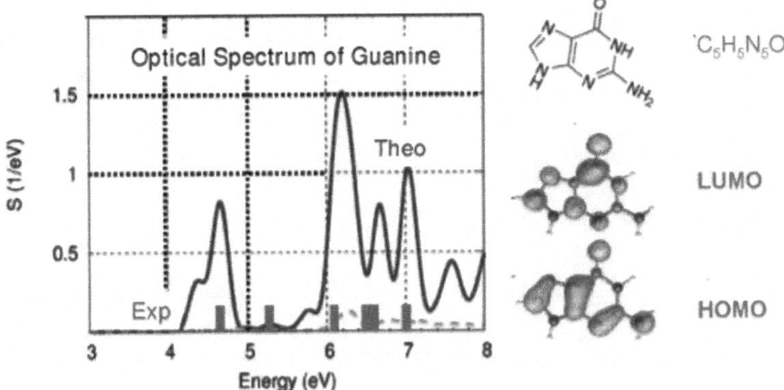

Fig. 1.16 Optical response of guanine ($C_5H_5N_5O$). Experimental results (*bars*) are compared to theoretical estimates obtained from linearized time-dependent DFT (Chap. 8). For completeness maps of electronic densities are plotted as inserts for the HOMO (highest occupied molecular orbital) and LUMO (lowest unoccupied molecular orbital) levels of guanine. Adapted from [107]

1.2.4.3 Level Density Parameter in Nuclei

The level density parameter provides a direct measure of the nature of least-bound (occupied and unoccupied) levels of the nucleon spectrum in nuclei. Given a quantum system one can characterize its spectrum by introducing the level density $\mathcal{D}(\epsilon)$ which is defined as the number of single-particle levels per unit energy $\mathcal{D}(\epsilon) = dn_{\text{level}}/d\epsilon$. For a discrete spectrum $\mathcal{D}$ is a sequence of individual peaks. For a continuum spectrum $\mathcal{D}$ becomes a smooth function. When levels are close to each other, like inside a shell, $\mathcal{D}$ becomes large, while in between two levels $\mathcal{D}$ vanishes. The level density can be explicitly measured in a variety of physical systems, for example, in solids. In nuclei one further focuses on the least-bound levels, close to the Fermi energy ϵ_F (Sect. 2.2). The level density parameter $a = \pi^2 \mathcal{D}(\epsilon_F)/6$ then provides a direct measure of the degree of occupation of levels near the ionization threshold (see also the precise definition of the associated Fermi energy ϵ_F in Sect. 2.2). The level density parameter has been measured in a systematic way in nuclei and is presented for illustration in Fig. 1.17. It exhibits a remarkable systematic trend, taking on the average a value well represented by the simple formula $a \simeq A/8$, where A is the total nucleon number, on which are superimposed some structures, especially strong minima. The sequence of minima is easy to identify and interpret, as it is simply the set of magic numbers. The effect is easy to understand. As already discussed above, nucleon shell closures are associated to a gain in stability which is reflected in particular in a larger separation between the last occupied and the first unoccupied level. At shell closure the level density is thus necessarily particularly small, whence the corresponding small values of a. The effect is especially large for doubly magic nuclei such as ^{208}Pb in which both neutron and proton numbers are magic.

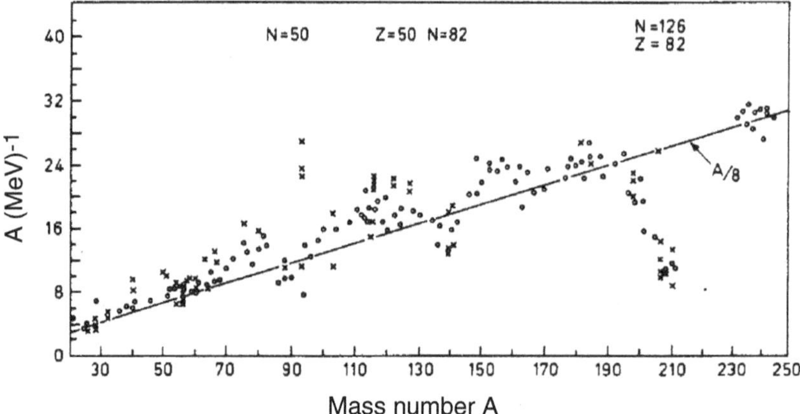

Fig. 1.17 Systematics of the level density parameter a in nuclei, as a function of nucleon number A. The smooth $A/8$ trend is plotted as a *solid line* to guide the eye. The experimental values exhibit fluctuations around the smooth trend and are attributed to nucleonic shell closures. The case of doubly magic ^{208}Pb is exemplary in this respect. Adapted from [17]

Surprisingly enough the general trend $a \simeq A/8$ is much more difficult to recover quantitatively by simple arguments. It is a well-known problem in nuclear theory and requires elaborate many-body techniques. Simple estimates can be attained, for example, with the help of the Fermi gas (Sect. 2) or with mean-field techniques (Chap. 5). But such approaches miss the value typically by a factor of 2 (Fermi gas) or by 50% (mean-field approaches). The level density parameter is thus by no means a trivial quantity to estimate and has led to numerous elaborate theoretical investigations. It is also important to note here that a plays a crucial role for analyzing many experimental nuclear data. Indeed when nuclei are excited, for example, in the course of a nuclear collision, the excitation energy is (as in any other systems) primarily stored in least-bound levels, especially for small to moderate perturbations. It is thus a key issue to know in detail how many levels are accessible to the system, and one can easily show that the statistical de-excitation of the excited system strongly depends on a, whence the key importance of evaluating it properly. This point is further investigated in Sect. 2.3.

1.3 Basic Theoretical Tools

After having introduced the various fermion systems of interest, both in specific and is generic ways, we now want to introduce some basic theoretical tools which will be used all over this book. A key question in quantum mechanics, and for many-fermion systems in particular, is the choice of a proper representation for the system under study. The standard approach in elementary quantum mechanics is based on a wave function representation of the quantum state of a particle. Many-fermion systems add the complication that their wave function is antisymmetric with respect to the exchange of any two particles, a condition which renders the actual writing of wave functions, even in the simplest cases somewhat cumbersome. Thus alternative approaches were developed, from which we will use two techniques both based on operators. The fermion operator formalism relies on a description in terms of creation and annihilation operators, similar to the well-known formalism used to describe the elementary harmonic oscillator. The other approach is based on density operators, which can be defined from wave functions but can also be used on their own. The latter are quite helpful, especially for formal manipulations, as they provide a compact representation of the system. They also allow a clean formulation of the semi-classical approximation which can be very useful especially in the case of large systems at high excitations. We briefly outline the basic content of these three approaches in the forthcoming sections and take the opportunity to define notations for the following chapters.

1.3.1 The Wave Function Approach

Starting from the usual Schrödinger picture, the a priori natural way to describe particles is by means of wave functions which will constitute the building blocks

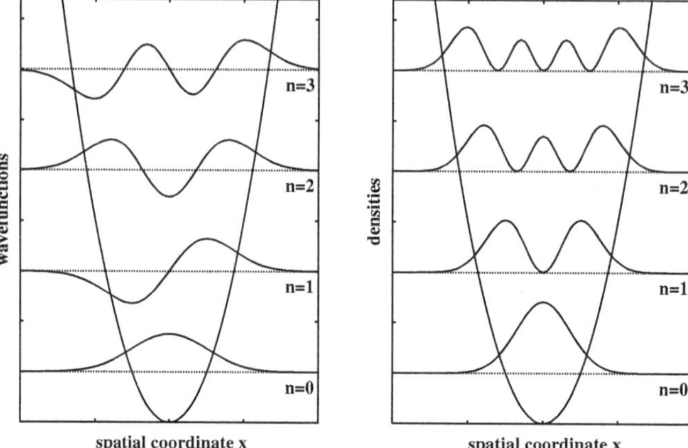

Fig. 1.18 Lowest four eigenfunctions (*left panels*) of the 1D harmonic oscillator and the corresponding density distributions

of the many-body state. For sake of simplicity, let us first start with a well-defined problem, namely a single-particle Hamiltonian. It exhibits a series of eigenstates of increasing eigenenergies. The point is illustrated in Fig. 1.18 in the case of a 1D harmonic oscillator (see also Chap. 3). In this figure we show both the eigenfunctions and the corresponding densities (square moduli of the wave functions). As long as only one particle is involved, a stationary state of the system is fully defined from one of these eigenstates and the nature of the particle (boson, fermion) is irrelevant, but the difficulty shows up as soon as we consider two or more identical particles.

Let us start with the simplest case of two independent particles of wave functions $\varphi_{\alpha_i}(i)\{i = 1, 2\}$ where α_i denotes a certain state in the single-particle spectrum and the "i" in the brackets summarizes space and spin dependencies. This already exhibits the crucial difficulties to be overcome. Furthermore assume that these two particles interact only with an external field. In terms of the Schrödinger equation this amounts to say that the total Hamiltonian of the system is just the sum of the separate Hamiltonians for each particle, each piece acting on its own particle only. Then the total energy of the system is just the sum of the two single-particle energies ε_{α_i}. The simplest choice for the two-body wave function is the product of the two single-particle wave functions $\Phi_{\alpha_1\alpha_2}(1, 2) = \varphi_{\alpha_1}(1)\varphi_{\alpha_2}(2)$. This choice would be correct for distinguishable particles, such as an electron plus a proton. It is not valid for indistinguishable fermions because the total wave function $\Phi(1, 2)$ has to be antisymmetric with respect to the exchange of particles 1 and 2. The simplest solution is to explicitly antisymmetrize by writing

$$\Phi(1, 2) = \frac{1}{\sqrt{2}}[\varphi_1(1)\varphi_2(2) - \varphi_1(2)\varphi_2(1)] = \frac{1}{\sqrt{2}}\begin{vmatrix}\varphi_1(1)\varphi_2(1)\\ \varphi_1(2)\varphi_2(2)\end{vmatrix} \tag{1.2}$$

which is a 2×2 determinant, up to a normalization factor $1/\sqrt{2}$. This wave function is called a Slater determinant and explicitly fulfills antisymmetrization requirements for fermions. It can be generalized to any number of fermions by using the determinantal form as

$$\Phi_{\alpha_1\ldots\alpha_N}(1,2,\ldots,N) = \frac{1}{\sqrt{N!}} \begin{vmatrix} \varphi_{\alpha_1}(1) & \varphi_{\alpha_2}(1) & \ldots & \varphi_{\alpha_N}(1) \\ \varphi_{\alpha_1}(2) & \varphi_{\alpha_2}(2) & \ldots & \varphi_{\alpha_N}(2) \\ & & \ldots & \\ \varphi_{\alpha_1}(N) & \varphi_{\alpha_2}(N) & \ldots & \varphi_{\alpha_N}(N) \end{vmatrix}$$
$$= \frac{1}{\sqrt{N!}} \mathcal{A}[\prod_{1\ldots N} \varphi_{\alpha_i}(i)] \tag{1.3}$$

where an even more compact formulation was introduced through the N-body antisymmetrizing operator $\mathcal{A}$ applied to the product of single-particle wave functions. The ground state of the system is the Slater determinant where the N lowest levels are filled successively, i.e., where $\alpha_i = i$. Slater determinants will play a key role in the following discussions. Still the picture is valid only for strictly independent particles, namely when neglecting interactions between particles. As we have seen in Sect. 1.1, most fermion systems can be described, at least to lowest order, by assuming that fermions do feel a common average potential in which they move more or less independently from each other. It thus a priori makes sense to consider Slater determinants as a relevant ansatz for the many-body wave functions.

In realistic situations interactions may have to be taken into account. To go beyond the simple Slater-determinant picture one can envision two general strategies. One is to refine the wave function itself, for example, by considering excited states built on a given Slater state and linear combinations thereof, i.e., dealing with a sum of Slater determinants $\Psi = \sum_{\alpha_1\ldots\alpha_N} \Phi_{\alpha_1\ldots\alpha_N} c_{\alpha_1\ldots\alpha_N}$ where the $c_{\alpha_1\ldots\alpha_N}$ are expansion coefficients. This is a typical strategy followed in quantum chemistry. Another strategy consists in implementing the correlations into the Hamiltonian (which thus becomes "effective") to preserve the simple nature of the Slater state. This is achieved by means of density functional theory (DFT) (Sect. 6.1) and widely used in electronic systems, to some extent also in nuclear physics. Both strategies have their virtues and defects and both have led to remarkable successes. The "effective Hamiltonian" approach will be used in several places in this book as it provides the simpler treatment. The treatment of correlations will be addressed briefly in Chap. 9. In any case, expressing quantities by means of explicit wave functions may soon become cumbersome when the number of particles grows too large. The technique of fermion operators which we discuss in the next section provides a substantial simplification of the handling, at least for most physically relevant operators.

1.3.2 Creation and Annihilation Operators

The fermion operator technique was developed in the late 1920s to provide a formal basis for the quantization of fields. It allows to treat in a simple way the (in fact quite generic) case of a particle inside an assembly whose state changes due either to an external perturbation or to an interaction with a neighboring particle. An intuitive insight into the fermion operator technique can be gained by remembering the algebraic treatment of the harmonic oscillator in terms of creation and annihilation operators [24].

The construction of fermion creation and annihilation operators starts from book-keeping Slater states in an occupation number representation. We start from a given complete basis set of single-particle states $\{|\alpha_i\rangle\}$ from which one will build the Slater states. In the previous section, we have specified a full state by the N occupied single-particle states $\alpha_1 \ldots \alpha_N$. It could equally well be characterized by attributing occupation $n_i = 0$ or $n_i = 1$ to every one of the states $|\alpha_i\rangle$. The fermion creator $\hat{a}^\dagger_\alpha$ is then defined as that operator which increases the occupancy n_α by one, and the annihilator $\hat{a}_\alpha$ correspondingly reduces $n_\alpha \longrightarrow n_\alpha - 1$. In order to preserve the fermionic nature of the particles these creation (and associated annihilation) operators have to fulfill the anticommutation relations

$$\{\hat{a}_\alpha, \hat{a}^\dagger_\beta\} = \hat{a}_\alpha \hat{a}^\dagger_\beta + \hat{a}_\beta \hat{a}^\dagger_\alpha = \delta_{\alpha,\beta} \quad , \quad \{\hat{a}_\alpha, \hat{a}_\beta\} = \{\hat{a}^\dagger_\alpha, \hat{a}^\dagger_\beta\} = 0 \quad . \tag{1.4}$$

These relations state that as long as α and β are different one can create or annihilate particles in the corresponding states independently from each other. When $\alpha = \beta$ the anti-commutators express the fermion statistics, e.g., they imply that $\hat{a}^\dagger_\alpha \hat{a}^\dagger_\alpha = 0$ which means that one cannot create two particles in the same α state (Pauli principle).

The fermion operators change the number of particles in the state (one up or one down). Observables are usually related to operators which conserve the number of particles. The definition of the "particle-number" operator illustrates this. Let us first consider the operator $\hat{N}_\alpha = \hat{a}^\dagger_\alpha \hat{a}_\alpha$. When applied to a state with $n_\alpha = 0$, it will give 0 because applying $\hat{a}_\alpha$ tries to annihilate a particle in state α which is not present. On the other hand, if $n_\alpha = 1$ the $\hat{a}_\alpha$ will "empty" this state and the subsequent action of $\hat{a}^\dagger_\alpha$ repopulates it. The net result of the action of $\hat{N}_\alpha$ is thus, not surprisingly, 1 for $n_\alpha = 1$ and 0 otherwise. Summing up the $\hat{N}_\alpha$ defines the particle-number operator $\hat{N} = \sum_\alpha \hat{a}^\dagger_\alpha \hat{a}_\alpha$ which exactly counts the number of occupied states in the system and thus the number of particles N.

A similar $\hat{a}^\dagger_\alpha \hat{a}_\beta$ sequence also naturally enters the expression of any one-body operator. Indeed one can easily show that a one-body operator $\hat{K}$, like the kinetic energy operator, can be expanded as

$$\hat{K} = \sum_{\alpha,\beta} K_{\alpha\beta} \hat{a}^\dagger_\alpha \hat{a}_\beta \quad , \quad K_{\alpha\beta} = \langle\alpha|\hat{K}|\beta\rangle \quad . \tag{1.5}$$

Two-body operators, like typical interaction potentials, involve two states and thus are expanded in terms of pairs of fermion annihilators and creators, $\hat{a}^\dagger_\alpha \hat{a}^\dagger_\beta \hat{a}_\gamma \hat{a}_\delta$. The fermion operator formalism is a quite useful tool in the theoretical description of many-fermion systems. We shall use it in many places throughout this book. More details and derivations of some of properties of the operators are given in Appendix A.4.

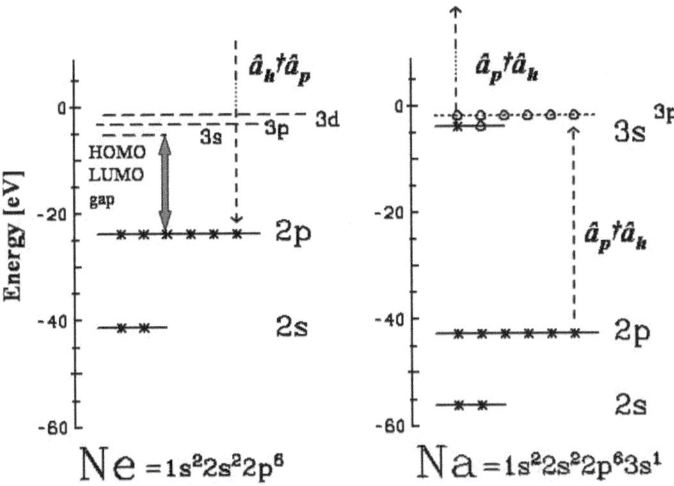

Fig. 1.19 Level schemes for the Ne and Na atoms. *Open circles* and/or *dashed lines* indicate empty states, *crosses* occupied states. The same energy scale is used for all the two atoms for better comparison. The spectroscopic notation is given for each atom. Also are indicated typical particle–hole transitions, also expressed through creation and annihilation operators and the HOMO–LUMO gap in Ne. In Na there exists no HOMO–LUMO gap as the 3s level is only half filled

The notion of creation and annihilation operators can be simply visualized in practical examples of many-body systems. Examples are shown in Fig. 1.19 in the case of two simple atoms, Na and Ar. The single-particle levels of both atoms are indicated on the figure (for the notation see Sect. 1.1.3). The electronic structure of the alkaline Na is $1s^2 2s^2 2p^6 3s$ and that of the rare gas Ne is $1s^2 2s^2 2p^6$. These two atoms also provide two nice examples of very different HOMO–LUMO gaps. The HOMO–LUMO gap is the energy difference between the highest occupied molecular orbital (HOMO) and the lowest unoccupied molecular orbital (LUMO). In a rare gas shell closure leads to a very large gap, while in the alkaline case it simply vanishes because the HOMO is not fully occupied (one more electron can still occupy the $3s$ level in Na). One usually calls *hole* (h) states as the ones initially occupied (to become unoccupied after action of the annihilation operator a_h) and which thus lie below the HOMO in the ground state. *Particle* (p) states are the unoccupied states (to become occupied by a particle after action of the creation operator $a_p^{\dagger}$) which lie above the LUMO. An excitation is produced by transitions from hole (occupied) to particle (unoccupied) levels. That is called a one-particle–one-hole $1ph$ excitation. It is described by the operator $\hat{a}_p^{\dagger}\hat{a}_h$ where p stands for the particle state and h for the hole. Any excitation of the system can then be formulated on the basis of such $1ph$ excitations. We shall often come back to these terms in the following. Typical $1ph$ excitations are illustrated in Fig. 1.19 in the simple case of Na and Ar and are indicated by up-oriented arrows. Conversely, de-excitation mechanisms can be formulated in terms of hole to particle transitions $\hat{a}_h^{\dagger}\hat{a}_p$ depopulating a particle level p (populated by a former excitation) to populate a hole level h (depopulated by a former excitation). These transitions are indicated graphically by down-oriented arrows.

1.3.3 Density Matrices

Density matrices provide a third, alternative, description of a system of identical fermions. The one-body density matrix can be defined, starting from a general (not necessarily Slater determinant) N-body wave function $\Psi(1,\ldots,N) = \Psi(x_1,\ldots,x_N)$, as

$$\varrho(x,x') = N\int \mathrm{d}x_2 \ldots \int \mathrm{d}x_N \Psi(x,x_2,\ldots,x_N)\Psi^*(x',x_2,\ldots,x_N), \tag{1.6}$$

which represents an average over all degrees of freedom but one. Note that the notation $x_i \equiv (\mathbf{r}_i,\sigma_i)$ is to be understood as covering all degrees of freedom including spin. It should also be stressed that $\hat{\varrho}$ is an operator, and $\varrho(x,x')$ corresponds to its coordinate-space representation. This operator is Hermitian as can be seen by exchanging x and x'. The diagonal elements of the density matrix represent the local density of matter at a given point and the integral over them yields the total number of particles in the system ($\mathrm{Tr}\hat{\varrho} = N$). Where "Tr{...}" stands for the trace, the sum of diagonal elements of the operator. It reads, e.g., in coordinate space $\mathrm{Tr}\{\hat{\varrho}\} = \int \mathrm{d}^3r\, \hat{\varrho}(\mathbf{r},\mathbf{r})$.

If the many-body wave function $\Psi(x_1,x_2,\ldots,x_N)$ is chosen to be a Slater state (which is a frequent practical choice) built out of the single-particle wave functions $\varphi_i(\mathbf{r})$, the density matrix can be expressed in the simpler form

$$\varrho(\mathbf{r},\mathbf{r}') = \sum_{i=1}^{N} \varphi_i(\mathbf{r})\varphi_i^*(\mathbf{r}') \quad . \tag{1.7}$$

It is easy to show that $\hat{\varrho}$ in this case becomes a projector: $\hat{\varrho}^2 = \hat{\varrho}$, and then the density operator has eigenvalues 0 or 1, corresponding to the occupation numbers n_α. The expectation value of one-body operators also takes an especially simple form, namely

$$\langle\Psi|\hat{K}|\Psi\rangle = \mathrm{Tr}\{\hat{\varrho}\hat{K}\}, \tag{1.8}$$

One can also consider more general definitions in which the integration is performed on all but two (or three, or ...) particles thus leading to two-body, three-body, ... density matrices. The two-body density matrix is thus defined as

$$\varrho(x_1,x_2;x_1'x_2') = \frac{N(N-1)}{2}\int \mathrm{d}x_3 \ldots \int \mathrm{d}x_N \\ \Psi(x_1,x_2,x_3,\ldots,x_N)\Psi^*(x_1',x_2',x_3,\ldots,x_N), \tag{1.9}$$

which is antisymmetric with respect to the exchange $x_1 \leftrightarrow x_2$ and $x'_1 \leftrightarrow x'_2$ and hermitian. The one-body density matrix can be obtained from the two-body density matrix by integration

$$\int \mathrm{d}x_2 \varrho(x_1, x_2; x_1' x_2) = \frac{N-1}{2} \varrho(x_1, x_1') \quad . \tag{1.10}$$

Note that we are using the same symbol for one- and two-body matrices. The number of arguments serves to distinguish.

In the following, as we shall see, the one-body matrix will play a major role. We shall thus often omit the term "one-body," especially when dealing with the diagonal part of the local one-body density. The local one-body density distribution is especially interesting as it can be measured experimentally in certain systems. The nuclear charge density is for example measured by electron scattering. Analyzing the differential cross section for electron scattering, one can deduce the charge form factor $F(\mathbf{q})$ which is simply the Fourier transform of the charge density with respect to real space

$$F(\mathbf{q}) = \int \mathrm{d}^3 r \rho_c(\mathbf{r}) \exp(\mathrm{i}\mathbf{q} \cdot \mathbf{r}) \quad . \tag{1.11}$$

The local charge density distribution $\rho_c(\mathbf{r})$ can then be recovered by Fourier back transformation compared to theoretical predictions. A simple example is presented in Fig. 1.20 in the case of ^{16}O (other nuclei were shown in Fig. 1.3). In the case of the light nucleus ^{16}O the wave functions can be well represented by harmonic oscillator wave functions (assuming that the nuclear potential for that small system is very close to a harmonic oscillator, see Chap. 3). With its doubly magic structure (8 neutrons and 8 protons) ^{16}O is furthermore spherical and the problem can thus be reduced to a spherical harmonic oscillator. Only the first two levels (shells) are occupied having angular momentum 0 (2 protons with spin up and down) and 1 (6 protons for the three $m = 0, \pm 1$ degenerate sublevels). The proton density is then immediately obtained as

$$\rho_p(\mathbf{r}) = \frac{1}{4\pi}[2R_{00}^2 + 6R_{01}^2], \tag{1.12}$$

where the R_{nl} are radial wave functions for the spherical oscillator, see Appendix A.1.2. This leads to a very simple form of the density

$$\rho_p(\mathbf{r}) = \frac{2\alpha^3}{\pi^{3/2}}[1 + 2\alpha^2 r^2]\exp(-\alpha^2 r^2), \tag{1.13}$$

with the width parameter $\alpha = (m\omega/\hbar)^{1/2}$ (where m is the proton mass). By Fourier transform, which because of spherical symmetry takes a simple form, one immediately obtains the form factor as

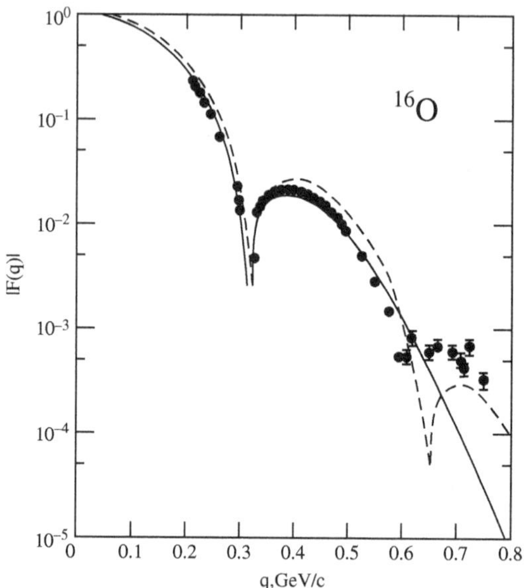

Fig. 1.20 The absolute value of the electromagnetic form factor of ^{16}O as a function of momentum transfer q. A comparison is done between experimental data points and a theoretical estimate in terms of the harmonic oscillator potential (see text for detail)

$$
\begin{aligned}
F(\mathbf{q}) &= \int \mathrm{d}^3 r \rho_c(\mathbf{r}) \exp(\mathrm{i}\mathbf{q}\cdot\mathbf{r}) = 4\pi \int_0^\infty \mathrm{d}r \rho_p(\mathbf{r}) \frac{\sin qr}{qr} r^2 \\
&= [8 - \frac{q^2}{\alpha^2}] \exp(-q^2/(4\alpha^2)),
\end{aligned}
\tag{1.14}
$$

whose absolute values are compared to the experimental ones in Fig. 1.20, assuming that the proton density ρ_p can be identified with the charge density ρ_c. Electron scattering in nuclei thus provides an almost direct access to nuclear density. Note that this technique has also been extensively used to deduce nuclear radii as presented in Fig. 1.2.

1.4 Concluding Remarks

As stated at the beginning of this chapter, fermions and fermion assemblies are everywhere around us. They constitute very different systems from the smallest ones like nuclei to the largest ones like stars. This means that the many-fermion problems covers an impressive range of distance and energy scales, not to mention time. Most of the four elementary interactions play a key role in these systems, even aside from the possibilities offered by artificial devices such as traps to tune the strength of the interaction itself. Still, in spite of such a large span of interacting systems we have seen that they share common trends. It is possible to define universal scales (and sometimes even scalings) which allow to consider on the same footing systems involving natural scales differing by several (even sometimes several tens of) orders of magnitude. This is the major justification for trying to understand and describe many-fermion systems by means of common universal tools and this is precisely the goal of this book. The major fermion constituents we shall consider in the following are nucleons and electrons with some excursions to atoms. The dominant interactions at work will then be the nuclear strong and Coulomb interactions, again with excursions, for example, to the gravitational interaction. As already mentioned our central theme will be the progressive account of interactions in the description of the system. We shall thus start with non-interacting particles to terminate with strongly interacting ones.

Chapter 2
The Fermi-Gas Model

The simplest quantum-mechanical systems are free particles. There is no spatial variation in the potential and the relevant Hamiltonian is given merely by the kinetic energy operator. Eigenstates are plane waves labeled by their eigenmomentum $\mathbf{k}$ with eigenenergies $\varepsilon_{\mathbf{k}} = \hbar^2\mathbf{k}^2/(2m)$. An interaction-free, continuous many-fermion system is called a Fermi gas. All plane-wave states are occupied up to a certain Fermi energy ε_F, or corresponding Fermi momentum k_{F}. The density of each plane wave is constant and so is the total density. This, in turn, means that interacting bulk matter when treated at the level of a mean-field approximation (Hartree–Fock or density functional approach, see Chaps. 5 and 6) still has plane waves as eigenstates for the single particle wavefunctions. This allows to use the Fermi gas as an instructive and powerful lowest-order approximation to a great variety of many-fermion systems. The most prominent example for a Fermi gas comes from bulk metals. An example from a quite different regime is given by compact stars (neutron star, white dwarf, see Sect. 1.1.8). A dilute gas is realized for large clouds of fermions in traps (see Sect. 1.1.8).

Bound finite fermion systems differ in that the occupied states reside in the regime of discrete spectra. Nonetheless, the Fermi gas is also a useful starting point for examining volume properties and orders of magnitude. For sufficiently large particle numbers, the spectrum becomes dense and can be approximated by a continuum (as will be discussed in Sect. 2.1). The complementing view comes from the spatial density distribution $\rho(\mathbf{r})$. A homogeneous gas has, by definition, a homogeneous density, ρ = constant. As an example, Fig. 1.3 shows nuclear charge densities for a broad range of nuclei. All four cases indicate a region of nearly constant density in the interior growing with increasing system size. This suggests that from the density point of view a Fermi-gas limit can provide a reasonable lowest-order approximation also for finite fermion systems (see also the examples in Sects. 2.1.2.2, 2.1.2.3, and 2.4).

This reasoning is applicable at least to all systems which have a nearly constant density in the interior. It holds for saturating systems, i.e., systems where the radius grows as $R \propto N^{1/3}$ (see e.g. Fig. 1.2) which applies to metallic nano-particles (metal clusters), atomic nuclei, or droplets of liquid ^{3}He. There are other systems whose radius does not behave $\propto N^{1/3}$ but which, nonetheless, have constant density like compact stars (their radius results from an interplay with gravitation, see Sect. 2.4),

J.A. Maruhn et al., *Simple Models of Many-Fermion Systems*,
DOI 10.1007/978-3-642-03839-6_2, © Springer-Verlag Berlin Heidelberg 2010

electrons in quantum dots (the radius is a design property), and fermions in a trap (the radius is tunable). All these systems can be well approximated by a Fermi gas. For a quick quantitative overview, let us recall Table 1.2 which summarizes key parameters for a variety of fermion systems, Wigner–Seitz radius r_s, Fermi momentum k_F, and Fermi energy ε_F. The table demonstrates the wide variety of physical dimensions for the systems of interest.

The Fermi gas also plays a crucial role in density functional theory. Many practical density functionals are derived by carefully computing the properties of the electron gas and transferring that, so to say piecewise, to an energy-density functional for inhomogeneous systems. That is the well-known local-density approximation (LDA), see Sect. 6.1.2 for the exchange functional and Sect. 6.5 about the Thomas–Fermi approximation for the kinetic energy functional.

One often encounters the notion of a "Fermi liquid" [85]. The difference to a Fermi gas consists in the role played by correlations. A Fermi gas applies if the interaction can be accounted for at a mean-field level. One speaks of a Fermi liquid if the two-body short-range interaction becomes dominant. Simple models are rare in that regime. A profound theory of finite droplets of Fermi liquids goes far beyond the scope of this book.

2.1 From Finite Box to Continuum States

2.1.1 Single-Fermion Wave Functions in General

The Fermi gas is a generic model for nearly homogeneous systems of independent fermions. Let us address first the features of a finite system. Each fermion is associated with a wave function $|\varphi_\alpha\rangle$. It is a state in the Hilbert space of one-particle wave functions. It is often used in coordinate-space representation $|\varphi_\alpha\rangle \longrightarrow \varphi_\alpha(\mathbf{r})$. This still hides the fact that fermions necessarily have spin and that the wave function is to carry a spinor part. To make it quite explicit we write it in detailed components

$$|\varphi_\alpha\rangle \equiv \varphi_\alpha(\mathbf{r}, \frac{1}{2})\chi_{\frac{1}{2}} + \varphi_\alpha(\mathbf{r}, -\frac{1}{2})\chi_{-\frac{1}{2}}.$$

We assume here and in the following that the fermions we consider are spin 1/2 which correspond to the vast majority of physically relevant cases (electrons, nucleons in particular ...). The components $\varphi_\alpha(\mathbf{r}, \nu)$ compose a complex field in $\mathbb{R}^3$. The two ν components together define a spinor. It is often advantageous to combine the coordinates into a super-vector $x = (\mathbf{r}, \nu)$. The integration and summation over the whole space is then abbreviated as $\sum_\nu \int d^3r \longrightarrow \int dx$. This can be summarized as

$$x = (\mathbf{r}, \nu), \quad \mathbf{r} \in \mathbb{R}^3, \quad \nu \in \{-\tfrac{1}{2}, \tfrac{1}{2}\}, \tag{2.1a}$$

$$|\varphi_\alpha\rangle = \int \mathrm{d}x |x\rangle \varphi_\alpha(x) = \sum_\nu \int \mathrm{d}^3 r \, |\mathbf{r}, \nu\rangle \varphi_\alpha(\mathbf{r}, \nu), \tag{2.1b}$$

$$\varphi_\alpha(x) = \langle x | \varphi_\alpha \rangle, \tag{2.1c}$$

$$\langle \varphi_\beta | \varphi_\alpha \rangle = \int \mathrm{d}x \, \varphi_\beta^*(x) \varphi_\alpha(x) = \sum_\nu \int \mathrm{d}^3 r \, \varphi_\beta^*(\mathbf{r}, \nu) \varphi_\alpha(\mathbf{r}, \nu). \tag{2.1d}$$

A formally sound connection is established by the Hilbert space basis of eigenstates $|\mathbf{r}, \nu\rangle = |x\rangle$ of space coordinate $\mathbf{r}$ and spin ν. It depends on the particular application whether the compact notation with x or the detailed form $\mathbf{r}$, ν is better suited.

The N-fermion state built from those occupied single-particle states is a Slater determinant as given in (1.3), for details see Appendix A.3. The notion of a determinant becomes impracticable for continuum states. One needs to find a more robust definition. That is achieved by employing the operators for creation or annihilation of a fermion in the state φ_α, i.e., the $\hat{a}_\alpha^\dagger$ or $\hat{a}_\alpha$, see Sect. 1.3.2 and Appendix A.4.

2.1.2 From Bound States to Plane Waves

2.1.2.1 1D Periodic Boundary Conditions – Plane Wave Limit

Here we discuss the handling of infinite systems by deriving their properties from a continuum limit of the corresponding finite system. We do that for the simpler case of one dimension (1D). We also ignore spin because it remains unchanged throughout all considerations. It allows to demonstrate all essential steps with simple sums and integrals. The generalization to realistic three dimensions (3D) is obvious but tedious.

The homogeneous 1D system is first limited to a finite box $0 \leq z \leq L$ with periodic boundary conditions. The normalized eigenstates thus become

$$\varphi_n(z) = \frac{1}{\sqrt{L}} \exp(\mathrm{i}k_n z), \quad k_n = \frac{2\pi}{L} n, \quad n = 0, \pm 1, \pm 2, \ldots . \tag{2.2}$$

The energies are $\varepsilon_n = \frac{\hbar^2}{2m} n^2$. The ground state is obtained by filling the lowest energies up to the given particle number N, which means $|n| \leq (N-1)/2$ (note that N must be odd to produce a unique ground state). The states are orthonormalized

$$\langle \varphi_n | \varphi_{n'} \rangle = \int_0^L \mathrm{d}z \, \varphi_n^*(z) \varphi_{n'}(z) = \int_0^L \frac{\mathrm{d}z}{L} \exp\left(\mathrm{i}(n - n') \frac{2\pi z}{L}\right) = \delta_{nn'}. \tag{2.3}$$

The density becomes

$$\rho_0 = \sum_{n=-n_\mathrm{F}}^{n_\mathrm{F}} \frac{\exp(\mathrm{i}k_n z)}{\sqrt{L}} \frac{\exp(-\mathrm{i}k_n z)}{\sqrt{L}} = \frac{N}{L}, \quad n_\mathrm{F} = \frac{N-1}{2}, \tag{2.4}$$

as it should be (remember that in 1D L plays the role of the volume V in 3D). We now view the summation as a discrete approximation to a continuous integral, i.e., $\int dk \ldots \longleftrightarrow \sum_n \Delta k \ldots$ and identify $\Delta k = k_{n+1} - k_n = 2\pi/L$. This suggests to rewrite (2.4) as

$$\rho_0 = \underbrace{\sum_{n=-n_F}^{n_F} \Delta k}_{} \frac{\exp(ik_n z)}{\sqrt{2\pi}} \frac{\exp(-ik_n z)}{\sqrt{2\pi}}, \qquad \Delta k = k_{n+1} - k_n = \frac{2\pi}{L}$$

$$\sum_n \Delta k \ldots \longrightarrow \int dk \ldots$$

$$= \int_{-k_F}^{k_F} dk \frac{\exp(ikz)}{\sqrt{2\pi}} \frac{\exp(-ikz)}{\sqrt{2\pi}}, \qquad k_F = \Delta k \, n_F = \tfrac{2\pi n_F}{L}$$

$$= \frac{1}{2\pi} \int_{-k_F}^{k_F} dk = \frac{N}{L}.$$

The integral trivially becomes k_F/π and allows to recover the appropriate N/L in the limit $N \gg 1$, while $N/L =$ constant. This shows that the factor $1/\sqrt{2\pi}$ provides the correct normalization of the continuum states, which are then

$$\varphi_k(z) = \frac{\exp(ikz)}{\sqrt{2\pi}}. \tag{2.5}$$

The continuum limit has an important consequence for the orthonormality relation. Form (2.3) applies only to a discrete spectrum because the Kronecker-δ, i.e., the δ_{nm}, is defined only for integer numbers, while the continuum limit produces the continuous label k. The wave functions now have to be "normalized" to a Dirac δ-distribution as

$$\langle \varphi_k | \varphi_{k'} \rangle = \int_{-\infty}^{\infty} dz \, \varphi_k^\dagger(z) \varphi_{k'}(z) = \int_{-\infty}^{\infty} dz \, \frac{\exp(i(k'-k)z)}{2\pi} = \delta(k'-k). \tag{2.6}$$

2.1.2.2 *N* Particles in a 1D Box – Limit for the Density $\rho(z)$

The periodic boundary conditions were particularly useful to explain the continuum limit in terms of plane waves. Realistic situations start from bound states and consider increasing system size. For simplicity, we stay in 1D and consider the Schrödinger equation in a square-well potential

$$\hat{H} = \frac{\hat{p}^2}{2m} + V(z),$$
$$V(z) = \begin{cases} 0 & \text{for} \quad 0 \le z \le L \\ \infty & \text{for} \quad z < 0 \text{ or } \&z > L \end{cases}, \quad L = \frac{N}{\rho_0}. \tag{2.7}$$

This potential produces bound states throughout, it is very well suited for the continuum limit because it has a constant value inside the infinite walls, so that it is a useful model for saturating many-fermion systems which develop constant potential bottoms in the interior for large system sizes (see the discussion at the beginning of Chap. 3). If the system were composed of N classical particles, the average density would be constant in space with $\rho = N/L$, which is also the density (2.4) in case of periodic boundary conditions. We take that as a guideline for the quantum-mechanical result. This is why we scale the length $L = N/\rho_0$ to simulate a saturating system which has a strong tendency to a constant equilibrium density ρ_0.

The eigenstates for the Hamiltonian (2.7) are [24]

$$\varphi_n(z) = \sqrt{\frac{2}{L}}\sin(k_n z), \quad k_n = n\Delta k, \quad \Delta k = \frac{\pi}{L}, \tag{2.8}$$

as can be seen from the boundary conditions $\varphi(0) = \varphi(L) = 0$. The total density of such a many-fermion system is the incoherent sum of the densities of the occupied single-particle states (see Appendix A.5.1). We use

$$\sum_{n=1}^{N} \sin^2(na) = \frac{N}{2} - \frac{\cos\left((N+1)a\right)\sin\left(Na\right)}{2\sin(a)}$$

to obtain

$$\rho(z) = \sum_{n=1}^{N} |\varphi_\alpha|^2 = \frac{2}{L}\sum_{n=1}^{N}\sin^2(n\Delta kz) = \underbrace{\frac{N}{L}}_{\rho_0} - \frac{\cos\left((N+1)\Delta kz\right)\sin\left(N\Delta kz\right)}{L\sin(\Delta kz)},$$

which we finally write as

$$\rho(z) = \rho_0\left[1 - \frac{1}{N}\frac{\cos\left((N+1)\Delta kz\right)\sin\left(N\Delta kz\right)}{\sin(\Delta kz)}\right]. \tag{2.9}$$

The resulting densities for three system sizes N are illustrated in Fig. 2.1. The convergence toward a homogeneous system with increasing N is clearly apparent. The deviation are largest at the boundaries because the step down to $\rho(0) = \rho(L) = 0$ has to be enforced. In the interior, however, there is soon a smooth pattern.

One can read Fig. 2.1 the reverse way and point out the irrepressible spatial fluctuations in the densities. This is a quantum-mechanical effect, often referred to as "shell fluctuations" [37]. The probability distributions of bound-state wave functions are necessarily inhomogeneous and excited states do have large fluctuations with zeroes, the more the higher the excitation. These fluctuations persist in the total densities, but in fact tend to compensate each other, the better the larger the particle number. This is why we see the nice approach to the continuum limit with constant

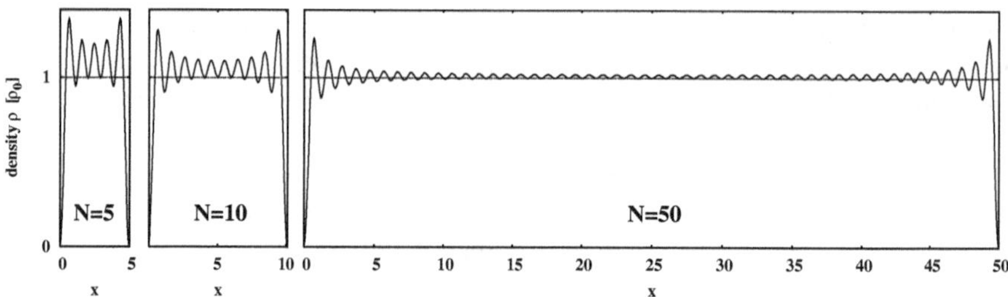

Fig. 2.1 Spatial density distributions for N-fermion states in the 1D box potential. The length of the box scales as $L \propto N$ to maintain the same average density. The *faint horizontal line* indicates the (classical and) continuum limit of equi-distribution

density. Consider, e.g., the nuclear charge densities in Fig. 1.3. Unlike the present rough model of particles in a box, they show an extended surface zone, but in the interior one can also spot small oscillations about a constant value, which are mainly due to these shell effects. The patterns also do show up in the density of the cluster Na_{339}^{+}, see the right-hand part of Fig. 2.4.

2.1.2.3 2D Periodic Box – Limit For Eigenvalues

In the last step of illustrating the continuum limit, we are now going to discuss the density of states and the associated level density (see Sect. 1.2.4.3). The starting point is the Fermi sphere of occupied single-particle states in a homogeneous fermion system. The Fermi "sphere" is literally true in a 3D system. For simplicity of graphical representation, we here choose a 2D system which then will yield a Fermi circle to describe the set of occupied states.

The homogeneous 2D system limited to a finite box $0 \leq x, y \leq L$ with periodic boundary conditions in both directions has solutions quite similar to the 1D case discussed in Sect. 2.1.2.1. They read

$$\varphi_{\mathbf{n}}(x, y) = \frac{1}{L} \exp(\mathrm{i}k_{n_x}x) \exp(\mathrm{i}k_{n_y}y), \tag{2.10a}$$

$$\mathbf{k}_{\mathbf{n}} = \frac{2\pi}{L}\mathbf{n} = \frac{2\pi}{L}(n_x, n_y), \quad n_x, n_y = 0, \pm 1, \pm 2, \ldots, \tag{2.10b}$$

$$\varepsilon_{\mathbf{n}} = \frac{\hbar^2 \mathbf{k}^2}{2m} = \frac{\hbar^2 (k_x^2 + k_y^2)}{2m}. \tag{2.10c}$$

We consider again systematically increasing the system size N while keeping the average density constant, which means $L = \sqrt{N/\rho_0}$ for that 2D system. The resulting spectrum of quantum numbers (k_x, k_y) is illustrated in Fig. 2.2. The single-particle energy (2.10c) is a function of $k = \sqrt{|\mathbf{k}|}$ only. The occupied states then reside within a circle whose radius is called the Fermi momentum k_{F}, as indicated in the figure. The selection of states does not so perfectly match the circle for small system sizes (leftmost panel), but becomes a smoother distribution and approaches

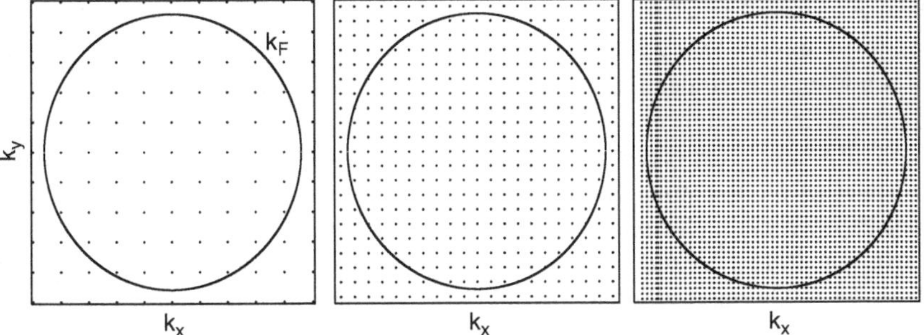

Fig. 2.2 The momenta (k_x, k_y) of the eigenvalues for the 2D box with periodic boundary conditions for different values of L ($\propto \sqrt{N}$), increasing from *left* to *right*

the circle quickly with increasing size (see Sect. 2.2.5 for the detailed computation of the density of states in a 3D Fermi gas.)

2.2 Basics: Density, Fermi Momentum, Fermi Energy

2.2.1 *Fermi Momentum, Fermi Energy, Density*

Consider the ground state of independent fermions with the symmetries of free space. "Independent" means that each particle can be associated with one single-particle wave function $\varphi_{\mathbf{k}\sigma}(\mathbf{r})$ and "fermion" implies that each single-particle state can be occupied only once. The "symmetries of free space" are invariance under translation and rotation, called together homogeneity of space. Homogeneous systems are necessarily infinite. This causes no technical problems but conceptual subtleties like dealing with continuum states. Translational and rotational symmetry apply to the Hamiltonian $\hat{h}$ which defines the single-particle wave functions, e.g., the self-consistent mean-field Hamiltonian as will be discussed in Chap. 5. For the moment we need only the symmetry property of $\hat{h}$. The most general translationally and rotationally invariant form is $\hat{h} = \sum_n a_n(\mathbf{p}^2)^n$ with constant coefficients a_n. The Fermi-gas model assumes the simplest form

$$\hat{h} = \frac{\mathbf{p}^2}{2m}. \tag{2.11}$$

In any case, plane waves, the eigenstates of the momentum operator $\hat{\mathbf{p}} = -\mathrm{i}\hbar\nabla$, will also be eigenstates of $\hat{h}$. We have obtained them for the 1D case in Sect. 2.1.2.1. The plane waves in 3D similarly read

$$\hat{h}\varphi_{\mathbf{k}\sigma} = \varepsilon_{\mathbf{k}}\varphi_{\mathbf{k}\sigma} \quad \longleftrightarrow \quad \varphi_{\mathbf{k}\sigma}(\mathbf{r}, \nu) = \frac{\exp(\mathrm{i}\mathbf{k}\cdot\mathbf{r})}{(2\pi)^{3/2}}\delta_{\sigma\nu}, \quad \varepsilon_k = \frac{\hbar^2}{2m}\mathbf{k}^2, \tag{2.12}$$

where $\sigma = \pm 1/2$ labels the spin and ν labels the components of the Pauli–Spinor. This form applies to a gas from a single-fermion species like the electron gas or neutron matter. Their degeneracy factor is two due to the two-spin orientations. That feature carries through to most formulae in the sequel. Symmetric nuclear matter would also have a dependence on an isospinor (which treats proton and neutron as two states of the nucleon distinguished by isospin $\pm\frac{1}{2}$) and consequently a degeneracy factor four, for details see Sect. 2.4. Note that the energies depend only on $k = |\mathbf{k}|$ which is a consequence of rotational invariance. The wave-vectors $\mathbf{k} \in \mathbb{R}^3$ constitute a continuum of quantum numbers. The orthonormality relation is the 3D generalization of (2.6), i.e.,

$$\langle \varphi_{\mathbf{k}\sigma} | \varphi_{\mathbf{k}'\sigma'} \rangle = \sum_{\nu} \int d^3 r \, \varphi^{\dagger}_{\mathbf{k}\sigma} \varphi_{\mathbf{k}'\sigma'} = \delta^3(\mathbf{k} - \mathbf{k}') \delta_{\sigma\sigma'} . \tag{2.13}$$

A many-fermion system occupies a large number of these states. The many-body ground state $|\Phi_0\rangle$ fills the lowest kinetic energies up to a Fermi momentum k_F, i.e.,

$$|\Phi_0\rangle \equiv \left\{ \varphi_{\mathbf{k}\sigma}, k = |\mathbf{k}| \leq k_\mathrm{F}, \sigma = \pm\tfrac{1}{2} \right\} . \tag{2.14}$$

The filling of occupied states may equally well be expressed in terms of the Fermi energy ε_F as

$$\varepsilon_k \leq \varepsilon_\mathrm{F} = \frac{\hbar^2}{2m} k_\mathrm{F}^2 . \tag{2.15}$$

Fermion creation and destruction operators, see Sect. 1.3.2 and Appendix A.4, allow a clear and compact definition of the ground state $|\Phi_0\rangle$ as

$$\begin{aligned} \hat{a}^{\dagger}_{\mathbf{k}\sigma} |\Phi_0\rangle &= 0 \;\; \text{for } k \leq k_\mathrm{F} \\ \hat{a}_{\mathbf{k}\sigma} |\Phi_0\rangle &= 0 \;\; \text{for } k > k_\mathrm{F} \end{aligned} \tag{2.16}$$

expressing formally that all states $k \leq k_\mathrm{F}$ are occupied.

The spatial density for independent-Fermion systems is generally (see Appendix A.5.1) $\rho(\mathbf{r}) = \sum_{\alpha \in \text{occup.}} \sum_{\nu} |\varphi_\alpha(\mathbf{r}, \nu)|^2$. For the present continuum of states, we have to replace the summation over discrete states by an integral over the continuum states. Thus

$$\begin{aligned} \rho(\mathbf{r}) &= \sum_{\sigma} \int_{k \leq k_\mathrm{F}} d^3 k \sum_{\nu} \varphi^{*}_{\mathbf{k}\sigma}(\mathbf{r}, \nu) \varphi_{\mathbf{k}\sigma}(\mathbf{r}, \nu) \\ &= \frac{1}{(2\pi)^3} \sum_{\sigma\nu} \int_{k \leq k_\mathrm{F}} d^3 k \, \delta_{\sigma\nu} \underbrace{\exp(-\mathrm{i}\mathbf{k}\cdot\mathbf{r}) \exp(\mathrm{i}\mathbf{k}\cdot\mathbf{r})}_{=1} \delta_{\sigma\nu} = \frac{2}{(2\pi)^3} \frac{4\pi}{3} k_\mathrm{F}^3 , \end{aligned}$$

which finally yields

$$\rho_0 = \rho(\mathbf{r}) = \frac{k_\mathrm{F}^3}{3\pi^2}. \tag{2.17}$$

The density is constant, i.e., it also obeys the symmetries of free space. From another aspect, the density is the number of particles per volume, $\rho_0 = N/V$. Both particle number and volume are infinite, but the density is finite. This provides a way to regulate the particle number in homogeneous systems, and (2.17) is often read in reverse. The density ρ_0 is given and determines the corresponding Fermi momentum as

$$k_\mathrm{F} = \left(3\pi^2\rho_0\right)^{1/3} \simeq 3.18\,\rho_0^{1/3}. \tag{2.18}$$

Consequently the Fermi energy behaves as $\varepsilon_\mathrm{F} \propto \rho_0^{2/3}$.

A further way to characterize the density of the system is the Wigner–Seitz radius r_s. It is the radius of the sphere whose volume fits just one particle, i.e.,

$$\frac{1}{\rho_0} = \frac{4\pi}{3}r_s^3 \quad \Longrightarrow \quad r_s = \left(\frac{9\pi}{4}\right)^{1/3}\frac{1}{k_\mathrm{F}} \simeq \frac{1.92}{k_\mathrm{F}} = \left(\frac{3}{4\pi\rho_0}\right)^{1/3}. \tag{2.19}$$

2.2.2 The One-Body Density Matrix

For finite N the one-body density matrix is defined as (see Appendix A.5)

$$\varrho(x, x') = \sum_{\alpha\in\mathrm{occup.}} \varphi_\alpha(x)\varphi_\alpha^\dagger(x').$$

It contains the same amount of information as the independent-fermion state $|\Phi_0\rangle$. It is instructive to evaluate the one-body density matrix for the Fermi gas as a model system. Identifying $\alpha \leftrightarrow (\mathbf{k}\sigma)$ this yields

$$\varrho(\mathbf{r}\nu, \mathbf{r}'\nu') = \sum_\sigma \int_{k\le k_\mathrm{F}} \frac{\mathrm{d}^3k}{(2\pi)^3}\delta_{\nu\sigma}\exp\left(\mathrm{i}\mathbf{k}\cdot(\mathbf{r}-\mathbf{r}')\right)\delta_{\nu'\sigma}.$$

The spin overlaps are trivially evaluated. The integration of the wave-vectors is performed in spherical coordinates $\mathbf{k} \equiv (k, \theta, \phi)$. We align the coordinate system along the direction of $\mathbf{r} - \mathbf{r}'$. Thus

$$\begin{aligned}
\varrho(\mathbf{r}\nu,\mathbf{r}'\nu') &= \frac{\delta_{\nu\nu'}}{(2\pi)^3}\int_0^{k_F}\mathrm{d}k\,k^2\int_{-1}^{1}\mathrm{d}(\cos\theta)\int_0^{2\pi}\mathrm{d}\phi\,\exp\Big(\mathrm{i}k\underbrace{|\mathbf{r}-\mathbf{r}'|}_{y}\cos(\theta)\Big)\\
&= \frac{\delta_{\nu\nu'}}{(2\pi)^2}\int_0^{k_F}\mathrm{d}k\,k^2\frac{\exp(\mathrm{i}ky)-\exp(-\mathrm{i}ky)}{\mathrm{i}ky}\\
&= \frac{\delta_{\nu\nu'}}{2\pi^2}\int_0^{k_F}\mathrm{d}k\,k\,\sin(ky) = \frac{\delta_{\nu\nu'}k_F^3}{2\pi^2}\underbrace{\frac{\sin(k_F y)-k_F y\cos(k_F y)}{(k_F y)^3}}_{j_1(k_F y)/(k_F y)},
\end{aligned}$$

where j_1 is the spherical Bessel function of first order [96]. The final result is

$$\varrho(\mathbf{r}\nu,\mathbf{r}'\nu') = \frac{\delta_{\nu\nu'}}{2}\rho_0\mathcal{J}(k_F|\mathbf{r}-\mathbf{r}'|), \quad \mathcal{J}(x) = \frac{3j_1(x)}{x}. \tag{2.20}$$

It is a rather simple universal function, depending only on $|\mathbf{r}-\mathbf{r}'|$, which expresses translational and rotational invariance. The shape of the function is shown in Fig. 2.3. Note that it scales with k_F^{-1} and so applies to all densities ρ_0. The larger ρ_0, the faster the decay in relative distance.

The spatial dependence can also be expressed in terms of the Wigner–Seitz radius (2.19) as $\mathcal{J}(1.92|\mathbf{r}-\mathbf{r}'|/r_s)$, which shows that exchange properties scale with r_s. The typical decay length of $\mathcal{J}$ is given by the first zero. That appears at a distance of $|\mathbf{r}-\mathbf{r}'| \approx 1.6\,r_s$.

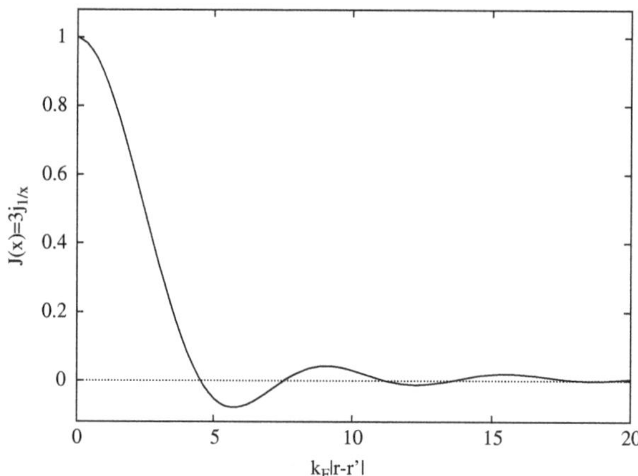

Fig. 2.3 The function $\mathcal{J}$ carrying the spatial dependence of the one-body density matrix (2.20) of the homogeneous Fermi gas

2.2.3 A Model for the One-Body Density Matrix in Finite Systems

The one-body density matrix is a rather powerful quantity carrying a large deal of information about a system. For example, it allows to compute all one-body observables directly (Sect. 1.3.3). It thus provides a rather detailed view of the state of a system and, correspondingly, a critical analyzing instrument for the quality of the Fermi-gas approximation. As a test case, we consider the valence electron cloud of the metal cluster Na_{339}^+. The ionic background is simplified in terms of the jellium approximation (1.1). A DFT calculation (see Sect. 6.1) is performed and yields the one-body density matrix $\varrho(\mathbf{r},\mathbf{r}')$ for the finite system, which is then the benchmark result for the fully detailed description of the system.

The Fermi-gas model yields a compact expression of the one-body density matrix with a very specific and pronounced dependence on the difference coordinate $|\mathbf{r}-\mathbf{r}'|$, see (2.20) and Fig. 2.3. The pattern in a finite system close to a Fermi gas should look similar in that difference coordinate while the local density has to reproduce the finite spatial distribution. That suggests a simple model for the one-body density matrix

$$\varrho^{(\mathrm{LDA})}(\mathbf{r}\nu,\mathbf{r}'\nu') = \rho(\overline{\mathbf{r}})\frac{\delta_{\nu\nu'}}{2}\mathcal{J}\left(k_F(\overline{\mathbf{r}})|\mathbf{r}-\mathbf{r}'|\right), \tag{2.21a}$$

$$k_F(\overline{\mathbf{r}}) = \left(3\pi^2\rho(\overline{\mathbf{r}})\right)^{1/3}, \tag{2.21b}$$

$$\overline{\mathbf{r}} = \frac{1}{2}\left(\mathbf{r}+\mathbf{r}'\right), \tag{2.21c}$$

which employs a local Fermi momentum $k_F(\overline{\mathbf{r}})$ deduced from the local density according to (2.17). It is called a *Local Density Approximation (LDA)* for the one-body density matrix. (An application of LDA to the energy is at the heart of DFT as we will see in Sect. 6.1.)

Figure 2.4 shows a comparison between the exact one-body density and the LDA (2.21). We look at the pattern in the relative coordinate $|\mathbf{r}-\mathbf{r}'|$ for a series of average positions $\overline{\mathbf{r}}$ and normalize to the value at $|\mathbf{r}-\mathbf{r}'| = 0$. There is nice agreement between the exact pattern and the LDA for most reference radii. The largest deviation is seen at the center of the cluster. A quick glance at the local density distribution on the right side of Fig. 2.4 shows a large deviation from the average local density just at the center. This is an effect of spatial shell fluctuations as discussed in Sect. 2.1.2.2, which here happen to accumulate particularly at the center. The mismatch of the LDA model in the lowest left panel of Fig. 2.4 is then understandable considering that going along growing $|\mathbf{r}-\mathbf{r}'|$ one always runs into a regime of higher densities than those at the reference point. On the other hand, the center occupies only a very small fraction of the total volume and thus Fig. 2.4 indicates that LDA is fairly well justified on the average. Note also that the density matrix falls off properly to zero outside the cluster.

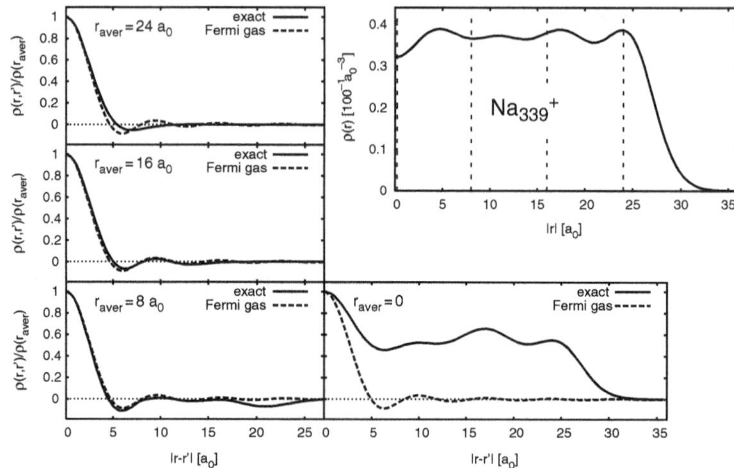

Fig. 2.4 *Upper right panel*: The local electron-density distribution of the cluster Na_{339}^+. The ionic background is described in the spherical jellium model (i.e. (1.1) with deformation $\delta = 0$). The electronic wave functions were computed with density functional theory (DFT) in local density approximation, see Sect. 6.1. The *vertical dashed lines* indicate the reference radii for the plots in the *left* and the *lower right panels*. *Left* and *lower right panels*: The one-body density matrices $\varrho(\mathbf{r}, \mathbf{r}')$ are normalized to the local density at $\rho(r_{\text{aver}})$ (where $r_{\text{aver}} = \frac{1}{2}(\mathbf{r}+\mathbf{r}')$) along the difference coordinate $|\mathbf{r} - \mathbf{r}'|$ for a variety of average radii r_{aver} as indicated. The *full line* shows the result from the DFT calculation of the finite system and the *dashed line* the density matrix in Fermi-gas approximation (2.21)

2.2.4 The Two-Body Density

The two-body density matrix ϱ_2 for a Slater state can be expressed completely through the one-body density matrix ϱ as (see Appendix A.5)

$$\varrho_2(x_1, x_2; x_1', x_2') = \varrho(x_1; x_1')\varrho(x_2; x_2') - \varrho(x_1; x_2')\varrho(x_2; x_1').$$

We consider its diagonal element, the local two-body density $\rho_2(x_1, x_2) = \varrho_2(x_1, x_2; x_1, x_2)$ which in terms of the one-body density matrix becomes

$$\rho_2(x_1, x_2) = \varrho(x_1; x_1)\varrho(x_2; x_2) - \varrho(x_1; x_2)\varrho(x_2; x_1). \tag{2.22}$$

It represents the probability to find one particle at x_1 and at the same time another particle at x_2. It thus characterizes the spatial correlations between particle 1 and particle 2 for the various spin relations between ν_1 and ν_2. To evaluate the local two-body density for the Fermi gas, we use the result (2.20) for the one-body density matrix and very quickly obtain

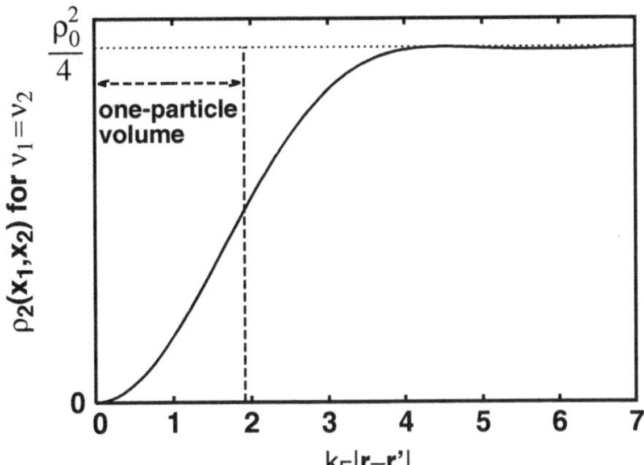

Fig. 2.5 The two-body density $\rho_2(x_1, x_2)$ as given in (2.23) for the case of aligned spin $\nu_1 = \nu_2$ as function of $k_F|\mathbf{r}_1 - \mathbf{r}_2|$. The *vertical dashed line* indicates the radius of a sphere covering one fermion

$$\rho_2(x_1, x_2) = \frac{\rho_0^2}{4}\left[1 - \delta_{\nu_1\nu_2}\mathcal{J}^2(k_F|\mathbf{r}_1 - \mathbf{r}_2|)\right] , \ \mathcal{J}(y) = \frac{3j_1(y)}{y}. \tag{2.23}$$

The result sensitively depends on the spin of the two particles. The case that they have the same spin is shown in Fig. 2.5. There is a deep hole around distance zero which is the coordinate-space appearance of the Pauli principle: two identical fermions cannot occupy the same position. For larger r, the function approaches quickly the value $(\rho_0/2)^2$ which means that the two particles become independent (no more "Pauli correlated"). The excluded volume around $r = 0$ is often called the exchange hole [27]. The vertical dashed line in Fig. 2.5 indicates the phase space volume of a sphere covering one fermion. This one-particle equivalent radius cuts almost precisely the half-value of the correlation function. This corresponds to the fact that the "exchange hole" excludes precisely one fermion. The local two-body density becomes $(\rho_0/2)^2$ constantly at all r if the two particles have different spin. For then they are distinguishable and thus independent everywhere.

2.2.5 The Kinetic Energy

The kinetic energy E_{kin} can be computed with the same steps as were used for the density in the previous subsection. The plane waves are eigenstates of $\hat{p}^2$ with $\hat{p}^2\varphi_{\mathbf{k}\sigma} = \hbar^2k^2\varphi_{\mathbf{k}\sigma}$. Thus we have just to modify the weight in the k-integration by a factor $\hbar^2k^2/(2m)$. This proceeds as

$$
\begin{aligned}
E_{\text{kin}} &= \int \mathrm{d}^3r \, \frac{2}{(2\pi)^3} \int_{k \le k_\mathrm{F}} \mathrm{d}^3k \, \frac{\hbar^2 k^2}{2m} \\
&= \frac{\hbar^2}{2m} \int \mathrm{d}^3r \, \frac{2}{(2\pi)^3} \underbrace{\int_0^{k_\mathrm{F}} \mathrm{d}k \, k^4}_{k_\mathrm{F}^5/5} \underbrace{\int_0^{\pi} \mathrm{d}\cos(\theta) \int_0^{2\pi} \mathrm{d}\phi}_{=\int \mathrm{d}\Omega = 4\pi} \\
&= \frac{\hbar^2}{2m} \underbrace{\int \mathrm{d}^3r}_{\mathcal{V}} \, \frac{k_\mathrm{F}^3}{3\pi^2} \frac{3}{5} k_\mathrm{F}^2 = \frac{\hbar^2}{2m} \frac{3}{5} k_\mathrm{F}^2 \rho_0 \mathcal{V},
\end{aligned}
$$

where $\mathcal{V} = \int \mathrm{d}^3r$ stands for the volume of the system. It is infinite and so is the kinetic energy. Thus it is more appropriate to discuss the energy density, which is a finite quantity. In fact, one usually prefers to specify the energy per particle which is related to the energy per volume through the density $\rho_0 = N/\mathcal{V}$. This then yields the finite result

$$
\frac{E_{\text{kin}}}{N} = \frac{E_{\text{kin}}}{\mathcal{V}} \frac{\mathcal{V}}{N} = \frac{E_{\text{kin}}}{\mathcal{V}} \frac{1}{\rho_0} = \frac{\hbar^2}{2m} \frac{3}{5} k_\mathrm{F}^2 = \frac{3}{5} \varepsilon_\mathrm{F}. \tag{2.24}
$$

The kinetic energy per particle grows $\propto k_\mathrm{F}^2$. The result as such is also quite interesting as it gives a typical estimate of the kinetic energy of a particle in a Fermi gas ($3\epsilon_F/5$). This again confirms the key importance of the Fermi energy to provide a relevant scale in simple fermion systems.

It is interesting to reformulate the kinetic energy in terms of system density using relation (2.18). That yields

$$
\frac{E_{\text{kin}}}{N} = \frac{\hbar^2}{2m} \frac{3}{5} (3\pi^2)^{2/3} \rho_0^{2/3}.
$$

It explicitly shows the growth of the kinetic energy with system density. Let us consider now a change the volume for fixed particle number which amounts to a change of density. This defines a kinetic pressure

$$
P_{\text{kin}} = \frac{\partial E_{\text{kin}}}{\partial \mathcal{V}} = -\rho_0^2 \frac{\partial}{\partial \rho_0} \frac{E_{\text{kin}}}{N} = \frac{\hbar^2}{2m} \frac{2}{5} (3\pi^2)^{2/3} \rho_0^{5/3}. \tag{2.25}
$$

This is the Pauli pressure opposing compression of the system because it is easier to accommodate the Pauli principle if the particles are farther apart.

2.2.6 The Level Density

In the computation of the kinetic energy the angular integration was trivial while the crucial integral ran over the absolute value of the momentum $k = |\mathbf{k}|$. For that very common situation, it is instructive and advantageous to replace the momentum integration by one over the energy. This substitution uses ε as

$$\varepsilon = \frac{\hbar^2 k^2}{2m_\mathrm{e}} \quad \Longrightarrow \quad \mathrm{d}\varepsilon = \frac{\hbar^2}{m_\mathrm{e}} k\,\mathrm{d}k \quad \Longrightarrow \quad \mathrm{d}k = \frac{\sqrt{m}}{\hbar}\frac{\mathrm{d}\varepsilon}{\sqrt{\varepsilon}} \tag{2.26}$$

to yield the kinetic energy as

$$E_\mathrm{kin} = \underbrace{\int \mathrm{d}^3 r}_{\mathcal{V}} \underbrace{\frac{2}{(2\pi)^3} \underbrace{\int \mathrm{d}\Omega}_{4\pi} \int_0^{\varepsilon_\mathrm{F}} \mathrm{d}\varepsilon \frac{2m^{3/2}}{\hbar^3} \sqrt{\varepsilon}}_{\int_0^{\varepsilon_\mathrm{F}} \mathrm{d}\varepsilon \mathcal{D}(\varepsilon)}\, \varepsilon = \mathcal{V}\frac{3}{5}\varepsilon_\mathrm{F}\rho_0. \tag{2.27}$$

The volume element for energy integration can be expressed through one physical quantity, the density of states $\mathcal{D}(\varepsilon)$, which then becomes useful for all expectation values of functions of energy alone. Assume that we consider an observable $f(\varepsilon)$ which is a function of energy. The expectation value can then be written as

$$\frac{\overline{f}}{\mathcal{V}} = \int_0^{\varepsilon_\mathrm{F}} \mathrm{d}\varepsilon\, \mathcal{D}(\varepsilon)\, f(\varepsilon), \quad \mathcal{D}(\varepsilon) = \frac{2m^{3/2}}{\hbar^3 \pi^3}\sqrt{\varepsilon}. \tag{2.28a}$$

The simplest examples are the particle number N for $f = 1$, and the kinetic energy E_kin for $f = \varepsilon$ and given in (2.27). The density of states can be rewritten in several different forms. Recall (2.17) connecting ρ with k_F and (2.15) connecting k_F with ε_F. Combining these two equations yields

$$\mathcal{D}(\varepsilon) = \frac{3}{4}\frac{N}{\mathcal{V}}\frac{\sqrt{\varepsilon}}{\varepsilon_\mathrm{F}^{3/2}} = \frac{3}{4}\rho_0\frac{\sqrt{\varepsilon}}{\varepsilon_\mathrm{F}^{3/2}}, \quad \mathcal{V}\mathcal{D}(\varepsilon) = \frac{3}{4}\frac{N}{\varepsilon_\mathrm{F}^{3/2}}\sqrt{\varepsilon}. \tag{2.28b}$$

It is important to note that this expression for the density of states holds for a Fermi gas in three dimensions. For a general dimension D, one has $\mathcal{D} \propto \varepsilon^{D/2-1}$.

2.3 Fermi Gas at Finite Temperature

One of the interests of a description in terms of level density is that temperature effects can easily be included in the picture. To that end, one introduces occupations numbers $n(\varepsilon)$ which regulate the probability to find a particle at single-particle energy ε. Particle number and kinetic energy can then be written as

$$N = \mathcal{V}\int_0^\infty d\varepsilon\, \mathcal{D}(\varepsilon)n(\varepsilon), \quad E = \mathcal{V}\int_0^\infty d\varepsilon\, \mathcal{D}(\varepsilon)n(\varepsilon)\varepsilon. \tag{2.29}$$

The occupation number $n(\varepsilon)$ for a Fermi gas at zero temperature is just a step function

$$n(\varepsilon) = \vartheta(\varepsilon - \varepsilon_F) = n_{T=0}(\varepsilon). \tag{2.30}$$

Switching to a finite temperature is achieved through replacing the step function for zero-temperature occupation numbers (2.30) by a smooth Fermi distribution

$$\vartheta(\varepsilon - \varepsilon_F) \quad \longrightarrow \quad n_T(\varepsilon) = \frac{1}{1 + \frac{\varepsilon - \mu(T)}{k_B T}}, \tag{2.31}$$

where $\mu = \mu(T)$ is the chemical potential which has to be adjusted such that the first integral in (2.29) reproduces the desired particle number. At temperature zero, the chemical potential merges into the Fermi energy, $\mu(T=0) = \varepsilon_F$. $k_B = 0.8617 \times 10^{-4}$ eV/K is the Boltzmann factor, so that it also often called the (temperature-dependent) Fermi energy.

The Fermi gas at finite temperature thus involves Fermi occupation numbers, which render most calculations non-analytical, except in the case of sufficiently small temperatures. Let us explore that case. Note first that the term "small temperature" deserves some explanations. In a Fermi system the temperature, as defined in classical kinetic theory, for example, in the case of perfect gases, has to be understood with respect to the scale determined by the Fermi energy ε_F or rather the associated temperature $T_F = \varepsilon_F/k_B$. Even at zero temperature, the Pauli exclusion principle implies a non-vanishing kinetic energy due to the motion of the fermions in the various occupied single-fermion states. This scales with Fermi temperature T_F, and the notion of low or high temperature is then to be defined with respect to T_F. We will confine the discussions to the case of low temperatures $T \ll T_F$.

The aim of the calculation in this subsection is to compute the thermal excitation energy E^* stored in the system and relate it to the temperature T. This amounts to evaluating the specific heat of the system. The excitation energy is defined as $E^*(T) = E(T) - E(T=0)$ where the energy in the Fermi gas is purely kinetic and reads

$$E(T) = \mathcal{V}\int_0^\infty d\varepsilon\, \mathcal{D}(\varepsilon)n_T(\varepsilon)\varepsilon = \int_0^\infty d\varepsilon\, \mathcal{V}\mathcal{D}(\varepsilon)\frac{1}{1 + \frac{\varepsilon - \mu(T)}{k_B T}}\varepsilon g(\varepsilon), \tag{2.32}$$

where the Fermi distribution (2.31) defines the thermal occupation weights. It has to be kept in mind that the Fermi energy $\mu(T)$ depends on temperature. It is determined from the condition that the distribution reproduces the given total number of particles N as

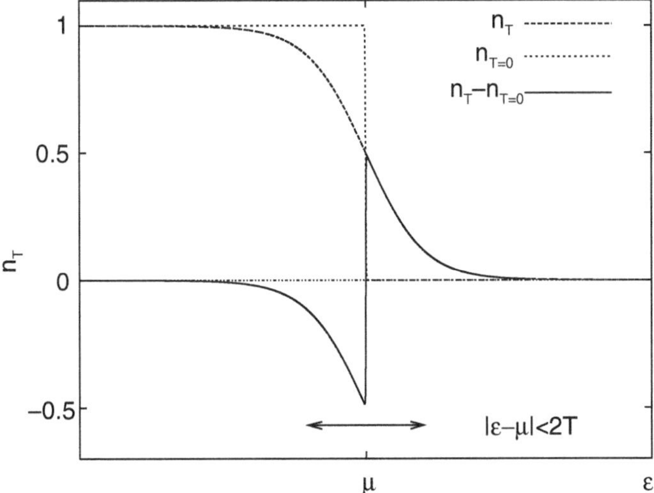

Fig. 2.6 The Fermi distribution n_T at finite temperature (*dashed*), the distribution at $T = 0$ (*dotted*) and their difference (*solid*)

$$N = \int_0^\infty \mathrm{d}\varepsilon \, n_T(\varepsilon) g(\varepsilon). \tag{2.33}$$

The condition (2.33) and the integral (2.32) for the total energy cannot be expressed in closed form, but the case of small temperatures ($T \ll T_\mathrm{F}$) still can be worked out in detail. Consider the change in occupation relative to the ground state at $T = 0$. The situation is sketched in Fig. 2.6. The difference

$$n_T - n_{T=0} = \frac{1}{1 + \frac{\varepsilon - \mu(T)}{k_\mathrm{B} T}} - \vartheta(\varepsilon_\mathrm{F} - \varepsilon)$$

is concentrated in a region around μ which is as narrow as T is small (remember that for small T we have $\varepsilon_\mathrm{F} \approx \mu$). That difference is thus a rapidly changing function, whereas all other ingredients in the integral vary smoothly. The idea is then to approximate the smooth parts by a Taylor expansion. We do that in one stroke for the normalization integral (2.33) and for the energy (2.32) by considering a functional

$$I(T)[f] = \int_0^\infty \mathrm{d}\varepsilon \, n_T(\varepsilon) \, f(\varepsilon), \quad f(\varepsilon) = \begin{cases} f = \mathcal{V}\mathcal{D}(\varepsilon) & \text{for } N \\ f = \varepsilon \mathcal{V}\mathcal{D}(\varepsilon) & \text{for } E \end{cases}. \tag{2.34}$$

The Taylor expansion will be applied to the smooth function $f(\varepsilon) \approx f(\mu) + (\varepsilon - \mu) f'(\mu)$. The steps read

$$I(T)[f] = \underbrace{\int_0^\infty \mathrm{d}\varepsilon n_{T=0}(\varepsilon)\, f(\varepsilon)}_{\int_0^\infty \mathrm{d}\varepsilon\, f(\varepsilon) = I_0} + \int_0^\infty \mathrm{d}\varepsilon\, [n_T - n_{T=0}]\; f(\varepsilon)$$

$$\approx I_0 + \underbrace{f(\mu)\int_0^\infty \mathrm{d}\varepsilon\, [n_T - n_{T=0}]}_{=0} + f'(\mu)\int_0^\infty \mathrm{d}\varepsilon\, [n_T - n_{T=0}]\,(\varepsilon - \mu).$$

Substituting $(\varepsilon - \mu)/T = x$ the remaining integral becomes

$$I - I_0 \approx f'(\mu)T^2 \int_0^\infty \mathrm{d}x \left[\frac{1}{1+e^x} - \vartheta(-x)\right] = 2f'(\mu)T^2 \underbrace{\int_0^\infty \mathrm{d}x \frac{1}{1+e^x}}_{\pi^2/12},$$

from which finally

$$\int_0^\infty \mathrm{d}\varepsilon\, n_T\, f(\varepsilon) \overset{T\to 0}{\longrightarrow} \int_0^\mu \mathrm{d}\varepsilon f(\varepsilon) + \frac{\pi^2 T^2}{6} f'(\mu). \tag{2.35}$$

This relation is evaluated first to determine $\mu(T)$ in lowest order. According to (2.34) we identify $f = \mathcal{D}(\varepsilon)$ and, furthermore, expand $\mu(T) = \varepsilon_\mathrm{F} + \delta\mu(T)$. This yields

$$N \approx \underbrace{\int_0^{\varepsilon_\mathrm{F}} \mathrm{d}\varepsilon\, \mathcal{V}\mathcal{D}(\varepsilon)}_{=N} + \underbrace{\int_{\varepsilon_\mathrm{F}}^{\varepsilon_\mathrm{F}+\delta\mu} \mathrm{d}\varepsilon\, \mathcal{V}\mathcal{D}(\varepsilon)}_{\delta\mu\mathcal{V}\mathcal{D}(\varepsilon_\mathrm{F})} + \frac{\pi^2}{6} T^2 \mathcal{V}\mathcal{D}'(\varepsilon_\mathrm{F}),$$

and thus

$$\delta\mu = -T^2 \frac{\pi^2}{6} \frac{\mathcal{D}'(\varepsilon_\mathrm{F})}{\mathcal{D}(\varepsilon_\mathrm{F})}. \tag{2.36}$$

One can now use the above expression for $\mu(T)$ in the expansion of the energy (2.32) for the limit of small temperature. Here, we identify $f = \varepsilon\mathcal{D}(\varepsilon)$ and insert it into expansion (2.35). This yields

$$E(T) \approx \underbrace{\int_0^{\varepsilon_\mathrm{F}} \mathrm{d}\varepsilon\, \varepsilon\mathcal{V}\mathcal{D}(\varepsilon)}_{E(T=0)} + \int_{\varepsilon_\mathrm{F}}^{\varepsilon_\mathrm{F}+\delta\mu} \mathrm{d}\varepsilon\, \varepsilon\mathcal{V}\mathcal{D}(\varepsilon) + \frac{\pi^2}{6} T^2 \mathcal{V} \frac{\mathrm{d}(\varepsilon\mathcal{D}(\varepsilon))}{\mathrm{d}\varepsilon}\Big|_{\varepsilon_\mathrm{F}}$$

$$\approx E(T=0) + \underbrace{\delta\mu\varepsilon_\mathrm{F}\mathcal{V}\mathcal{D}(\varepsilon_\mathrm{F}) + \frac{\pi^2}{6} T^2 \varepsilon_\mathrm{F}\mathcal{V}\mathcal{D}'(\varepsilon_\mathrm{F})}_{=0} + \frac{\pi^2}{6} T^2 \mathcal{V}\mathcal{D}(\varepsilon_\mathrm{F}).$$

The first term of this latter equation is nothing but the energy of the system at zero temperature. The excitation energy is thus directly obtained as

$$E^*(T) = E(T) - E(T=0) \approx \frac{\pi^2}{6}\mathcal{V}\mathcal{D}(\varepsilon_F))T^2. \tag{2.37}$$

It is to be noted that starting from $T = 0$ the energy initially grows quadratically with the temperature. The specific heat $c = \partial_T E^*$ of a Fermi gas thus begins with a linear growth and becomes zero in the limit $T \longrightarrow 0$.

Formula (2.37) for the excitation energy is still very general as it allows to insert different expressions for the level density $\mathcal{D}$. (For an example, see the discussion in Sect. 1.2.4.3.) The standard application is a Fermi gas in three dimensions for which the level density (2.28b) applies. Inserting that relation for the level density yields the excitation energy E^*, and the specific heat c, as

$$E^*(T) = \frac{\pi^2}{8}\frac{N}{\varepsilon_F}T^2, \quad c = \frac{\mathrm{d}E^*}{\mathrm{d}T} = \frac{\pi^2}{4}\frac{N}{\varepsilon_F}T. \tag{2.38}$$

This simple relation can serve as a quick estimate for a great variety of 3D systems when inserting the Fermi energies as given in Table 1.2. A typical example of application is the level density parameter in nuclei (Sect. 1.2.4.3) and its use for studying statistical nuclear deexcitation. It holds for low temperatures $T \ll T_F$. There are systems, however, for which the temperature range of applicability is also limited from below. A transition to a BCS condensate (see Sect. 9.4) changes the pattern below a critical temperature T_{BCS}. Small many-fermion systems can also experience sizeable shell effects. Estimate (2.38) then still holds averaged over system sizes and becomes valid for any N as soon as T is of the order of the shell gap which, in turn, shrinks with $N^{-1/3}$.

2.4 Fermi Gas in Stars: The Example of White Dwarfs

One of the probably most famous examples of a Fermi gas is provided by the electron gas in white dwarfs, for a brief introduction see Sect. 1.1.8. Chandrasekhar showed that in order to reach stability such a self-gravitating object should have a mass smaller than the Chandrasekhar mass M_{Ch} which is about 1.4 solar masses [77]. We give a short outline of the derivation below. Let us denote the total number of electrons in the star by N_e and the stellar radius by R.

2.4.1 "Low" Densities

For a particle density $\rho \sim 0.5 \times 10^{30}\,\mathrm{cm}^{-3} = 0.5 \times 10^5\,\mathrm{a}_0^{-3}$ (Sect. 1.1.8) and Fermi energy of order $\varepsilon_F \sim 50$ keV, the electrons still move at non-relativistic velocities, or momenta $\hbar k \ll m_e c$. The total kinetic energy is then given by (2.24). We express the Fermi momentum therein by the Wigner–Seitz radius using (2.19), and the star radius as $r_s N^{1/3}$. For the kinetic energy this yields

$$E_{\rm kin} = \frac{3}{5} N_e \frac{\hbar^2 k_{\rm F}^2}{2m_e} = \frac{3}{5}\left(\frac{9\pi}{4}\right)^{2/3} \frac{\hbar^2}{2m_e}\frac{N_e}{r_s^2} = \frac{3}{5}\left(\frac{9\pi}{4}\right)^{2/3} \frac{\hbar^2}{2m_e}\frac{N_e^{5/3}}{R^2}. \tag{2.39}$$

The electron charge is neutralized by the positive proton charges. Thus the Coulomb energy becomes negligible and the potential energy is purely gravitational. The mass of the star mostly stems from the nucleon masses. Assuming equal number of protons and neutrons, the total number of nucleons becomes $2N_e$. We, furthermore, ignore the small mass difference between protons and neutrons. The potential energy is then the gravitational energy of a homogeneous sphere with radius R and mass $2N_e m_p$ which reads

$$E_{\rm pot} = -\frac{3}{5} G \frac{(2N_e m_p)^2}{R} = -\frac{12}{5} G \frac{m_p^2}{R} N_e^2, \tag{2.40}$$

where m_p is the proton (or neutron) mass and $G = 6.67 \times 10^{-11}$ m^3/(kg s^2) is the gravitational constant. The total energy of the star then becomes

$$E = E_{\rm kin} + E_{\rm pot} = \frac{3}{5}\left(\frac{9\pi}{4}\right)^{2/3} \frac{\hbar^2}{2m_e}\frac{N_e^{5/3}}{R^2} - \frac{12}{5} G \frac{m_p^2}{R} N_e^2.$$

It should be minimized with respect to the radius R at constant mass M, namely constant N_e. This leads to

$$R_{\rm eq} = \frac{1}{2}\left(\frac{9\pi}{4}\right)^{2/3} \frac{\hbar^2}{2m_e}\frac{(2m_p)^{1/3}}{G m_p^2} M^{-1/3}, \tag{2.41}$$

expressed as a function of the total mass $M = 2m_p N_e$. Taking for M the solar mass $M_\odot = 2\,10^{30}$ kg leads to a radius $R \sim 8 \times 10^6$ m in agreement with observations.

2.4.2 Stability in the Relativistic Domain

The stability radius (2.41) shrinks with increasing total mass M. Decreasing R means increasing density and consequently increasing Fermi momentum. The non-relativistic procedure, as outlined in the previous section, provides a stable solution for every M, or R, respectively. With increasing M, however, one quickly reaches a regime where the assumption of non-relativistic momenta does not hold anymore. In this section we want to discuss an upper limit for the mass M, i.e., a lower limit for the radius R. To that end, we consider the relativistic case. The electronic kinetic energy (properly subtracting the electron rest mass) then becomes

$$\begin{aligned}
E_{\text{kin}} &= \frac{2\mathcal{V}}{(2\pi)^3}\int_0^{k_F} \mathrm{d}^3k \left[\sqrt{(\hbar c k)^2 + (mc^2)^2} - mc^2\right] \\
&\approx \frac{2\mathcal{V}4\pi}{(2\pi)^3}\int_0^{k_F} \mathrm{d}k\, k^2 \left[\hbar c k - mc^2 + \frac{m^2c^3}{2\hbar k}\right] \\
&= \frac{\mathcal{V}}{4\pi^2}\hbar c k_F^4 - \frac{\mathcal{V}}{3\pi^2} mc^2 k_F^3 + \frac{\mathcal{V}}{4\pi^2}\frac{m^2c^3}{\hbar}k_F^2 \\
&= \frac{3}{4}N_e \hbar c k_F - N_e mc^2 + \frac{3}{4}N_e\frac{m^2c^3}{\hbar k_F} \\
&= \frac{3}{4}\left(\frac{9\pi}{4}\right)^{2/3}\hbar c\frac{N_e^{4/3}}{R} - N_e mc^2 + \frac{3}{4}\left(\frac{4}{9\pi}\right)^{2/3} N_e^{2/3}\frac{m^2c^3}{\hbar}R, \qquad (2.42)
\end{aligned}$$

where we have used that the volume $\mathcal{V}$ is directly linked to the electron number by the relation $\mathcal{V}k_F^3/(3\pi^2) = N_e$ and the relation $k_F = (4/(9\pi))^{1/3}N_e^{1/3}/R$. It is to be noted that the integration starts from a regime of very small values of k where the high-momentum expansion is grossly wrong. Looking at the whole integral, this region is extremely small (k^2 weight) and the error made therein is outweighed by the huge region of relativistic k, thus validating the above approximation. The potential energy remains as given before in (2.40). The leading term of the kinetic energy now has the same trend $\propto R^{-1}$ as the potential energy. We thus write the total energy

$$E = \frac{C}{R} - N_e mc^2 + \frac{3}{4}\left(\frac{4}{9\pi}\right)^{2/3} N_e^{2/3}\frac{m^2c^3}{\hbar}R, \qquad (2.43a)$$

$$C = \frac{3}{4}\left(\frac{9\pi}{4}\right)^{2/3}\hbar c N_e^{4/3} - \frac{12}{5}G m_p^2 N_e^2. \qquad (2.43b)$$

It has a stable minimum if $C > 0$. For $C < 0$, however, the minimal energy corresponds to a singularity $R \longrightarrow 0$ which means that the white dwarf would be unstable and implode. The critical point lies at $C = 0$. It corresponds to a critical particle number $N_{e,c}$ and critical mass $M_{\text{Ch}} = 2m_p N_{e,c}$, which are given by

$$N_{e,c} = \left(\frac{5}{16}\left(\frac{9\pi}{4}\right)^{2/3}\frac{\hbar c}{G m_p^2}\right)^{3/2}, \quad M_{\text{Ch}} = \left(\frac{5}{16}\left(\frac{9\pi}{2}\right)^{2/3}\frac{\hbar c}{G m_p^{4/3}}\right)^{3/2}. \qquad (2.44)$$

The critical mass M_{Ch} is known as the *Chandrasekhar mass* [77]. The star becomes unstable if $M > M_{\text{Ch}}$. On the contrary for $M < M_{\text{Ch}}$ an equilibrium is possible. The Chandrasekhar mass thus fixes the maximum mass a self-gravitating object may possess. Our simple-minded calculation leads to $M_{\text{Ch}} \simeq 1.7 M_\odot$ ($N_{e,c} \sim 1.2\,10^{57}$) while more elaborate calculations give a slightly smaller value of $M_{\text{Ch}} \simeq 1.45 M_\odot$ [77].

2.5 Coulomb Energy of a Charged Fermi Gas

We now go one step beyond the Fermi-gas model and consider a homogeneous electron gas including the repulsive Coulomb interaction between the electrons. This is, e.g., a model for the valence electrons in bulk metal where the positive ionic background has been smoothed to a homogeneous jellium density (1.1). The Hamiltonian of that many-body system is the sum of kinetic energy and Coulomb interaction:

$$\hat{H} = \sum_{i=1}^{N} \frac{\hat{p}_i^2}{2m_\mathrm{e}} + \frac{1}{2} \sum_{i \neq j=1}^{N} \frac{e^2}{|\mathbf{r}_i - \mathbf{r}_j|} . \tag{2.45}$$

For a finite system of independent particles the energy then reads

$$E = E_\mathrm{kin} + E_\mathrm{pot,dir} + E_\mathrm{pot,ex} . \tag{2.46a}$$

The kinetic energy was discussed and evaluated in Sect. 2.2.5. The potential energy consists of two terms,

$$\begin{aligned} E_\mathrm{pot,dir} &= \frac{1}{2} \sum_{\alpha\beta} \int \mathrm{d}x\, \mathrm{d}x'\, \varphi_\alpha^*(x)\varphi_\beta^*(x') \frac{e^2}{|\mathbf{r}-\mathbf{r}'|} \varphi_\alpha(x)\varphi_\beta(x') \\ &= \frac{1}{2} \int \mathrm{d}^3r\, \mathrm{d}^3r'\, \rho(\mathbf{r}) \frac{e^2}{|\mathbf{r}-\mathbf{r}'|} \rho(\mathbf{r}') , \end{aligned} \tag{2.46b}$$

$$\begin{aligned} E_\mathrm{pot,ex} &= -\frac{1}{2} \sum_{\alpha\beta} \int \mathrm{d}x\, \mathrm{d}x'\, \varphi_\alpha^*(x)\varphi_\beta^*(x') \frac{e^2}{|\mathbf{r}-\mathbf{r}'|} \varphi_\beta(x)\varphi_\alpha(x') \\ &= -\frac{1}{2} \sum_{\nu\nu'} \int \mathrm{d}^3r\, \mathrm{d}^3r'\, \delta_{\nu\nu'} \varrho(\mathbf{r}\nu, \mathbf{r}'\nu') \frac{e^2}{|\mathbf{r}-\mathbf{r}'|} \varrho(\mathbf{r}'\nu', \mathbf{r}\nu). \end{aligned} \tag{2.46c}$$

The direct part of the Coulomb energy, $E_\mathrm{pot,dir}$, diverges dramatically. Even when considering the energy per volume (or per particle), there remains a quadratic divergence, as the long-range Coulomb force does not allow infinite amounts of charge. The average electron charge needs to be compensated by an equally dense positively charged background with $\rho_\mathrm{back} = \rho_0 = \text{constant}$, since the slightest amount of finite charge density would again lead to a divergent Coulomb energy. Actually, an electron gas in the degenerate regime ($T \ll \varepsilon_\mathrm{F}$) is realized by the valence electrons in a metal. Here, the positive metal ions serve as neutralizing background. It is not homogeneous. But one often ignores the detailed ionic structure and smooths the total positive charge to a homogeneous density distribution, often called the jellium approximation, see (1.1). It is justified by the argument that the electrons have long wavelengths which cannot resolve the ionic details anyway and thus deliver nearly constant density. After the approximation $\rho_{\text{back}} = \rho_0$, we have the quite trivial result that the total direct Coulomb energy is zero because the total charge density $(\rho_{\text{back}} - \rho_0)$ vanishes everywhere.

The exchange energy (2.46c) requires quite a lengthy calculation. On the other hand, it serves as a welcome example to practice a moderately lengthy formal calculation in detail. Before carrying on, we first summarize a few integrals which will be needed in the following:

$$\int \mathrm{d}^3r \frac{\exp(\mathrm{i}\mathbf{k}\cdot\mathbf{r})}{r} = \frac{4\pi}{k^2}, \tag{2.47a}$$

$$\int \mathrm{d}x\,(ax+b)^{-1} = \frac{\log|ax+b|}{a}, \tag{2.47b}$$

$$\int \mathrm{d}k'\,k' \log\left|\frac{k+k'}{k-k'}\right| = kk' + \frac{1}{2}(k'^2 - k^2)\log\left|\frac{k+k'}{k-k'}\right| \tag{2.47c}$$

$$\int \mathrm{d}k\,k^3 \log\left|\frac{k_F+k}{k_F-k}\right| = \frac{1}{4}(k^4 - k_F^4)\log\left|\frac{k_F+k}{k_F-k}\right| + \frac{1}{2}k_F^2k^2 + \frac{1}{6}k_Fk^3. \tag{2.47d}$$

Inserting the plane-wave eigenstates (2.12) and performing the continuum limit $\sum_\alpha \longrightarrow \sum_\sigma \int \mathrm{d}^3k$ yields

$$E_{\text{pot,ex}} = e^2 \sum_{\sigma\sigma'} \int \frac{\mathrm{d}^3r\mathrm{d}^3r'\mathrm{d}^3k\mathrm{d}^3k'}{(2\pi)^6} \underbrace{\chi_\sigma^+ \chi_{\sigma'}}_{\delta_{\sigma\sigma'}} \chi_{\sigma'}^+ \chi_\sigma \frac{\exp \mathrm{i}(\mathbf{k}-\mathbf{k}')\cdot(\mathbf{r}-\mathbf{r}')}{|\mathbf{r}-\mathbf{r}'|}.$$

It is advantageous to transform to center of mass and relative coordinates as

$$\mathbf{R} = \frac{1}{2}(\mathbf{r}+\mathbf{r}'), \quad \tilde{\mathbf{r}} = \mathbf{r}-\mathbf{r}'.$$

The integration volume element changes as $\mathrm{d}^3r\mathrm{d}^3r' \longrightarrow \mathrm{d}^3R\mathrm{d}^3\tilde{r}$. We are now in shape to go through the steps quickly, using the integrals of (2.47). In a first step, spatial integration is performed:

$$E_{\text{pot,ex}} = -\frac{e^2}{(2\pi)^6} \underbrace{\int \mathrm{d}^3R}_{\mathcal{V}} \int \mathrm{d}^3k\mathrm{d}^3k' \underbrace{\int \mathrm{d}^3\tilde{r}\, \frac{\exp \mathrm{i}(\mathbf{k}-\mathbf{k}')\cdot\tilde{\mathbf{r}}}{\tilde{r}}}_{(2.47\text{a})}$$
$$= -\frac{4\pi e^2}{(2\pi)^6}\mathcal{V} \int \mathrm{d}^3k\mathrm{d}^3k' \frac{1}{k^2 + k'^2 - 2\mathbf{k}\cdot\mathbf{k}'}.$$

Again, it is advantageous switching from Cartesian coordinates for the wave vectors to spherical ones. The choice of the polar axis is free and a proper choice can save a huge amount of work. First, we observe that the integrand depends only on three quantities, the absolute values of the wave vectors k as well as k' and the angle between the two vectors $\cos(\theta) = \mathbf{k}\cdot\mathbf{k}'/(kk')$. After the d^3k' integration and the change to spherical coordinates $\mathrm{d}\cos(\theta')\,\mathrm{d}\phi'\,\mathrm{d}k'k'^2$, the integration over d^3k may be attacked for the given $\mathbf{k}$. The choice of the axis for the spherical coordinates of $\mathbf{k}$ is still free. We choose the z-axis parallel to the vector $\mathbf{k}'$. The polar angle θ in that frame is precisely the angle between $\mathbf{k}$ and $\mathbf{k}'$ as given above. The azimuthal angle ϕ thus does not appear in the integrand because the relative distance $|\mathbf{k}-\mathbf{k}'|$ is invariant under azimuthal rotations. The angles θ' and ϕ' also do not appear anywhere in the integrand. This simplifies the integration greatly. The trivial part is $\int_{-1}^{1}\mathrm{d}\cos(\theta')\int_0^{2\pi}\mathrm{d}\phi'\int_0^{2\pi}\mathrm{d}\phi = 2(2\pi)^2$. There remain only three non-trivial integrals

$$\begin{aligned}
E_{\text{pot,ex}} &= -\frac{32\pi^3e^2}{(2\pi)^6}\mathcal{V}\int_0^{k_F}\mathrm{d}k\,k^2\int_0^{k_F}\mathrm{d}k'k'^2\int_{-1}^{+1}\mathrm{d}(\cos\theta)\frac{1}{k^2+k'^2-2kk'\cos\theta} \\
&\qquad\text{use (2.47b)} \\
&= -\frac{e^2}{2\pi^3}\mathcal{V}\int_0^{k_F}\mathrm{d}k\,k\int_0^{k_F}\mathrm{d}k'k'\log\left|\frac{k+k'}{k-k'}\right| \\
&\qquad\text{use (2.47c)} \\
&= -\frac{e^2}{2\pi^3}\mathcal{V}\int_0^{k_F}\mathrm{d}k\left(k^2k_F+kk_F^2\frac{1}{2}\log\left|\frac{k+k_F}{k-k_F}\right|-k^3\frac{1}{2}\log\left|\frac{k+k_F}{k-k_F}\right|\right) \\
&\qquad\text{use (2.47d)} \\
&= -\frac{e^2k_F^4}{2\pi^3}\mathcal{V}\left(\frac{1}{3}+\frac{1}{2}-\frac{1}{4}-\frac{1}{12}\right) = -\frac{2e^2k_F^4}{(2\pi)^3}\mathcal{V}\quad.
\end{aligned}$$

This energy also grows to infinity with $\mathcal{V}$ like the kinetic energy, so that it is preferable similarly to consider the finite energy per particle

$$\frac{E_{\text{pot,ex}}}{N} = \frac{E_{\text{pot,ex}}}{\mathcal{V}}\frac{1}{\rho_0} = -\frac{2e^2k_{\mathrm{F}}}{(2\pi)^3}\,. \tag{2.48}$$

This grows $\propto k_{\mathrm{F}}$, i.e., more slowly than the kinetic energy.

After all, the total energy becomes

$$\frac{E}{N} = \frac{E_{\text{kin}}}{N}+\frac{E_{\text{pot,ex}}}{N} = \frac{\hbar^2}{2m_{\mathrm{e}}}\frac{3}{5}k_{\mathrm{F}}^2-\frac{3e^2}{4\pi}k_{\mathrm{F}} = 0.6\,\frac{\hbar^2}{2m_{\mathrm{e}}}\,k_{\mathrm{F}}^2-0.239\,e^2\,k_{\mathrm{F}}\,. \tag{2.49}$$

The electron–electron interaction is repulsive while the exchange term has the opposite sign and so becomes attractive. That is plausible. But a quick glance at the total energy (2.49) gives the puzzling impression that all binding comes from the electron–electron interaction which is usually considered to be repulsive. Note that the direct term is the one responsible for repulsion and we have seen that it is huge, in fact, insurmountably infinite. It is, however, fully counterweighted by the equally huge attractive contribution from the external positively charged background. The

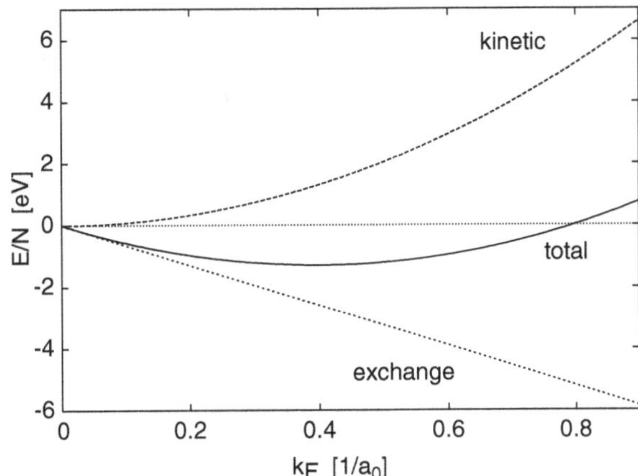

Fig. 2.7 The binding energy per particle for the Fermi gas of electrons, (2.49), and its two constituents, the kinetic energy and the exchange energy

full compensation of the leading forces leaves the small exchange effect on the final tip in the balance and that small contribution provides the binding in the Fermi-gas model, which is counterbalanced by the Fermi pressure from the kinetic energy. The trends are sketched in Fig. 2.7. The linear growth of (negative) exchange energy dominates for small Fermi momenta k_F. It is overruled by the quadratic trend of the repulsive kinetic energy for increasing k_F. The turnover produces a unique minimum which establishes the equilibrium state of the electron gas, having the energy $E/N = -1.295$ eV at the Fermi momentum $k_\mathrm{F} = 0.4/\mathrm{a}_0$ corresponding to $r_s = 4.8$ a_0. It is remarkable how well this number fits typical Wigner–Seitz radii of metals, e.g., $r_s = 3$ a_0 for Ag, $r_s = 4$ a_0 for Na, or $r_s = 5$ a_0 for K. Of course, the ionic structure, adding core polarizability and core repulsion, determines the final detailed value of r_s, but the basic balance seems to be determined already by the electron gas, a simple model with surprisingly realistic aspects.

2.6 Concluding Remarks

The Fermi-gas approximation, simple though it is, was shown to be surprisingly successful in understanding many features of quite distinct systems at least in a semi-quantitative way. The reason was that many systems are characterized by a relatively constant density and weak correlations, which makes the approximation by a gas of non-interacting particles trapped in a constant potential quite adequate. Through the exploration of the thermal behavior, relativistic formulation, and the addition of gravitational and Coulomb interactions we have shown the tremendous flexibility that also characterizes the model. For a first basic understanding of fermion systems, the Fermi-gas model remains an approach of choice.

The next chapters will show how to introduce more realism by discarding the constant potential using either a phenomenological potential adapted to specific systems or a self-consistent one calculated from the single-particle wave functions themselves. In both cases, the many-body state is still a wave function built out of single-particle states that are filled up to Fermi energy, so that many qualitative properties of the Fermi-gas model remain valid.

Chapter 3
Particles in an External Field

In the previous chapter, it was shown how the Fermi-gas model can be a useful lowest-order description even for a finite system. It provides, e.g., correct volume properties, scaling, or level densities, but neglects crucial quantum effects, particularly quantum shell structure, related to the spatial confinement of a finite system. A rough confinement is given by an infinite potential barrier enclosing a flat potential region. More realistic potential wells have finite depth and a smooth transition zone at the surface. There may also be contributions depending on spin or a coupling to the wave functions in other more complicated ways. Each of these features influences the single-particle levels and thus the shell structure considerably.

All these effects are included, of course, in self-consistent models where the fermion–fermion interaction is taken into account at least at the mean-field level. These approaches will be discussed in Chaps. 5 and 6. They reach a high degree of descriptive power, but are usually rather complicated and computationally expensive. Thus for an intermediate level of description simple mean-field potentials have been developed, which are appropriate for describing some system properties and leave a few free parameters for fine tuning. This sort of empirically guessed mean-field potential is often called a *(phenomenological) shell model*, see, e.g., [9]. The aim is to choose, within the intuitively reasonable possibilities, a realistic potential which is also easy to use, ideally providing analytical solutions. The simplest and most widely used model potential is the harmonic oscillator which thus will cover the largest portion of this chapter.

In such phenomenological models it is possible to also prescribe the *shape* of the system. This is crucial because finite fermion systems often have non-spherical ground states because of the shell structure. This will be illustrated later in Sects. 3.2.4 and 3.2.5.

3.1 The Variety of Shell-Model Potentials

3.1.1 Saturating Fermion Systems

A first demonstration of the principal types of potentials is given in Fig. 3.1 for a Na cluster as a typical test case for a saturating fermion system. The fully self-consistent

J.A. Maruhn et al., *Simple Models of Many-Fermion Systems*,
DOI 10.1007/978-3-642-03839-6_3, © Springer-Verlag Berlin Heidelberg 2010

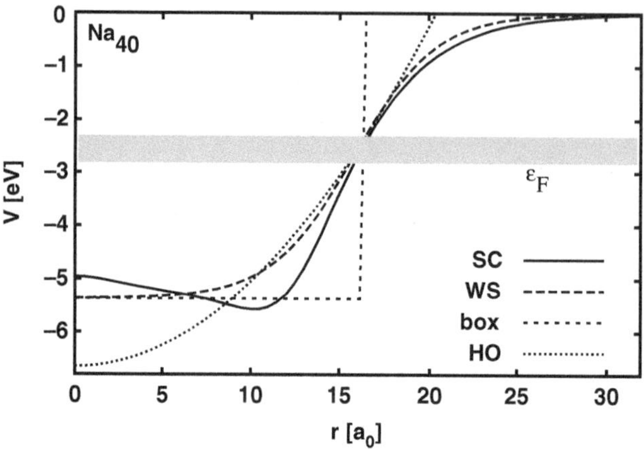

Fig. 3.1 One-body potentials for a Na_{20} cluster. The benchmark is the self-consistent potential from a DFT calculation denoted as "SC". The approximate choices are: "WS" = Woods–Saxon, "box" = spherical square well, "HO" = harmonic oscillator. The Fermi energy ϵ_F (depending on energy) is indicated by a *grey horizontal stripe*

potential was obtained numerically with DFT, see Sect. 6.1, and the positive background was described in jellium approximation, (1.1). It is compared with three widely used shell-model potentials as defined in Table 3.1. Closest to the typical mean-field potential shape is the Woods–Saxon potential with its three parameters depth V_{WS}, radius R_{WS}, and width σ_{WS}. Since it is also a good approximation to the nuclear mean field, the Woods–Saxon potential has been widely used in many fields of physics. Analytical solutions, however, do not exist, or only approximately [35]. Therefore, one often resorts to even simpler models, the spherical box (also called square well) potential (denoted "box" in Fig. 3.1) or the harmonic oscillator (denoted "HO"). The latter is particularly useful as it has simple analytical solutions. We will in this chapter mainly discuss the harmonic oscillator for this reason.

The comparison in Fig. 3.1 shows that some model potentials are not flexible enough to follow the given mean field everywhere. Fortunately, that is not really necessary. For most purposes, it suffices to reproduce the situation in the active zone around the Fermi energy as sketched in Fig. 3.1. In fact, the model parameters are usually adjusted to reproduce the single-particle levels around the Fermi energy because the level sequence here is the crucial feature which decides shell effects

Table 3.1 Three typical spherically symmetric model potentials

Woods–Saxon	: $V_{WS}(r) = \dfrac{V_{WS}}{1 + \exp((r - R_{WS})/\sigma_{WS})}$
Square well	: $V_{box}(r) = \begin{cases} -V_{box} & \text{for } r < R_{box} \\ +\infty & \text{for } r \geq R_{box} \end{cases}$
Harmonic oscillator	: $V_{HO}(r) = \dfrac{m}{2}\omega^2 r^2 - V_{HO,0}$

like magic numbers or deformations. The self-consistent potential (denoted "SC") in Fig. 3.1 falls in between the two simplest potentials (box and HO). The (almost) flat bottom of the SC potential becomes more extended for larger systems, thus making the spherical box preferable there, while smaller systems come much closer to the harmonic oscillator.

The situation as sketched in Fig. 3.1 applies to all saturating systems, nuclei, metal clusters, and ^{3}He droplets. It is also valid for quantum dots, whose dimensions are tunable, but where the density distribution follows the same pattern (for given overall radius) and it is often realistic for atomic clouds in a trap. For saturating systems the central density is always about the same, independent of the number of particles and the system's shape. Thus the volume grows $\propto N$, where N is the total particle number. This leads to the crucial condition of *volume conservation*, which is a condition for the geometric size parameters in the potential, see Sect. 3.2.

3.1.2 Atomic Systems

The non-scalable systems, atoms and molecules, require quite different potentials which are not so easy to model in simple terms. An attempt to reproduce the atomic mean field is, for example, the Hulthén potential

$$V(r) = -\frac{Ze^2}{a}\frac{1}{\mathrm{e}^{r/a} - 1},$$

which becomes proportional to Ze^2/r for $r \longrightarrow 0$ and simulates screening for $r \longrightarrow \infty$. There exist analytical solutions for angular momentum $l = 0$ [35], but unfortunately only approximate ones for other l so that it is of limited use. The difficulty to develop simple model potentials for atoms becomes even more dramatic for all-electron models of molecules. Thus the typical approximation schemes in the realm of atoms and molecules are variational methods with simple basis wave functions. These methods range from nice instructive schematic models like the tight-binding model up to quite elaborate quantum chemical calculations [84]. We will address the tight-binding approach in Chap. 4. More elaborate methods are beyond the scope of this book.

For atoms, there is an alternative path of approximations where the aim is to reduce the many-body problem of all electrons to an effective one- (or few-) body problem for the valence electron(s). This leads to the pseudo-potential strategy [101]. It is illustrated in Fig. 3.2 for the case of the Na atom, having a $1s$, $2s$, $2p$ core and one valence electron in the $3s$ shell. The left part shows part of the level scheme from an all-electron calculation (the $1s$ state lies too deep to fit into the plot). The energetic separation between the core levels and the valence state is large, so that one expects the core electrons to stay rather inert in the low-energy dynamics around the valence shell. The pseudo-potential is now constructed such that the solution of the one-electron problem yields a spectrum which is similar to the excitation spectrum of the atom near the valence shell and such that the intermediate and long-range radial extension of the valence state is also reproduced. Such a situation is shown

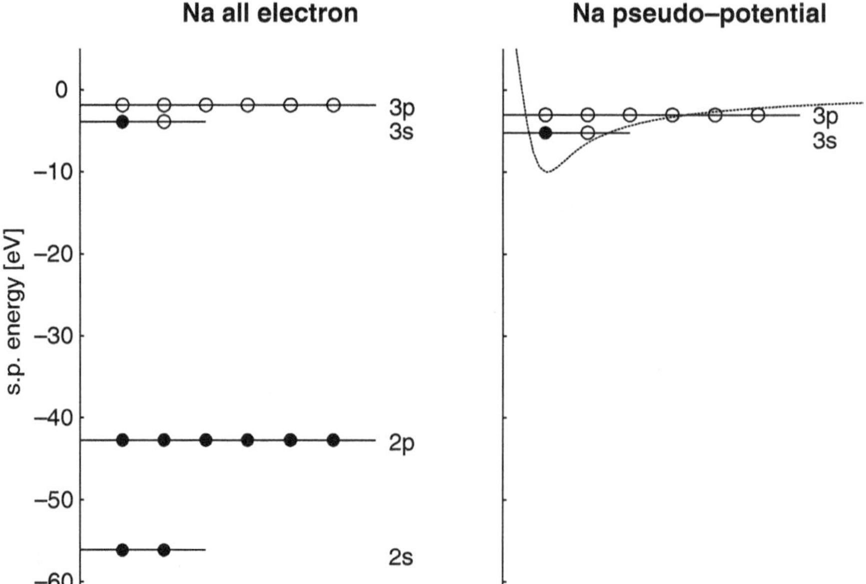

Fig. 3.2 The single-electron energies of the Na atom from an all-electron DFT calculation (*left*) and from a one-electron pseudo-potential (shape indicated by dotted line) calculation (*right*). *Full circles* indicate occupied and *open circles* unoccupied states

in the right part of Fig. 3.2. Pseudo-potentials are calibrated carefully to atomic properties and then extensively used for calculating electronic states and bonding of molecules. Although they simplify quantum-chemical calculations enormously, pseudo-potentials are rarely simple enough to allow for an analytical solution. They thus do not belong to the sort of simple models which we are discussing in this book. We will nevertheless deal with the simplest model for an atomic transition (e.g., the $3s \longrightarrow 3p$ one in Fig. 3.2), namely a two-level model which is isomorphous to a spin $1/2$ algebra, see Chap. 7.

3.2 The Harmonic Oscillator in the Various Subfields

The technically simplest case is the harmonic oscillator, so that developments start with an oscillator model wherever possible. It turns out that a great variety of physical situations allow this. We here briefly summarize the major applications. Detailed solutions and results will be outlined in the next section.

Volume conservation is a key feature of saturating systems. It shows up first at the level of the spatial density distribution which is basically constant up the boundaries where it drops to zero rather quickly. Density is N/V which means that the enclosed volume grows proportional to system size N. The relation $R^3 \propto N \propto V$ then holds for the radius of spherically symmetric systems. Deformed systems can be characterized by three different extensions R_x, R_y, R_z. Again the constancy of density imposes $R_x R_y R_z \propto V =$ constant for a given particle number N. When using a model Hamiltonian as we do here, we are not given the density initially but a potential. The surface of an N-fermion system can then approximately be given by the equipotential line at the Fermi surface. Thus one takes the condition that the Fermi energy is independent of particle number and shape for a saturating system. This condition implies that the oscillator frequency follows a trend $\omega \propto \epsilon_F/N^{1/3}$ for a 3D system, and $\omega \propto \epsilon_F/\sqrt{N}$ in 2D. These properties will be obeyed in all the following examples.

3.2.1 The Harmonic Oscillator in Two and Three Dimensions

The general form of the harmonic oscillator Hamiltonian for the 3D case is

$$\hat{h} = \frac{\hat{\mathbf{p}}^2}{2m} + \tfrac{1}{2}m\left(\omega_x^2 x^2 + \omega_y^2 y^2 + \omega_z^2 z^2\right). \tag{3.1}$$

The 2D case can be obtained trivially by just omitting the contribution of the z-coordinate and momentum.

There are several interesting limiting cases: if all the ω's are equal, the oscillator has spherical or, in the 2D case, circular symmetry. If two ω's are equal in the 3D case, the oscillator is axially symmetric. If all ω's are different, the oscillator becomes triaxial. The deviation of the oscillator from spherical or circular symmetry is determined by the ratios of the different ω's, their absolute size is dictated by volume conservation. The spectrum can be formulated in different ways depending on the symmetry (details are given in Appendix A.1).

The single-particle spectrum of the pure oscillator is easy to write down in the Cartesian basis, it is just the sum of the contributions of 1D oscillators in each coordinate direction,

$$\varepsilon_{n_x n_y n_z} = \hbar\omega_x(n_x + \tfrac{1}{2}) + \hbar\omega_y(n_y + \tfrac{1}{2}) + \hbar\omega_z(n_z + \tfrac{1}{2}), \tag{3.2}$$

where again the 2D case can be obtained by omitting the contribution of the z-direction. This form is particularly useful for truly triaxial deformation. More symmetric situations allow alternative representations which display symmetry quantum numbers explicitly.

For the spherical oscillator with $\omega_x = \omega_y = \omega_z = \omega$, the spectrum can be expressed also in a spherical basis with good angular momentum quantum numbers (see Appendix A.1):

$$\varepsilon_{nlm} = \hbar\omega(n + \tfrac{3}{2}), \; n = 0, 1, \ldots, \; l = n, n-2, \ldots \geq 0, \; m = -l, \ldots, +l. \tag{3.3}$$

Note that the energy does not depend on the value m of the z-component of angular momentum. It shows the typical $2l+1$ fold degeneracy of all central-field problems. Moreover, there is a degeneracy covering different l values which is specific for the harmonic oscillator.

The deformed axially symmetric oscillator has $\omega_x = \omega_y = \omega_r$. The z-component of angular momentum then remains a good quantum number m. The practical solution is performed in cylindrical coordinates and the spectrum becomes

$$\begin{aligned} \varepsilon_{n_r m n_z} &= \hbar\omega_r(2n+|m|+1) + \hbar\omega_z(n_z+\tfrac{1}{2}) \,, \\ \omega_r &= \frac{R_x}{R_0}\omega \,, \quad \omega_z = \frac{R_z}{R_0}\omega \,, \\ n &= 0, 1, \ldots \,, \; m = 0, \pm 1, \ldots \,, \; n_z = 0, 1, \ldots \end{aligned} \tag{3.4}$$

These spectra provide the central part of the discussions of oscillator models in the various applications, which introduce some differences via additional contributions to the potential and also through the notation used for the coefficients. We now give a brief overview of these applications before going into a more detailed discussion.

3.2.2 The 3D Oscillator Atomic Traps

The confining potentials for atoms in a trap are to a very good approximation harmonic such that the 3D oscillator Hamiltonian (3.1) is a standard starting point for theoretical investigations in that field [14, 38]. The oscillator frequencies $\omega_{x,y,z}$ are a design property and can be varied in a large range. We will exploit that in connection with density functional theory in Sect. 6.5.2.

The values of ω typically range between several kHz and a few MHz. The oscillators are usually deformed, mostly axially, i.e., $R_z \neq R_x = R_y$. The trapped systems differ from self-bound systems in that the oscillator frequency is adjustable by design and not determined by an internal systems property (like the ϵ_F in (3.6)). Moreover, shell effects are not the issue in trapped clouds of atoms. One rather uses the tunable conditions to explore the different regimes of correlations from basically free atoms to pairing correlations (see Sect. 9.3.1) and condensates [14].

3.2.3 The 2D Oscillator for Quantum Dots

The mean-field potential for electrons in quantum dots is often described by a 2D oscillator, which is usually written as [87]

$$\hat{h}_{\text{HO,2D}} = \frac{\hat{p}_x^2 + \hat{p}_y^2}{2m^*} + \frac{m^*}{2}\omega^2 \left(\frac{R_0^2}{R_x^2}x^2 + \frac{R_0^2}{R_y^2}y^2 \right), \quad R_x R_y = R_0^2, \tag{3.5}$$

$$\omega = \frac{e^2}{\epsilon m^* r_s^3 \sqrt{N}}, \quad r_s \approx 1.3\text{–}1.4\text{a}_B^* ,$$

where m^* is the effective mass of the electrons moving in the semiconductor material of the dot [3], ϵ its dielectric constant, a_B^* the corresponding effective Bohr radius for the elementary exciton, and r_s = the Wigner–Seitz radius of the confined electron cloud (Sect. 1.1.6). A possible deformation can be described by setting the relevant extensions $R_x \neq R_y$ while, of course, obeying volume conservation $R_x R_y = R_0^2$. The choice $R_x = R_y$ corresponds to a circular potential.

Two comments justifying this great simplification are in order. First, the electron cloud is, in principle, a 3D object. The extension in the third dimension is, however, much smaller than in x and y, so that the wave functions are all in the (oscillator) ground state with respect to the z-direction and this dimension becomes irrelevant. Second, the external confining potential walls in the semiconductor are rarely so nicely harmonic nor are they circular. It is the electron–electron interaction which smoothes the landscape such that the self-consistent mean field comes close to the oscillator ideal, at least for not too large electron numbers [60].

Finally, we ought to mention that such planar situations are realized not only in quantum dots but also for electrons in small metal clusters deposited on a surface with large interface interaction like, e.g., the combination of Na clusters on a NaCl surface [55]. All discussions below on the sequence of magic numbers and on deformation also apply to that case. There is an intriguing difference to quantum dots, though. The deformation is defined by construction for the dot, while the deposited cluster has more freedom to a self-adjustment of shape (similar to free clusters) within the bounds set by the strong interface interaction from the surface, i.e., in the two dimensions parallel to the surface. This holds, as discussed at the end of Sect. 3.3.1, for Na clusters at room temperature. Ionic effects also have to be considered for very cold and/or very large clusters.

3.2.4 The Clemenger–Nilsson Model for Metal Clusters

The mean field of metal clusters is often parametrized in terms of a 3D oscillator written as

$$\hat{h}_{\text{HO,3D}} = \frac{\hat{\mathbf{p}}^2}{2m} + \frac{m}{2}\omega^2 \left(\frac{R_0^2}{R_x^2}x^2 + \frac{R_0^2}{R_y^2}y^2 + \frac{R_0^2}{R_z^2}z^2 \right), \; R_x R_y R_z = R_0^3, \tag{3.6}$$

$$\omega \approx \epsilon_{\text{F}} N^{-1/3} \Big[1 + \frac{\delta R_{\text{sp}}}{R_0} \Big]^{-2},$$

where ϵ_F is the Fermi energy of the bulk material and the term with δR_{sp} corrects for the electronic spill out in small systems. The spill-out effect reflects the fact that the electronic density is more spread-out than the ionic one and thus literally spills out of the ionic density. The R_i allow to give the three directions different extensions (triaxial oscillator), and $R_x R_y R_z = R_0^3$ maintains volume conservation. The choice of ω describes the proper scaling with particle number, but with a spill-out correction as mentioned.

The strictly harmonic potential becomes increasingly unrealistic for larger systems (see the discussion of Fig. 3.1). One can stay close to the beneficial simplicity of the harmonic oscillator by a corrective term

$$\hat{h}_{\text{CN}} = \hat{h}_{\text{HO,3D}} - D\frac{\omega}{\hbar}\hat{\mathbf{l}}^2, \tag{3.7}$$

where $\hat{\mathbf{l}}$ is the operator of orbital angular momentum. In cluster physics this is called the *Clemenger–Nilsson model* [117, 59]. The term $\propto \hat{\mathbf{l}}^2$ shifts the states with larger angular momentum to lower energy, simulating this feature from a finite well. A typical choice for the dimensionless parameter D lies around 0.05, depending somewhat on the intended system size. At first glance, it appears complicated to correct by an involved operator rather than by a simple local term, but note that the term is trivially evaluated in the spherical oscillator where angular momentum becomes a good quantum number and the shift reduces to $-Dl(l+1)$ without the need for modifying the wave functions.

3.2.5 The Nilsson Model for Nuclei

The model potential (3.1) goes back, in fact, to the nuclear oscillator model called the *Nilsson model* [18, 42]. Here, a strong spin–orbit term has to be added because that is compulsory to explain nuclear shell structure. Thus we augment the 3D oscillator model (3.6) including the $\hat{\mathbf{l}}^2$ correction (3.7) and a strong spin–orbit interaction, yielding the Nilsson model Hamiltonian

$$\hat{h}_{\text{Nil}} = \hat{h}_{\text{HO,3D}} - \kappa\frac{\omega}{\hbar}\left(2\hat{\mathbf{l}}\cdot\hat{\mathbf{s}} + \mu\hat{\mathbf{l}}^2\right) \tag{3.8}$$

$$\hbar\omega = \frac{41\,\text{MeV}}{A^{1/3}}\,, \qquad \kappa \approx 0.05\,, \qquad \mu \approx 0.3\,. \tag{3.9}$$

The values for κ and μ are fitted to experimental data and vary somewhat with the size region considered. It ought to be mentioned also that there exist some variants of the $\hat{\mathbf{l}}^2$ correction in the literature.

3.2.6 Comparison of Spectra

We can now compare the spectra for the various cases discussed. It should be noted that the spherical representation allows a closed analytical solution even for the modified oscillators in the Nilsson (3.8) and Clemenger–Nilsson models (3.7). The cylindrical form is often employed for strongly deformed nuclei, see Sect. 3.3.3.2. It becomes particularly useful for a circular quantum dot, in which case one merely skips the z-contribution from (3.4), see Sect. 3.3.2.

Figure 3.3 gives an overview of the single-particle levels for the various shell-model potentials discussed above, all for the case of spherical symmetry (or circular symmetry for the 2D oscillator). Details of dealing with the corrective terms $\hat{\mathbf{l}}^2$ and $\hat{\mathbf{l}} \cdot \hat{\mathbf{s}}$ will be discussed with the applications later on. The multiplicity of a given level is indicated by the length of each level line. It is of course larger the higher the symmetry of the system (3D versus 2D for example). The figure also shows the magic numbers obtained by filling all states from below (including spin degeneracy) according to the Pauli principle up to the indicated shell closure. It is obvious that the 2D structure, e.g., of a quantum dot has a quite different sequence of magic numbers.

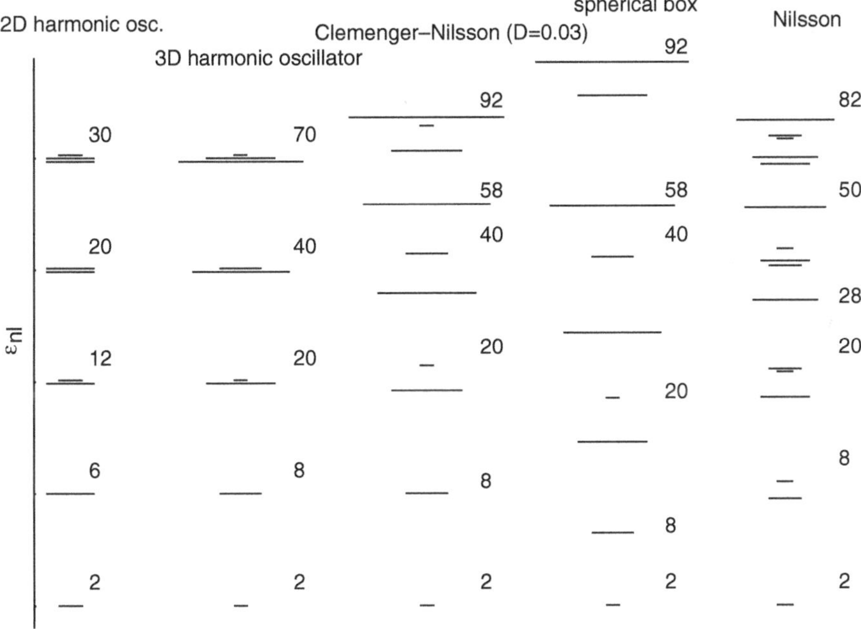

Fig. 3.3 Comparison of the single-particle spectra and magic numbers for various model potentials: *leftmost* = 2D oscillator, second from *left* = pure 3D oscillator (3.1), *middle* = the full Clemenger–Nilsson model (3.6) including the corrective term (3.7), second from *right* = spherical box potential (Table 3.1 and Appendix A.1.4), *right* = nuclear Nilsson model (3.8). The lengths of the lines are proportional to the degeneracy of the corresponding level, i. e., indicate how many particles can occupy each level

The three cases in the middle all represent an option for 3D Fermi systems without spin–orbit force such as metal clusters. The three potentials produce comparable sequences up to $N = 20$, possibly up to $N = 40$. Above that we find a different sequence of magic numbers for the pure oscillator as compared to the box and $\hat{\mathbf{l}}^2$ corrected oscillator (Clemenger–Nilsson). Self-consistent calculations confirm the sequence 2, 8, 20, 40, 58, 92 (which corresponds to the experimental one in simple metal clusters with one valence electron per atom such as Na, Cs, K, ...) and have spectra which lie between the Clemenger–Nilsson model and the spherical box. This result is plausible in view of Fig. 3.1, which showed that larger systems will have a broad flat bottom in the mean-field potential, which can hardly be simulated by a purely harmonic potential.

The rightmost example in Fig. 3.3 shows the sequence for the Nilsson model discussed in Sect. 3.2.5. The decisive influence of the strong spin–orbit splitting is clearly visible. The series of magic numbers above $N = 20$ changes from 40, 58, 92 in the Clemenger–Nilsson model to the typical nuclear values 28, 50, 82, 126. More details for these different oscillator models will be outlined in the following sections discussing a few selected applications from various subfields.

3.3 Applications

3.3.1 Metal Clusters

As the technically simplest example, we start with the case of metal clusters in the Clemenger–Nilsson model. We consider only small clusters for which the simple pure-oscillator form (3.1) is still appropriate. The aim is to demonstrate how shell structure can determine the global shape of the system. To keep geometry simple, we consider only axially symmetric shapes, i.e., $R_x = R_y \neq R_z$, which leaves only one parameter characterizing deformation. It is customary to use the distortion parameter [117]

$$\eta = 2\frac{R_z - R_x}{R_z + R_x} = 2\frac{R_z/R_x - 1}{R_z/R_x + 1}\,, \quad \frac{R_x}{R_0} = \left(\frac{2-\eta}{2+\eta}\right)^{1/3}\,, \quad \frac{R_z}{R_0} = \frac{R_0^2}{R_x^2}\,, \tag{3.10}$$

where volume conservation has been used to express the R_i in terms of η. The single-particle energies then become

$$\frac{\varepsilon_{n_x n_y n_z}(\eta)}{\hbar\omega_0} = (n_x + n_y + 1)\left(\frac{2+\eta}{2-\eta}\right)^{1/3} + \left(n_z + \tfrac{1}{2}\right)\left(\frac{2+\eta}{2-\eta}\right)^{-2/3}\,. \tag{3.11a}$$

A first rough estimate for the total energy is obtained simply from the sum of the single-particle energies running over the occupied states, i.e.,

$$E(\eta) = \sum_{n_x n_y n_z \in \text{occupied}} \varepsilon_{n_x n_y n_z}(\eta) \,. \tag{3.11b}$$

The occupied states are chosen in standard fashion by filling the energetically lowest single-particle states.

The left panel of Fig. 3.4 shows the sequence of single-particle levels (3.11a) as a function of deformation. Each level can be occupied by two electrons (spin up and spin down). Let us first consider the spherical shape ($\eta = 0$). Two electrons fill the lowest shell in a unique fashion. Adding the next electron leads to an ambiguous situation as there are six energetically equivalent levels available which could be occupied. Adding two electrons already yields 30 possible configurations. It is only at $N = 8$ that a unique filling is reached again. That is then a particularly stable configuration, called a magic electron number. The next full shell is reached at $N = 20$. The ambiguous situation in between the shell closures is resolved by deformation. The spherically degenerate levels develop differently with η. This dissolves spherical shell closures and produces some gaps at other places. For $N = 4$ we find a reasonably large gap at a prolate deformation of $\eta = 0.6$, for $N = 6$ at an oblate one of $\eta = -0.7$. Further shell fillings are indicated in the plot.

So far, the argument for the preferred shape was to achieve a maximal shell gap. That is closely related to a minimum in the total energy. The right panel of Fig. 3.4

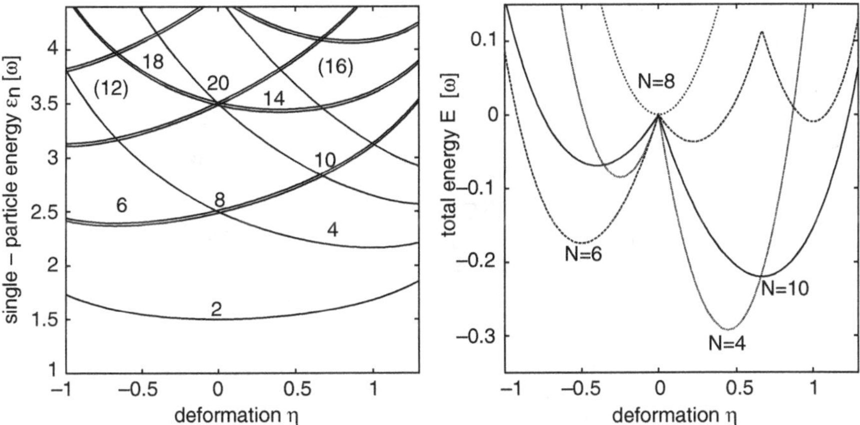

Fig. 3.4 *Left panel*: The level sequence (3.11a) for the axially symmetric, deformed harmonic oscillator drawn versus the distortion parameter η as defined in (3.10). The numbers in the shell gaps indicate the numbers of electrons reached when filling up to the last level below that number. The cases $N = 16$ and $N = 12$ are given in brackets because both these systems actually prefer a triaxial deformation. *Right panel*: The total energies (3.11b) for various electron numbers N as indicated computed for the axially symmetric, deformed harmonic oscillator drawn versus the distortion parameter

shows the total energy (3.11b) obtained as the sum of occupied single-particle energies. This is a rather rough approach as we will see in Chap. 5, but it suffices for the present qualitative discussion. To cover all systems, the energy is adjusted to zero at spherical shape, so that the curves show the relative energies, i.e., the energy changes through deformation. The minima for each N are found indeed at about the same deformations as predicted from looking at the gaps in the left panel.

Figure 3.5 compares the ground-state deformations estimated this way with experimental data for positively charged Na clusters up to $N = 40$. The experimental deformation is deduced via its effect on the Mie plasmon resonance (Sect. 1.1.5), because the resonance peaks for modes in x-, y-, and z-direction depend on the extension of the system in the corresponding direction and this, in turn, allows to determine the global quadrupole deformation [59, 117, 97, 89]. The agreement is excellent and it is noteworthy that this is achieved by the extremely simple 3D oscillator model.

The good agreement with experimental results suggests that the electronic shell dominates the cluster shape and that the underlying ionic configuration seems to willingly follow the electronic driving forces. That is corroborated by mean-field calculations of the electron cloud (see Sect. 6.1) including detailed ionic structure. Figure 3.6 shows the computed ionic shapes for three small Na clusters. These very small systems may be expected to experience strong perturbations through ionic structures, but the result shows that the shapes in the average conform nicely with the predictions of the Clemenger–Nilsson model, strongly oblate Na_6, almost spherical Na_8, and strongly prolate Na_{10}. The point is that the electron cloud in metal clusters extends over all ions and screens the ion–ion interaction to a large extent; consequently the electron cloud explores the shell effects which are given by the global quadrupole deformation and the weakly interacting ions move into the optimum positions as dictated by the electronic cloud. A word of caution is in place here. The

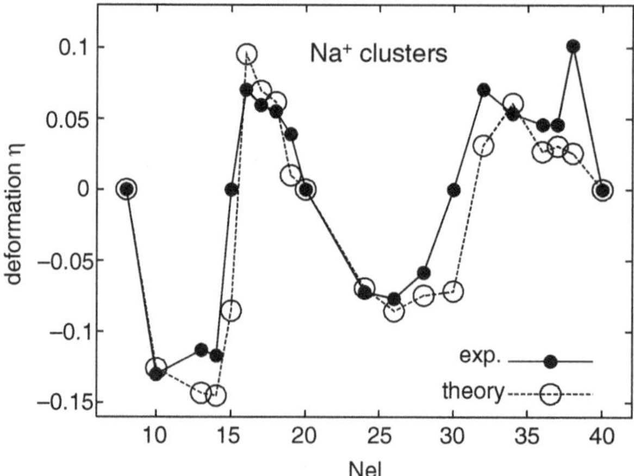

Fig. 3.5 Predictions of the ground-state deformation parameter η in Na^+ clusters as deduced from the Clemenger–Nilsson model with experimental data from [97]

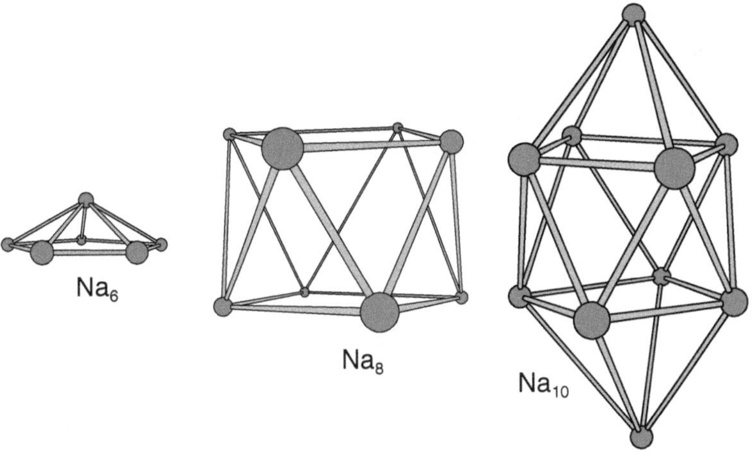

Fig. 3.6 Detailed ionic configurations for Na_6 (*left*), Na_8 (*middle*), and Na_{10} (*right*)

self-adjustment of cluster shape by electronic shell effects holds only if ionic effects on the shape can be ignored. This is approximately valid for simple metal clusters (alkalies) above melting temperature, e.g., for Na at room temperature [19, 89].

3.3.2 Quantum Dot in a Magnetic Field

Electrons in quantum dots are confined to two dimensions and can be well described by a 2D oscillator potential [87]. The level sequence of the 2D oscillator and associated magic electron numbers were shown in Fig. 3.3. Shell closures and magic numbers play an important role for the properties of the quantum dot (addition energies which are similar to ionization potential, conductivity). A particularly appealing feature of quantum dots is that their shell structure can be easily modified by external magnetic fields which, in turn, allows to modify their physical properties at will.

The Hamiltonian of a 2D oscillator in a homogeneous magnetic field perpendicular to the dot's plane, namely along the z-axis (axis of heterostructure confinement, see Sect. 1.1.6), while the harmonic field is along x and y reads

$$\hat{H} = \frac{1}{2m^*}\left(\hat{\mathbf{p}} - \frac{e}{c}\mathbf{A}\right)^2 + \frac{m^*}{2}\omega^2 r^2 \,, \quad \mathbf{A} = \frac{B}{2}(-y, x) \,, \quad r^2 = x^2 + y^2 \,,$$

where m^* is the effective electron mass of the given semiconductor material in the dot and B the strength of the magnetic field. Evaluating the kinetic term yields

$$\hat{H} = \frac{\hat{\mathbf{p}}^2}{2m^*} + \frac{m^*}{2}\left(\omega^2 + \frac{\omega_B^2}{4}\right) r^2 + \frac{\omega_B}{2}\hat{l}_z \,, \quad \omega_B = \frac{eB}{m^* c} \,, \tag{3.12}$$

where ω_B is the cyclotron frequency in the field B. The first two terms of this Hamiltonian are equivalent to the radial part of the oscillator Hamiltonian in axially symmetric representation, see (3.4) without the z-direction. The spectrum is as given in (3.4) when ignoring the dependence on n_z, complemented by the cyclotron term. It can be classified in terms of the angular momentum projection m and the radial oscillator quantum number n and becomes

$$\frac{\varepsilon_{nm}}{\hbar} = \sqrt{\omega^2 + \frac{\omega_B^2}{4}}(2n+|m|+1) + \frac{\omega_B}{2}m\,, \quad n = 0, 1, \ldots\,, \ m = 0, \pm 1, \ldots\,. \tag{3.13}$$

The left part of Fig. 3.7 shows the dependence of the spectrum on the cyclotron frequency ω_B. One clearly sees the bunching in shells for zero magnetic field with the magic numbers as already shown in (3.3). The magnetic field first makes the spectrum more diffuse and so destroys the shells. However, there appear new shell gaps at certain values of the magnetic field. They yield different magic numbers as

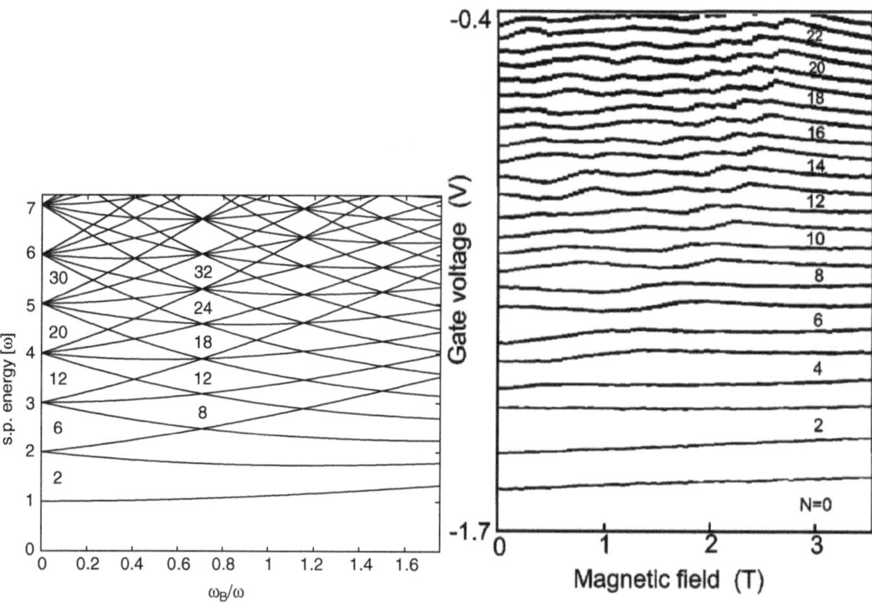

Fig. 3.7 *Left panel*: The single-particle spectrum (3.13) of the 2D oscillator in a magnetic field as a function of the cyclotron frequency ω_B. The magic numbers at shell closures are indicated for zero field and at the first occurrence for finite magnetic field. *Right panel*: Experimental results for the peaks of the Coulomb blockade (in terms of the corresponding gate voltage) as function of external magnetic field [104]. The electron number associated with each peak is indicated

indicated for the first spot of reappearing gaps. The shell structure can be measured by electron-addition energies, the analogue of ionization potentials in atoms. It also shows up in the conductance of a dot. Due to the quantization of electron number in the dot, there are well-defined conductance peaks at certain voltages and a strong suppression elsewhere (called Coulomb blockade [87]). The peaks are related to the energy levels. The right part of Fig. 3.7 presents experimental results on the effect of the magnetic field on conductivity of the dot. It shows the peak voltages, i.e., the voltages associated with a peak in conductivity (seen as function of voltage). Looking at zero magnetic field, we can spot small gaps between the peak voltages at $N = 2, 6, 12$. In the region of about 1 T we find the gaps at the sequence 2, 4, 8, 12. Both results are in accordance with the numbers shown in the left panel.

3.3.3 Nuclei

3.3.3.1 The Spherical Oscillator Model

In this section, we look at the spherical harmonic oscillator and its application to nuclear shell structure. The dependence of the oscillator parameter on the total nucleon number A is $\omega \propto A^{-1/3}$ because of volume conservation. A standard choice is

$$\hbar\omega = \frac{41\,\mathrm{MeV}}{A^{1/3}} ,$$

where the numerical factor is adjusted to reproduce the average shell spacing near the Fermi energy. The harmonic oscillator shows strong shell structure because of its high degeneracy. Using the single-particle energies in the labeling (3.3) and subtracting the contribution from the term $\propto \hat{\mathbf{l}}^2 = l(l+1)\hbar^2$ yields the spectrum as shown in Fig. 3.8 on the left-hand side. The different orbital angular momenta contained in each oscillator shell are already split up because of the $\mathbf{l}^2$-term. The letters $s, p, d, f, g, \ldots$ indicate $l = 0, 1, 2, 3, 4, \ldots$, following the familiar notation from atomic physics (see Sect. 1.1.3). Multiplying by a factor of 2 for spin, we find the degeneracies as indicated there (round brackets), and adding them up yields the magic numbers also indicated (square brackets). They do not agree with what is found in experiments. One needs to take care of the spin–orbit coupling in the Hamiltonian (3.8). Now the projections of the orbital angular momentum $\hat{l}_z$ and the spin $\hat{s}_z$ are no longer good quantum numbers and must be coupled to total angular momentum j. According to Appendix A.2, the triangle rule allows two values for the total angular momentum: $j = l \pm \frac{1}{2}$. If the harmonic oscillator wave functions are used in spherical coordinates, the spin–orbit coupling term then is immediately diagonal and its value can be computed from

$$\hat{\mathbf{l}} \cdot \hat{\mathbf{s}} = \tfrac{1}{2}\left[\left(\hat{\mathbf{l}} + \hat{\mathbf{s}}\right)^2 - \hat{\mathbf{l}}^2 - \hat{\mathbf{s}}^2\right] = \tfrac{1}{2}\hbar^2[\,j(j+1) - l(l+1) - s(s+1)]\ .$$

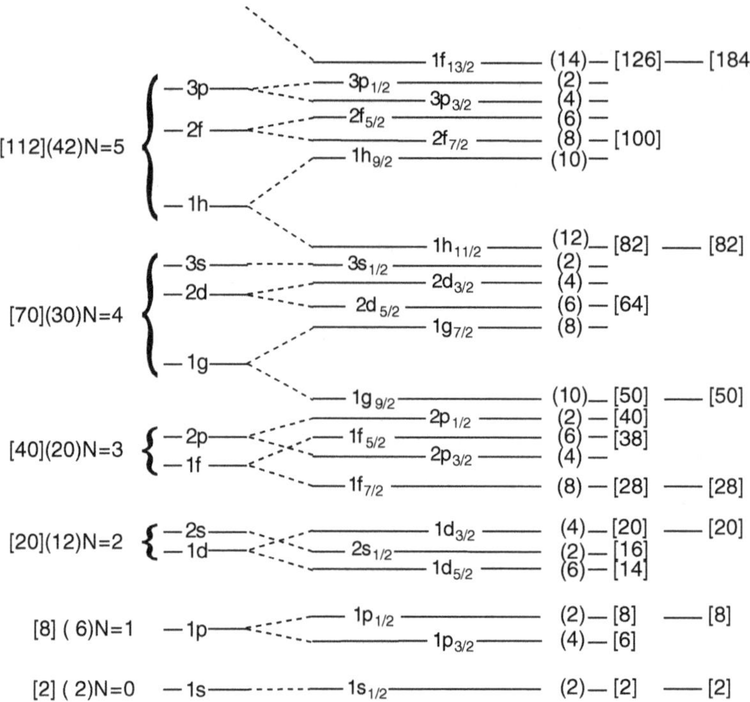

Fig. 3.8 Level scheme for a harmonic oscillator with spin-orbit coupling. The states of the pure harmonic oscillator plotted on the *left* split up to form the structure on the *right*. Both *left* and *right* the number of particles possible in each level is shown in parentheses and the total number up to that level in brackets

Of interest is the splitting of the two levels with $j = l \pm \frac{1}{2}$. It is given by

$$E_{j=l+1/2} - E_{j=l-1/2} = -2\kappa\hbar\omega\left(l + \tfrac{1}{2}\right) \quad .$$

This leads to the structures shown in the right part of Fig. 3.8. The spin–orbit coupling splits the levels depending on total angular momentum; a label like $2f_{7/2}$ corresponds to the second f-state coupled to $j = 7/2$, the sign of the splitting being such that the highest total angular momentum state becomes lowest in energy.

Clearly the inclusion of spin–orbit coupling rearranges the levels such that new magic numbers appear. These are indicated on the right and now agree with what is seen experimentally in nuclei. If the number of protons or neutrons is one of the magic values, the nucleus turns out to be especially stable; more specifically, it is characterized by

- a larger total binding energy of the nucleus,
- a larger energy required to separate a single nucleon,

- a higher energy of the lowest excited states, and
- a large number of isotopes or isotones with the same magic number for protons (neutrons)

(all of these in comparison to neighboring nuclei in the table of nuclides). The lower magic numbers are the same for protons and neutrons, namely 2, 8, 20, 28, 50, and 82, whereas the next number, 126, is established experimentally only for neutrons. Theoretically one would expect additional magic numbers near 114, 120, or 126 for protons and 184 for neutrons (the exact prediction depends on the theory) leading to *superheavy nuclei*, but these have not been confirmed in experiment, although there are hints of an increase of lifetimes in the heaviest elements observed up to now [49].

Aside from the magic numbers, the properties of the single-particle states themselves are also accessible to experiment through *pickup and stripping reactions*, which allow the determination of the binding energy and angular momentum of the particles near the Fermi level. The experiments of this type tend to agree with the predictions of the model near closed shells.

3.3.3.2 The Nilsson Model for Deformed Nuclei

The generalization of the phenomenological shell model to deformed shapes in the nuclear case was first given by S. G. Nilsson [75], so this version is often referred to as the *Nilsson model*. The Hamiltonian of the Nilsson model is a combination of the pure harmonic oscillator of (3.1) with a spin–orbit coupling and an $\hat{\mathbf{l}}^2$-correction added, so that the major difference to the spherical version is just allowing deformation in the pure-oscillator part. Most applications are done with axial symmetry, so that the Hamiltonian may be diagonalized in the basis of the harmonic oscillator using either spherical or cylindrical coordinates depending on the application. In spherical coordinates the spin–orbit and $\hat{\mathbf{l}}^2$ terms are diagonal, but the harmonic oscillator potential couples orbital angular momenta differing by ± 2. In cylindrical coordinates, on the other hand, the deformed oscillator potential is diagonal and the angular momentum terms must be diagonalized numerically. In any case, with modern computing resources neither is a problem.

It is worthwhile to study the quantum numbers resulting from both approaches. Consider first the spherical oscillator without spin effects. The energies from the oscillator potential in the spherical basis are given by (3.3) in terms of orbital angular momentum l. In the cylindrical basis the oscillator energies and quantum numbers are given by (3.4). Spin–orbit coupling, as mentioned, makes these quantum numbers only approximately applicable: there are no j and l quantum numbers, but only the projection j_z, often denoted as Ω in the Nilsson model.

For the spherical shape the levels will be grouped according to the principal quantum number N (with the splitting by the spin–orbit force then determined through the total angular momentum j), but the behavior with deformation depends on how much of the excitation is in the z-direction. For prolate deformation (one long axis and two small equivalent ones), the potential becomes shallower in this direction, and the energy contributed by n_z excitations decreases. The cylindrical quantum numbers are thus helpful in understanding the splitting for small deformations even near the spherical shape, while the spin–orbit properties are of course better discussed with spherical quantum numbers. For very large deformations, on the other hand, the influence of the spin–orbit and $\hat{\mathbf{l}}^2$ terms becomes less important and one may classify the levels according to the cylindrical quantum numbers. It has thus become customary to label the single-particle levels with the set $\Omega^\pi [N n_z m]$. The projection of total angular momentum Ω and the parity π are good quantum numbers while N, n_z, and m are only approximate by looking near the spherical state or, for m, at large deformation. Figure 3.9 shows the levels of the Nilsson model (3.8) as functions of deformation η. The trends are similar to the cluster example in Fig. 3.4, however, here including a strong spin–orbit term. The spherical level structure at zero deformation and the emergence of the asymptotic Ω^π quantum number can be recognized clearly.

The systematic behavior of the levels is made more complicated by *avoided energy-level crossings*. As a general rule, levels with the same (exact) quantum numbers should never cross if they are plotted as functions of a single parameter. This was first noticed by von Neumann and Wigner [1, 62], who investigated how the number of degrees of freedom of a general Hermitian matrix is reduced if the matrix is constrained to have two equal eigenvalues; they found that it is in general not sufficient to vary a single parameter to reduce the matrix to this special case. Thus a degeneracy should always be caused by a symmetry which produces additional quantum numbers to distinguish the states.

If two levels with the same quantum numbers get close to each other, they are instead repelled. In the Nilsson diagram of Fig. 3.9 the effect is clearly observable, for example, for the $1/2^-$ level coming from the $f_{5/2}$ spherical multiplet below magic number 40. Going toward negative deformations it is first repelled by the $1/2^-$-level from the $p_{1/2}$ state above and then by the one from the $p_{3/2}$ below. It is not, however, forbidden from crossing the $9/2^+$ coming from above.

The level structure is actually accessible in experiment through the angular momentum and parity of the nucleus. For even–even nuclei, pairing (see Sect. 9.3) lets all the nucleon pairs in the $\pm\Omega$ level pairs add up to a total angular momentum of zero, but in odd nuclei the angular momentum and parity of the state the single unpaired nucleon occupies determine the properties of the ground state. Thus at a fixed deformation one counts up the level scheme for protons or neutrons to find the level with the odd particle. Ideally the spin of the nucleus is then given by $I = \Omega$ and the parity is given by the parity of the single-particle state. In practice usually several levels are close together at that deformation: in this case one of them yields the ground-state properties, while the other quantum number combinations can be found among the low-energy excitations.

Let us finally mention that one can easily also define versions of the Woods–Saxon and other realistic potentials containing similar deformation properties, although these are less amenable to computation.

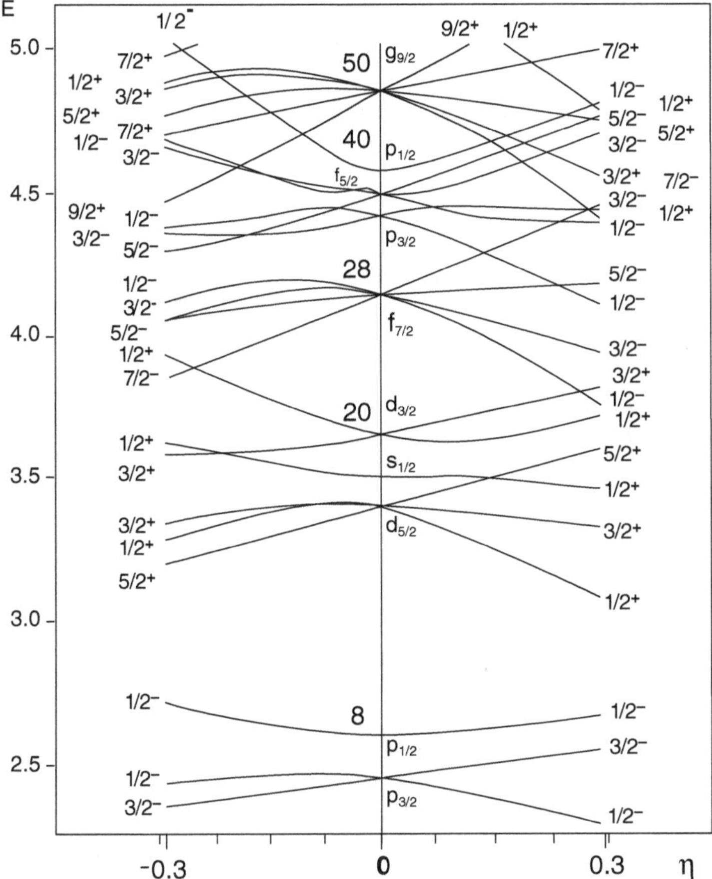

Fig. 3.9 Lowest part of the level diagram (Nilsson diagram) for the deformed shell model (3.8). The single-particle energies (in units of $\hbar\bar{\omega}_0$) are plotted as functions of the deformation parameter η as defined in (3.10). Note that $\eta = 0$ corresponds to the *spherical shape* and $\eta = 0.4$ has the symmetry axis longer than the other two by a ratio of 1.4. The quantum numbers Ω^π for the individual levels and l_j for the *spherical ones* are indicated as are the magic numbers for the *spherical shape*

3.3.3.3 The Two-Center Shell Model

The Nilsson model discussed in the previous section is very powerful in predicting nuclear deformation and its consequences like rotational spectra. However, it is not capable of modeling nuclear fission, the breakup into two comparable fragments. For this purpose there exists another interesting modification of the harmonic oscillator potential, the *two-center potential*. In its simplest version it corresponds to two identical oscillator potentials displaced along the symmetry axis; in cylindrical coordinates this looks like

$$V(\rho, z) = \tfrac{1}{2} m \omega_\rho \rho^2 + \begin{cases} \tfrac{1}{2} m \omega_z^2 (z - z_0)^2, \ z > 0 \\ \tfrac{1}{2} m \omega_z^2 (z + z_0)^2, \ z < 0 \end{cases} .$$

The distance between the two centers is $2z_0$. The potential is constructed such that it is continuous at $z = 0$. Wave functions for this simple model can be obtained analytically [69].

It is easy to introduce variations by allowing different oscillator frequencies in the two parts and the different directions, thus describing fragments of unequal masses and even with deformations. This makes the calculation of the wave functions more

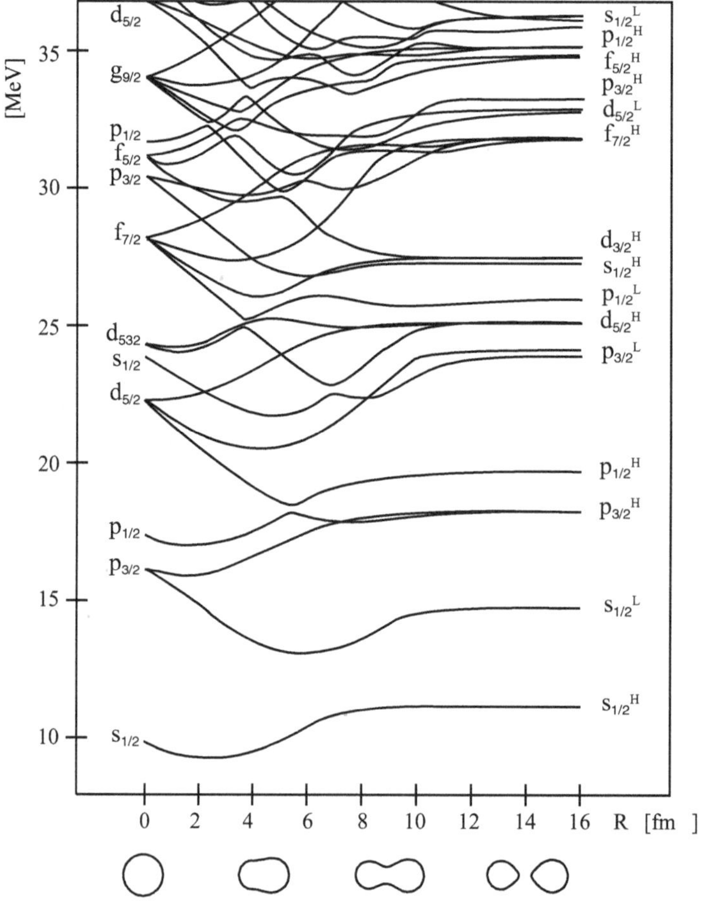

Fig. 3.10 Dependence of the single-particle energies on *center* distance in a two-center oscillator shell model. In this case the ratio of masses of the fragments is 1.4. The assignment of the levels to each fragment is shown by "L" and "H" for the Light and Heavy one; quantum numbers are indicated for this limiting case and for the combined spherical nucleus at $R = 0$. An indication of the shape of the system is given through the figures *below* the axis

complicated and also requires interpolating the potential to make it smooth [67]. As usual, the various oscillator frequencies have to fulfil volume conservation.

Figure 3.10 shows the dependence of the single-particle levels on the distance $R(= 2z_0)$ between the centers of the two fragments. To make the structure more transparent, an asymmetric situation was selected, which makes the emergence of the two spherical spectra with different oscillator frequency for the light and heavy nucleus at larger separation clearly visible. For $R = 0$ there is a spherical compound system with the appropriate spectrum of the spherical oscillator model. In many places gaps in the spectrum appear, which can lead to enhanced binding like in magic nuclei and are responsible for shell effects in the fission barriers.

3.4 Concluding Remarks

Simple external fields confining particles provide a very useful tool for investigation of the properties of many-fermion systems. While such potentials are constitutive of the system, for example, in quantum dots or traps, they are just a common approximation in self-bound systems such as nuclei or metal clusters. Still, even in these latter cases the self-consistent field binding fermions together can rather easily be approximated by a simple external potential. This allows more transparent calculations and gives access to a straightforward, sometimes even fully analytic, understanding of major shell closures. The use of such model systems is not restricted to spherical objects. Deformations can as well be inserted in the picture.

Among simple model potentials the harmonic oscillators play a special role as they allow analytical solutions. They are extremely useful in quantum dots (2D harmonic confinement) as well as nuclei, metal clusters, or traps (3D oscillator). We have also seen that the simple harmonic oscillator picture could be improved in several respects, especially what concerns angular momentum effects, and even provides a transparent understanding of the dynamics of fission in nuclei. Of course model potentials impose some restrictions, especially in the case of self-bound systems, in which they constitute at best a simple practical approximation to the true self-consistent potential. We will see in Chaps. 5 and 6 how such self-consistent fields can be constructed in a more microscopic way, but at the price of a strongly enhanced complexity.

Chapter 4
Approaches Based on Model Spaces

When considering a complex system from the quantum point of view the question of the form of the many-body wave function quickly arises. The most general form is an expansion into a couple of Slater states $\Phi_{\alpha_1\ldots\alpha_N}$ (1.3) as

$$\Psi(1,2,\ldots,N) = \sum_{\alpha_1\ldots\alpha_N} \Phi_{\alpha_1\ldots\alpha_N} c_{\alpha_1\ldots\alpha_N} . \tag{4.1}$$

Variation of $c_{\alpha_1\ldots\alpha_N}$ for a given (large) basis set of Slater states yields the configuration interaction (CI) methods which are widely used in quantum chemistry. Also varying the φ_{α_i} amounts to multi-configuration Hartree–Fock (MCHF). Although conceptually straightforward, the Hilbert space grows very quickly as the number of involved particles increases and may soon become huge. There are several solutions to deal with this problem, among them the neglect of interactions (Chap. 2), or a simplified account of the interactions through a common explicit potential (Chap. 3). Such approaches require a well-developed intuition for a given system in order to assure a good guess for the effective mean-field potential. This works fairly well for simple geometries (Chap. 3), but becomes increasingly tedious in quantum chemistry because each molecule can have a very different configuration requiring separate modeling. We will see later on that such complex situations can be handled in an unprejudiced manner when properly formulating a self-consistent mean-field theory (Chap. 5). Even if such a self-consistent approach represents a clear improvement over the introduction of schematic potentials, there remains as another crucial aspect the question of the Hilbert space, i.e., of the most appropriate expansion basis in (4.1). That question reappears notoriously at many levels of detail of the theory. It thus also makes sense to directly try an approximation on the side of the Hilbert space in order to simplify the problem. This is a strategy which has been largely developed in chemistry, and here we shall outline a few widely used concepts. The basic idea is simple. Keeping the Hamiltonian as simply built from elementary interactions between constituents of the system, one tries to express the electronic wave functions in a simple form using an adequate basis. In the case of molecular systems the building blocks are atoms and the associated atomic wave functions for electrons. One can then try to construct wave functions of molecular electrons building them from the ones of atomic electrons. The idea is elementary

J.A. Maruhn et al., *Simple Models of Many-Fermion Systems*,
DOI 10.1007/978-3-642-03839-6_4, © Springer-Verlag Berlin Heidelberg 2010

and has been used at many levels of sophistication in chemistry. We shall discuss it for standard simple cases in the following. Note that such an idea is not confined to chemistry and can be used in several situations, as soon as one can identify a reduced Hilbert space in which to express the wave functions of the constituents of the system. A typical example is the case of two-level systems which we shall also discuss in Chap. 7.

The use of atomic wave functions for building molecular ones is a standard method in chemistry. We shall discuss it starting from simple examples of dimer molecules in which the picture is especially transparent and then consider the case of more complex molecules where it turns out that it often becomes less accurate. We shall discuss two standard approaches along that line. We shall first discuss the *valence-bond* (VB) approach and then the well-known *linear combination of atomic orbitals* (LCAO) which leads to the simplified *tight-binding* approximations, such as, in particular, the *Hückel model*.

4.1 Valence-Bond Theory

The idea of valence-bond (VB) theory is to model binding in a molecule by associating the chemically active electrons in pairs. The corresponding wave function is shared between the two atoms joined by the bond and localizes predominantly along the bond. It is clearly a method which will make sense in systems in which binding occurs "between" atoms, namely essentially covalent systems. It can be used in particular in organic molecules. The VB approach by construction reduces the many-body problem to a set of two-body problems, whence the obvious simplification. Not surprisingly, we will see that its range of applicability is limited. In order to illustrate the method we start with the case of a dimer molecule, discuss then the general distinction of bond types, and finally proceed to a brief overview of VB for more complex molecules.

4.1.1 Dimer Molecules

We start from the simplest case of a dimer molecule with only one valence electron per atom, such as H_2. If the two atoms A and B are widely separated then the wave function of any electron (which would sit on its parent atom) can be written as a simple product. This reads for the spatial part

$$\Phi(\mathbf{r}_1, \mathbf{r}_2) = \varphi_A(\mathbf{r}_1)\varphi_B(\mathbf{r}_2) \tag{4.2}$$

if electron 1 is on atom A (centered at $\mathbf{r}_A$) and electron 2 on atom B (centered at $\mathbf{r}_B$). The wave function φ_A labels the wave function of the electron on atom A about origin $\mathbf{r}_A$ (the single occupied atomic level in this case), and similarly φ_B the wave function of the electron about origin $\mathbf{r}_B$ ($1s$ wavefunctions in the case of H_2).

The above expression becomes exact in the limit of infinitely separated atoms. But when atoms are close to each other the interactions make it only approximate and indistinguishability has to be taken into account. The actual spatial wave function should be written

$$\Phi(\mathbf{r}_1, \mathbf{r}_2) \propto \varphi_A(\mathbf{r}_1)\varphi_B(\mathbf{r}_2) \pm \varphi_A(\mathbf{r}_2)\varphi_B(\mathbf{r}_1). \tag{4.3}$$

The "+" sign applies if the spin-wave function is antisymmetric (singlet), the "−" if it is symmetric (triplet). For both combinations the total wave function is antisymmetric under particle exchange as it should be. The sign to be chosen is obtained by computing the total electronic energy of the system

$$\begin{aligned} E &= \frac{\int \psi^* H \psi \mathrm{d}^3 r}{\int \psi^* \psi \mathrm{d}^3 r} \\ H &= -\frac{\hbar^2}{2m}(\Delta_1 + \Delta_2) + V \\ V &= -e^2\left(\frac{1}{|\mathbf{r}_1 - \mathbf{r}_A|} + \frac{1}{|\mathbf{r}_2 - \mathbf{r}_A|} + \frac{1}{|\mathbf{r}_1 - \mathbf{r}_B|} + \frac{1}{|\mathbf{r}_2 - \mathbf{r}_B|}\right) \\ &\quad + e^2 \frac{1}{|\mathbf{r}_1 - \mathbf{r}_2|}). \end{aligned} \tag{4.4}$$

One then keeps the combination giving the smallest energy and this finally leads to the spatially symmetric (thus spin antisymmetric) form

$$\Phi \propto \varphi_A(1)\varphi_B(2) + \varphi_A(2)\varphi_B(1). \tag{4.5}$$

In other words, in VB theory, each pair of electrons binding together couple with anti-parallel spins in the singlet configuration.

The typical shape of the wave functions (4.3) is illustrated schematically in Fig. 4.1 for the widely used practical case of Gaussians, i.e., $\Phi_\pm \propto \exp(-(\mathbf{r}-\mathbf{d}/2)^2/(2\sigma^2)) \pm \exp(-(\mathbf{r}+\mathbf{d}/2)^2/(2\sigma^2))$ with distance $\mathbf{d}$ and width $\sigma = a_0$. The two separate atomic wave functions are clearly visible for the largest distance. The zone between the atoms fills up for the symmetric wave function with closer distances and reaches well-developed inter-atomic (covalent) localization at a typical equilibrium value $d = 2\,\mathrm{a}_0$ for the bond distance. The antisymmetric wave functions have a reduced density just between the atoms and thus are unfriendly to binding.

4.1.2 *σ*- and *π*-Orbitals

Valence-bond theory can be easily formulated for more complex homonuclear dimers. In this case the sequence of all atomic levels possibly participating in binding is considered, leading to the formation of a sequence of paired electrons ensuring binding. A few words on nomenclature are necessary here. In the case of

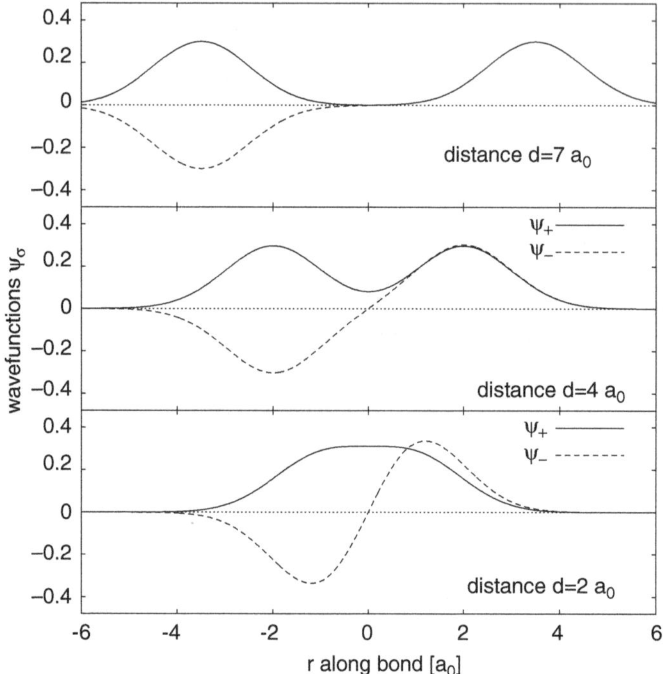

Fig. 4.1 Symmetric (ψ_+) and anti-symmetric (ψ_-) molecular wave functions (4.3) composed from two 1s Gaussians of width a_0 and at several distances as indicated

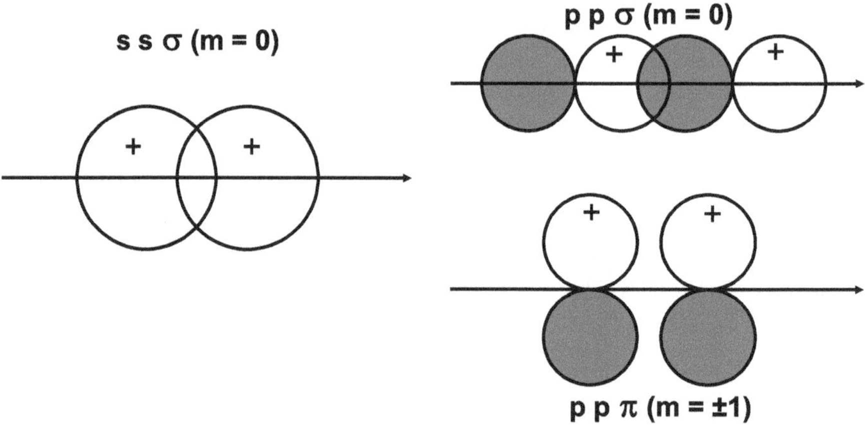

Fig. 4.2 Schematic illustration of the formation of σ and π orbitals. While s-orbitals can only lead to σ-orbitals, p ones may couple to form either a σ or a π-orbital, depending on the projections of angular momenta m on the interatomic axis (*horizontal axis* here). The signs label the actual signs of the wave functions

the hydrogen dimer the VB wave functions are constructed from $1s$ wave functions of each hydrogen atom. Each $1s$-orbital is spherically symmetric around its parent atom. When linked together in the VB wave function the two $1s$-orbitals will create an orbital which has of course lost its spherical symmetry but which remains symmetric under rotations about the axis connecting the two atoms. Such an orbital is called a σ-orbital as, when viewed along the internuclear axis, it resembles an s-orbital occupied by two electrons.

In the case of binding between more complex atoms, this axial symmetry may be lost. Let us, for example, consider atoms with p valence orbitals like Carbon or Nitrogen. These orbitals will again be associated together as a bonding pair but an atomic p-orbital does not have spherical symmetry. It only possesses axial symmetry. There are then two possible cases. If the symmetry axis of the p-orbitals coincides with the internuclear axis, axial symmetry is preserved and one will again speak of a σ bond. If they do not, the electronic clouds show a preferred direction within the plane perpendicular to the internuclear axis. This produces sort of a slightly sideward-twisted bridge between the two atoms. Such a bond is called a π bond as, again, when viewed along the internuclear axis it resembles a doubly occupied p-orbital. To illustrate the point, take the simple example of the N_2 dimer. Each Nitrogen atom has three valence electrons with corresponding atomic orbitals $2p_x$, $2p_y$ and $2p_z$. Taking conventionally the z-axis along the internuclear axis, we see that the two p_z orbitals will form a σ bond while the p_x and p_y-orbitals will form two complementary π bonds. This finally leads in that case to a triple $\sigma{-}\pi{-}\pi$ bond.

4.1.3 Polyatomic Molecules

The extension of VB theory to the case of polyatomic molecules is straightforward and yet, as we will see, also shows the limits of the method. The basic idea is again to group electrons pairwise between pairs of atoms to form σ or π bonds with corresponding product wave functions. Of course one now needs to decide which pairs to take into account and there is no systematic rule for this. As illustration let us regard the case of the water molecule H_2O. The atomic orbitals involved are the two $1s$ orbitals of the two hydrogen atoms and the two unpaired valence electrons of oxygen (remember that oxygen has electronic structure $1s^2 2s^2 2p_x^2 2p_y 2p_z$, and the $2p_y 2p_z$ electrons are the ones open for bonds). Following the spirit of the VB approach, one will thus pair the $1s$ hydrogen orbitals with the two $2p_y$ and $2p_z$ orbitals of oxygen, leading to two σ bonds. The net result is interesting because it automatically leads to a nonlinear structure which is compatible with reality. The first $1s$ bond ties to the $2p_y$ state and the second to $2p_z$ which means that there is an angle of 90° between these both bond directions. That is to be compared with the true bond angle between the two OH bonds in H_2O which is 104°, so that this prediction is at least qualitatively correct.

The difficulties of the VB method can be made even clearer when considering a more complex molecule such as CH_4, whose experimental structure is a perfectly symmetric pyramid. Indeed CH_4 is a typical example of a molecule in which a carbon atom is able to provide four electrons for binding with four atoms. The electronic structure of carbon is $1s^2 2s^2 2p_x 2p_y$ which a priori suggests two $2p$ valence electrons rather than four. In order to understand the mechanism of fourfold binding of carbon (as well as the threefold one actually) one is led to admit that inside a molecule there occurs a *promotion* mechanism, leading the carbon atom to have four unpaired electrons with a corresponding electronic structure $2s2p_x2p_y2p_z$. Of course such a promotion mechanism involves an energy loss as it corresponds to an excited state of the carbon atom, but one assumes that inside the molecule this energy loss is more than compensated by the energy gain due to the possibility of binding the thus created reservoir of unpaired electrons. It is interesting also to note that the promotion mechanism can be viewed as a simple mechanism to enlarge the Hilbert space from which binding is constructed by allowing for the occupation of one more atomic orbital.

The promotion mechanism still does not suffice to understand the highly symmetric structure of the CH_4 molecule. Indeed, following the standard VB scheme it should not lead to four equivalent bonds. Three σ bonds would stem from pairing Carbon $2p$ and Hydrogen $1s$-orbitals and one σ bond would correspond to the pairing of carbon $2s$ and hydrogen $1s$-orbitals, hence with a different symmetry. Strictly keeping to the idea of pairing atomic orbitals as such then leads to a major difficulty. The way out is to extend the simple VB picture of strict pairing of atomic orbitals to a linear combination thereof. The set of available atomic orbitals then forms a basis from which one can construct a set of equivalent orbitals. This procedure, which essentially amounts to a simple linear combination inside a reduced Hilbert space, is known as the hybridization procedure in VB theory and widely used to characterize binding mechanisms in molecules. In the specific case of CH_4 one will thus consider that the binding orbitals are to be constructed from the set of the four carbon orbitals $2s$, $2p_x$, $2p_y$, and $2p_z$. This is called a sp^3 hybridization. The four equivalent hybrid orbitals constructed from this set then take the form

$$h_1 = \psi_{2s} + \psi_{2p_x} + \psi_{2p_y} + \psi_{2p_z} \qquad h_2 = \psi_{2s} - \psi_{2p_x} - \psi_{2p_y} + \psi_{2p_z}$$
$$h_3 = \psi_{2s} - \psi_{2p_x} + \psi_{2p_y} - \psi_{2p_z} \qquad h_4 = \psi_{2s} + \psi_{2p_x} - \psi_{2p_y} - \psi_{2p_z}$$

and the associated VB wave functions are built from these hybrid orbitals as

$$\varphi_i = h_i(1)\psi_{\mathrm{H},i}(2) + h_i(2)\psi_{\mathrm{H},i}(1), \quad i = 1, 4,$$

where $\psi_{\mathrm{H},i}$ labels the $1s$-orbital of a given hydrogen atom i. The hybridization thus solves the paradox of inequivalent bonding by producing four equivalent hybrid wave functions. Note that the hybrid orbitals h_i are of course orthogonal to each other. Hybridization thus simply corresponds to a generalization of the original VB picture by considering as building blocks not the eigenstates of each atom individually, but the subspace spanned by these eigenstates as a whole, admitting linear combinations as better candidates for the formation of bonds.

The VB theory forms a basic approach in chemistry and is extremely interesting in many situations. Its foundations are simple but, as exemplified in the few examples above, it requires some specific adaptation as soon as one wants to consider slightly complex systems. There are of course systematic ways to construct hybrid orbitals, but we do not want to elaborate more on these methods here. We shall mostly remember for our purpose that binding is constructed pairwise on "well-chosen" orbitals, so that one deals with a dedicated reduced Hilbert space in which the choice of orbitals to be combined appears crucial. This also points out some intrinsic limitation in the use of the method. Without abandoning the idea of working in a reduced Hilbert space, we shall now consider an alternative approach in which the focus is not on individual bonds but rather on the electrons of the molecule as a whole. This is the viewpoint of molecular orbitals (MO) which in a practical form leads to the LCAO approach discussed in the next section.

4.2 Linear Combination of Atomic Orbitals (LCAO)

In the LCAO approach, binding is not described through individual pairs of electrons but rather by considering the binding problem globally, without a priori specifying the connection between a given electron and a bond. In a generalization of the hybridization picture, one now models all bonding-relevant states as superpositions of appropriate atomic orbitals, i.e.,

$$\varphi_\alpha(1) = \sum \psi_i(1) c_i, \tag{4.6}$$

where $\{\psi_i\}$ is a small set of relevant atomic orbitals. Referring back to the general ansatz (4.1), this corresponds to modeling the most relevant single-particle states φ_{α_i} in simple terms.

This approach is obviously less restrictive than the VB theory and should thus be applicable to a larger range of situations in multi-centered molecules (for simple dimer molecules LCAO is not very different from the VB approach). Although more flexible, LCAO also expresses the wave functions of the molecular electrons in a simplified Hilbert space. In contrast to the VB approach, which specifies bonds and thus the corresponding wave functions one by one, LCAO does not impose specific pairings and leaves the choice of optimal combinations to the variational principle. The question again comes up of what should be the "reservoir" of electronic wave functions from which to construct molecular orbitals. The selection here is quite simple: all the valence electrons of all atoms in the molecule are included in the problem. This means that the Hilbert space is constructed from the set of electronic valence wave functions and molecular wave functions are constructed as linear combinations of such "basis" wave functions. The term basis is here to be taken in a generalized sense as the atomic eigenstates are not necessarily orthogonal to each other which requires a (moderate) formal complication to deal with a diagonalization in non-orthogonal spaces. In terms of the most general ansatz (4.1), LCAO corresponds to picking the "active" single-particle wave functions φ_{α_i} and modeling them in terms of linear combinations of atomic orbitals. The restriction to a couple of relevant basis states simplifies the problem dramatically and it is hoped that the physically guided choice of that basis, nonetheless, allows a pertinent description. In order to illustrate how the LCAO method works, let us first consider the simple case of a dimer molecule.

4.2.1 LCAO in Dimer Molecules

4.2.1.1 The General Setup

We first consider the simple case of a dimer molecule formed of atoms A and B and restrict the analysis to two valence levels, one provided by each atom. We furthermore assume that the molecule is singly ionized, such as H_2^+, again for sake of simplicity. In the LCAO approximation it is assumed that the molecular orbital can be expressed in the form

$$\varphi = c_A \psi_A + c_B \psi_B, \tag{4.7}$$

where $\psi_{A,B}$ label the atomic orbitals (associated to atomic eigenvalues ϵ_A and ϵ_B) and the coefficients c_A, c_B are to be determined by a variational principle minimizing the total energy E. We are interested in a stationary problem and can thus take the wave functions as well as the coefficients c_A, c_B to be real. From the total Hamiltonian H

$$H = -\frac{\hbar^2}{2m}\Delta - \frac{e^2}{|\mathbf{r}-\mathbf{r}_A|} - \frac{e^2}{|\mathbf{r}-\mathbf{r}_B|}, \tag{4.8}$$

the energy $E = \int \psi^* H \psi \mathrm{d}^3 r / \int \psi^* \psi \mathrm{d}^3 r$ is computed. Note that the Hamiltonian (4.8) is just a simplified version of the VB one (4.5) due to the fact that we consider only one electron. Because the two orbitals $\varphi_{A,B}$ are not necessarily orthogonal to each other, the expression of the energy involves extra terms.

We first compute the denominator

$$\begin{aligned}\int \varphi^* \varphi \mathrm{d}^3 r &= \int |c_A \psi_A + c_B \psi_B|^2 \mathrm{d}^3 r \\ &= c_A^2 \int \psi_A^2 \mathrm{d}^3 r + c_B^2 \int \psi_B^2 \mathrm{d}^3 r + 2 c_A c_B \int \psi_A \psi_B \mathrm{d}^3 r \\ &= c_A^2 + c_B^2 + 2 c_A c_B S,\end{aligned} \tag{4.9}$$

where the fact was used that the atomic orbitals are normalized to unity and we have introduced what is called the *overlap integral*

$$S = \int \psi_A \psi_B \mathrm{d}^3 r, \tag{4.10}$$

which is nonzero because the atomic orbitals are not necessarily orthogonal. It is actually a welcome feature as, if the overlap of the two orbitals were strictly zero, that would imply that the two orbitals are not connected to each other and thus hardly capable of establishing a common bond. The integral over the Hamiltonian also involves an overlap integral. Explicitly it becomes

$$\int \varphi^* H \varphi \mathrm{d}^3 r = c_A^2 \int \psi_A H \psi_A \mathrm{d}^3 r + c_B^2 \int \psi_B H \psi_B \mathrm{d}^3 r$$
$$+2 c_A c_B \int \psi_A H \psi_B \mathrm{d}^3 r \quad . \tag{4.11}$$

Introducing simpler conventional notation, one can finally rewrite the total energy as

$$E = \frac{\int \varphi^* H \varphi \mathrm{d}^3 r}{\int \varphi^* \varphi \mathrm{d}^3 r} = \frac{c_A^2 \alpha_A + c_B^2 \alpha_B + 2 c_A c_B \beta}{c_A^2 + c_B^2 + 2 c_A c_B S} \tag{4.12}$$
$$\alpha_{A(,B)} = \int \psi_{A(,B)} H \psi_{A(,B)} \mathrm{d}^3 r, \quad \beta = \int \psi_A H \psi_B \mathrm{d}^3 r.$$

Note that the integrals $\alpha_{A,B}$ are nothing but the original eigenenergies of the atomic orbitals complemented by the overlap between the wave function of a given atom and the potential from the other one:

$$\alpha_X = \epsilon_X + \int \psi_X \frac{e^2}{|\mathbf{r} - \mathbf{r}_Y|} \psi_Y \qquad (X, Y) = (A, B) \quad \text{or} \quad (B, A). \tag{4.13}$$

It is then an easy task to find a stationary point of E with respect to c_A and c_B. Let us write it explicitly for c_A, the second equation being simply obtained by exchanging c_A and c_B. One obtains successively:

$$\begin{aligned} \frac{\partial E}{\partial c_A} &= \frac{2 c_A \alpha_A + 2 c_B \beta}{c_A^2 + c_B^2 + 2 c_A c_B S} \\ &- \frac{(2 c_A + 2 c_B S)(c_A^2 \alpha_A + c_B^2 \alpha_B + 2 c_A c_B \beta)}{(c_A^2 + c_B^2 + 2 c_A c_B S)^2} \\ &= 2 \frac{((c_A \alpha_A + c_B \beta) - (c_A + c_B S) E)(c_A^2 + c_B^2 + 2 c_A c_B S)}{(c_A^2 + c_B^2 + 2 c_A c_B S)^2}, \end{aligned} \tag{4.14}$$

by reintroducing the total energy E. The overlap integral S is bound by 1 so that the factor $c_A^2 + c_B^2 + 2 c_A c_B S$ is necessarily strictly positive and nonzero, as already pointed out. The two conditions $\partial E / \partial c_A = \partial E / \partial c_B = 0$ then finally lead to the two equations

$$\begin{aligned} c_A(\alpha_A - E) + c_B(\beta - E S) &= 0 \\ c_A(\beta - E S) + c_B(\alpha_B - E) &= 0, \end{aligned} \tag{4.15}$$

which can have a nontrivial solution only if the determinant of this linear system vanishes. This leads to the secular equation

$$\begin{vmatrix} \alpha_A - E & \beta - ES \\ \beta - ES & \alpha_B - E \end{vmatrix} = (\alpha_A - E)(\alpha_B - E) - (\beta - ES)^2 = 0. \tag{4.16}$$

The roots of the secular equation provide the two possible values of the energy $E^{\pm}$, from which one can deduce the values of $c_A^{\pm}$ and $c_B^{\pm}$. They are determined up to a constant, which can be determined from the normalization of the wave function φ (4.7) to unity.

4.2.1.2 Symmetric Dimer

In the general case the determination of E and c_A, c_B is a bit cumbersome. There are fortunately enough cases which allow a simplification, namely symmetric dimers where by construction ψ_A and ψ_B are identical but located at different space points. The energy integrals α_A and α_B are correspondingly equal, $\alpha_A = \alpha_B = \alpha$. The secular equation then reduces to

$$(\alpha - E)^2 - (\beta - ES)^2 = 0, \tag{4.17}$$

with obvious solutions

$$E^+ = \frac{\alpha + \beta}{1 + S}, \quad c_A^+ = \frac{1}{\sqrt{2(1+S)}}, \quad c_A^+ = c_B^+,$$
$$E^- = \frac{\alpha - \beta}{1 - S}, \quad c_A^- = \frac{1}{\sqrt{2(1-S)}}, \quad c_A^- = -c_B^- .$$

As in the VB model of the symmetric dimer the lowest energy solution associated to E_+ corresponds to the symmetric bonding orbital with a wave function which may be written as

$$\varphi^+ = \frac{1}{\sqrt{2(1+S)}}(\psi_A + \psi_B). \tag{4.18}$$

The overlap integral S as well as the bonding integral β are schematically represented for this symmetric case in the bottom panel of Fig. 4.3 as a function of distance d between the two atoms. In the upper panel of the figure are plotted the bonding and antibonding energies $E^{\pm}$, complemented by the additional (trivial) term e^2/d (remember the negative values of ϵ_A, ϵ_B and correspondingly of α_A, α_B at sufficiently large distance d). At short distance the overlap integral tends toward 1 and at large distance both the S and β integrals vanish, while the bonding and antibonding energies approach $E^{\pm} \rightarrow \alpha$: the dimer becomes a collection of two independent atoms. The minimum in (E^+) at finite distance corresponds to the dimer bond length, namely the equilibrium distance d_b between the two-bound atoms

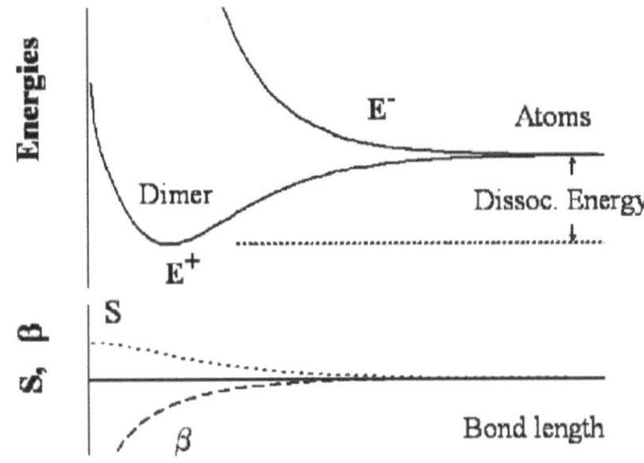

Fig. 4.3 Schematic plot (*arbitrary scales*) of the overlap (S) and bond (β) integrals (*bottom part*) and of the energies of the bonding (E^+) and antibonding (E^-) levels (complemented by ionic repulsion e^2/d) as functions of the distance d between two identical atoms

forming a dimer. The difference $(E^+(d_b) - E^+(\infty))$ then represents the dissociation energy of the dimer.

4.2.1.3 Vanishing Overlap

The second simple case is attained when the overlap integral can be neglected. For then the secular equation takes the simpler form

$$(\alpha_A - E)(\alpha_B - E) - \beta^2 = 0, \tag{4.19}$$

with the obvious solutions

$$E^+ = \frac{\alpha_A + \alpha_B}{2} - \sqrt{\frac{(\alpha_A - \alpha_B)^2}{4} + \beta^2},$$
$$E^- = \frac{\alpha_A + \alpha_B}{2} + \sqrt{\frac{(\alpha_A - \alpha_B)^2}{4} + \beta^2},$$
$$c_A^\pm = \frac{\alpha_B - E^\pm}{\sqrt{(\alpha_B - E^\pm)^2 + \beta^2}}, \quad c_B^\pm = \frac{\beta}{E^\pm - \alpha_B} c_A^\pm,$$

where again the bonding orbital corresponds to E_+. It is interesting to consider even more limiting cases. For symmetric dimers, for instance, $\alpha_A = \alpha_B = \alpha$, so that simply $E_+ = \alpha - \beta$ and $E_- = \alpha + \beta$, and also $c_A = \pm c_B$ is easily seen. The other extreme is a highly asymmetric molecule for which $|\alpha_A| \gg |\alpha_B|$ and $|\alpha_A| \gg |\beta|$ (remember that the α's and β are negative). In this case $E_+ \sim \alpha_A$ with $c_A \sim 1$ and $c_B \sim 0$. This is the typical situation encountered in a strongly ionic dimer in which one atom (here A) exhibits a much more strongly bound level than the other atom (here B). The molecular level is then almost equivalent to the deeply bound level in atom A. The electron is almost fully localized at site A, while site B remains empty, thus becoming positively charged.

4.2.2 A Quick Look at Neutral Dimers

The above discussion started out from a charged dimer as a very simple one-electron system. The results, being mostly qualitative, can easily be carried over to neutral dimers. Two electrons in spin-singlet state can occupy the same spatial state. A state with both electrons in the bonding state describes the typical covalent binding of H_2 or similar molecules. If the two electrons are brought into a spin-triplet state (aligned spins), this forces one electron to go into the anti-bonding state and thus renders the molecules unstable against dissociation. A He dimer has four electrons. Only two of them can occupy the bonding state. The other two necessarily occupy the anti-bonding state. The He dimer (as well as those of all other rare gases) is thus unstable in that zeroth-order consideration. In fact, the faint He binding comes from the van der Waals force which is a correlation effect, see Sect. 9.1. These qualitative estimates are corroborated by the results from a detailed Hartree–Fock calculation shown in Fig. 4.4. The energy from binding for the H_2 molecule and the practically non-binding situation for He_2 are clearly apparent. One also sees that the HF approximation underestimates the binding of H_2. This indicates that some correlation effects are needed for fine tuning, see Sect. 6.1 and Chap. 9.

The case of ionic binding also becomes more realistic for systems which are neutral in the whole. Coming back to the discussion at the end of the previous section, one assumes that the binding of atom A is sufficiently strong for both electrons, the one from A and the other one originally from B. For then the atom A becomes a negative ion in the dimer and atom B a positive one. Once the electron has been caught

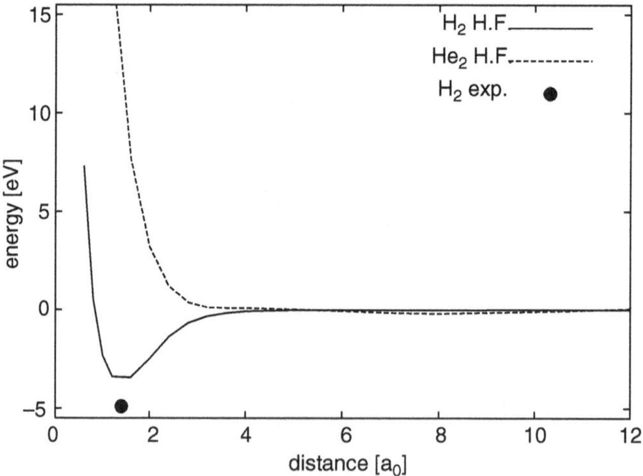

Fig. 4.4 Binding energy versus bond distance for the H_2 and He_2 dimer. The energies are scaled to the asymptotic value of two separate atoms. They are calculated in mean-field approximation at the level of Hartree–Fock, see Chap. 5. A stationary solution is produced for each fixed molecular distance (Born–Oppenheimer picture). The experimental binding of the H_2 dimer is indicated by a *fillet dot*

by atom A, the molecular binding is dominated by the Coulomb force between anion A and cation B. That is a typical situation in molecules formed from alkalies and halogens. In practice, however, the ionization stage does not reach unit charges.

4.2.3 From Dimer to Large Molecules

The extension of the LCAO to polyatomic molecules is straightforward. That is a great advantage in comparison to the VB approach, which required specific analysis of intervening atomic orbitals. In the LCAO one can immediately assume that the molecular orbital will be built from the whole set of atomic orbitals chosen to describe the system. It is important to realize here that the choice of active orbitals is of course a crucial aspect. In Sect. 4.2.1, we deliberately considered atoms with only one active (valence orbital). In the general case this restriction is not justified, nor it is necessary, and one has to carefully analyze which atomic orbitals should be included in the description of the system. That is the place where experience and intuition come into play. After all, the LCAO wave functions are expressed as the superpositions (4.6). To take again the example of water, the set of atomic wave functions φ_i is composed of the two $1s$ hydrogen orbitals plus the four $2s$, $2p_x$, $2p_y$, $2p_z$ oxygen orbitals, both doubled for spin. The total number of active states is then 12, to be occupied by 8 electrons.

Although the LCAO is in principle applicable to any polyatomic molecule, the expense grows rapidly with system size. One has thus been led to consider simplifications of the scheme in order to treat larger systems and to understand binding mechanisms for complex molecules in simple terms. A widely used simplification is the Hückel approximation [4]. It will be discussed in the next section.

4.3 The Hückel Approximation

4.3.1 Conjugated Molecules

In order to illustrate the Hückel approximation we shall consider two standard examples, the one of conjugated molecules and the limiting case of bulk solid. Remind that the conjugated molecules are organic compounds which exhibit an alternation of single and double bonds along a chain of carbon atoms. Typical simple examples are ethylene C_2H_4 and butadiene C_4H_6, see Fig. 4.5. Note that already for such seemingly simple molecules a direct application of LCAO would lead to a rather involved model with 12 active electrons for C_2H_4 and 22 for C_4H_6. The idea of Hückel is to simplify the problem by considering only π-orbitals, while the σ orbitals are treated separately and assumed to fix the geometry of the system. It is, so to say, a way to reduce the set of valence electrons, or in other words, again a way to reduce the Hilbert space in which to construct the LCAO. Furthermore one assumes that all carbon atoms can be treated identically, which means that the Coulomb integrals for the atomic states α_i are all set equal. The problem then finally reduces

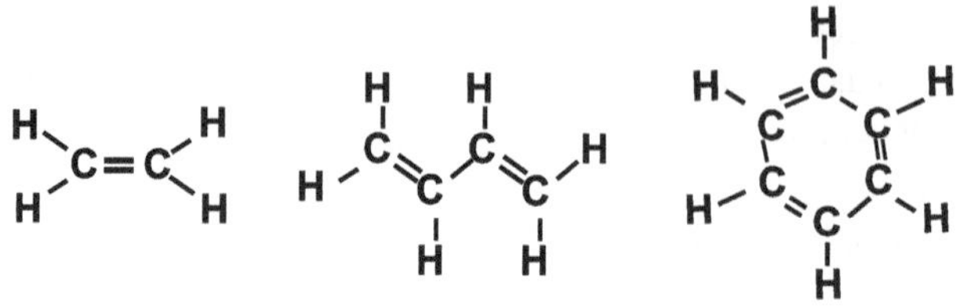

Ethylene C_2H_4 Butadiene C_4H_6 Benzene C_6H_6

Fig. 4.5 Structures of typical small conjugated molecules illustrating Hückel calculations

to determining two π-orbitals in C_2H_4 and four of them in C_4H_6. The symmetry of C_2H_4 finally reduces it to a problem strictly equivalent to a symmetric dimer, but now involving only π-orbitals.

In the case of butadiene, the situation remains more complicated even with the above simplifications. We deal here with four active electrons and even if the α_i are all set equal, there is still a collection of different values for β (the Coulomb integral between the atoms) and the overlap integrals (actually one per atom pair!). To be more explicit, we denote by A, B, C, D the set of four carbon atoms, each providing one (and the same, denoted φ) atomic orbital. The LCAO wave function then reads

$$\varphi = c_A\psi(A) + c_B\psi(B) + c_C\psi(C) + c_D\psi(D) \tag{4.20}$$

and the secular equation becomes

$$\begin{vmatrix} \alpha - E & \beta_{AB} - ES_{AB} & \beta_{AC} - ES_{AC} & \beta_{AD} - ES_{AD} \\ \beta_{BA} - ES_{BA} & \alpha - E & \beta_{BC} - ES_{BC} & \beta_{BD} - ES_{BD} \\ \beta_{CA} - ES_{CA} & \beta_{CB} - ES_{CB} & \alpha - E & \beta_{CD} - ES_{CD} \\ \beta_{DA} - ES_{DA} & \beta_{DB} - ES_{DB} & \beta_{DC} - ES_{DC} & \alpha - E \end{vmatrix} = 0.$$

Of course such an equation can be solved numerically, but if an analytic solution is desired, additional Hückel approximations are needed to reduce the complexity even more. These approximations are

i) to set all overlap integrals S_{ij} to zero;
ii) to set all β_{ij} integrals to zero except for nearest neighbor atoms; and
iii) to take all remaining β_{ij} integrals equal to β.

The secular equation then takes the much more forgiving form:

$$\begin{vmatrix} \alpha - E & \beta & 0 & 0 \\ \beta & \alpha - E & \beta & 0 \\ 0 & \beta & \alpha - E & \beta \\ 0 & 0 & \beta & \alpha - E \end{vmatrix} = 0,$$

which can easily be expanded to yield a quadratic equation $(\alpha-E)^4-3(\alpha-E)^2\beta^2+\beta^4=0$ with the four roots

$$E=\alpha\pm[(3\pm\sqrt{5})/2]^{1/2}\beta=\begin{cases}\alpha\pm 1.62\beta\\ \alpha\pm 0.62\beta\end{cases}.$$

The Hückel method is obviously quite powerful and can be applied to many systems. A particularly appealing example is the Benzene molecule with its cyclic structure C_6H_6. In the Hückel approximation six electrons are treated. The cyclic structure of the molecule leads to non-vanishing β values in the upper-right and lower-left corners of the determinant and the eigenvalue problem takes the form

$$\begin{pmatrix}\alpha-E & \beta & 0 & 0 & 0 & \beta\\ \beta & \alpha-E & \beta & 0 & 0 & 0\\ 0 & \beta & \alpha-E & \beta & 0 & 0\\ 0 & 0 & \beta & \alpha-E & \beta & 0\\ 0 & 0 & 0 & \beta & \alpha-E & \beta\\ \beta & 0 & 0 & 0 & \beta & \alpha-E\end{pmatrix}\begin{pmatrix}c_0\\ c_1\\ c_2\\ c_3\\ c_4\\ c_5\end{pmatrix}=0. \tag{4.21}$$

At first glance, a 6×6 eigenvalue problem is considered intractable analytically. However, the molecule has a symmetry which helps toward a solution: a rotation by one position along the ring (i.e., about 60°). The wave function should not change by such a rotation $c_n \longrightarrow c_{n+1 \mod 6}$, except for a phase factor. This condition can only be fulfilled by the ansatz

$$c_n^{(\nu)}=\frac{1}{\sqrt{6}}\exp\left(i\nu n\frac{2\pi}{6}\right),\qquad \nu\in\{0,\pm1,\pm2,3\}. \tag{4.22}$$

The ansatz is strictly periodic under $n \longrightarrow n+6$ (if the ring is recovered exactly) and the global phase for the basic symmetry transformation $c_n \longrightarrow c_{n+1 \mod 6}$ is $\exp(i\nu\pi/3)$. The six possible different phase factors are labeled by ν which is the symmetry quantum number for the solution. The normalization factor is obvious. The eigenenergies are easily obtained by inserting this ansatz into the above linear equation. They become

$$E_\nu=\alpha+2\beta\cos\left(\nu\frac{\pi}{3}\right)=\begin{cases}\alpha-2\beta \text{ for } \nu=3\\ \alpha-\beta \text{ for } \nu=\pm2\\ \alpha+\beta \text{ for } \nu=\pm1\\ \alpha+2\beta \text{ for } \nu=0\end{cases}, \tag{4.23}$$

where it is to be noted that $\beta<0$, so that $\nu=0$ corresponds to the ground state and $\nu=3$ to the highest excited state. The states $|\nu|\le 1$ have energies $E_\nu<\alpha$ and are

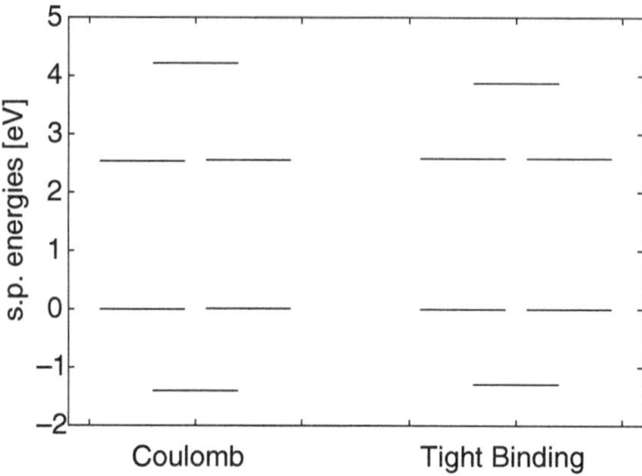

Fig. 4.6 *Left*: The spectrum of a ring of six unit charges distributed symmetrically along a ring of radius 4 a_0. The energies have been shifted to place the second state at zero. *Right*: A comparable tight-binding spectrum is also shown where the gap 2β has been fitted

thus bonding states. The complementing set with $|\nu| \geq 2$ are anti-bonding states. Considering spin, each state can be filled with two electrons. Thus we can conclude that the sixfold cyclic structure is stable for up to six electrons in a π bond while it becomes destabilized whenever a donor drives the electron number above that value. A favorable situation is filling with six electrons for which the HOMO–LUMO gap becomes 2β.

In order to test the relations for a sixfold symmetry, consider a simplest test system, an external field constituted by the Coulomb field of six point charges along a ring of radius 4 a_0. The first six eigenenergies are shown in the left part of Fig. 4.6. One immediately recognizes a pattern similar to the spectrum (4.23). For full comparison, we have adjusted the model parameter 2β to the gap of the computed spectrum and α to reproduce the zero level. The result is shown on the right side. There are, of course, deviations in quantitative detail, but the overall pattern is very much the same.

4.3.2 Bulk Solid

The Hückel approximation can also be simply applied to the case of bulk solid. Although bulk materials are not a central issue in this book, they provide an interesting illustration of the Hückel approximation, which is usually called the tight-binding (TB) approximation in this context [3]. As seen above (Sect. 4.2.1), binding between two atoms leads to the creation of one bonding and one antibonding orbital. The case of butadiene with four centers has two bonding and two anti-bonding orbitals. The cyclic structure with six centers comes up to three and three. If one now imagines progressively enlarging the chain/ring of atoms the process will successively lead to a series of nearby levels which ultimately lead to the creation

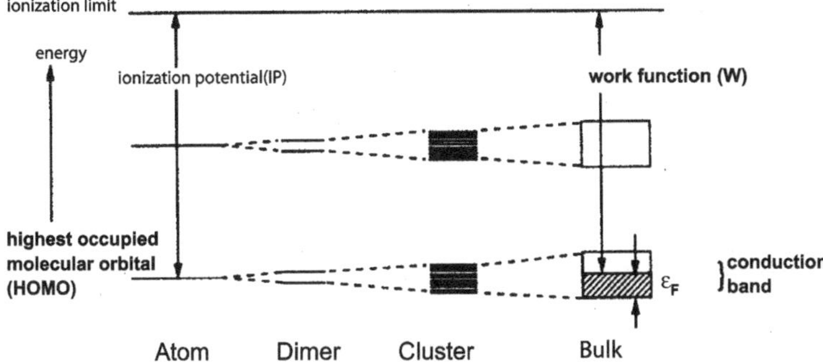

Fig. 4.7 Sketch of the evolution of the single-electron spectrum between atom and bulk, with clusters as an intermediate case. The figure points out the appearance of bands in extended materials. In realistic cases the band "width" does depend on system size (as long as it is finite) at odds with the simplest treatment in the text. We nevertheless keep in the sketch the more realistic size dependence of the band width

of continuous bands. The mechanism is illustrated in Fig. 4.7 and can be readily analyzed in the framework of the Hückel approximation.

The TB, or Hückel, approximation in crystals takes care of the concept of neighborhood in a given crystal symmetry [3] which has a strong influence on band structure. For a schematic demonstration, we consider here the simplest linear symmetry of a 1D "crystal" and approach the "bulk" limit by considering an N fold ring letting $N \longrightarrow \infty$. The secular equation for the N-ring is a straightforward extension of (4.21). The ansatz for the solution is similar to the form (4.22) with 6 replaced by N and the spectrum becomes simply (for even N)

$$E_\nu = \alpha + 2\beta \cos\left(\nu \frac{2\pi}{N}\right), \quad \nu \in \{-N/2+1, \ldots, 0, \ldots N/2\}.$$

The above formula is quite enlightening. As the ring becomes very large, namely $N \to \infty$, nearby levels come closer and closer to each other leading to the formation of a continuous band. The difference between the lowest and highest energy remains constant with a band width of

$$\Delta E = E_N - E_0 = 2\beta \left(\cos(\frac{2N\pi}{2N}) - cos(0)\right) = 4\beta. \tag{4.24}$$

The level spacing within the band shrinks as N^{-1}. In the spirit of molecular binding, the lowest energy level of the band provides the most bound level, while the highest level corresponds to the most anti-bonding level. The bands can be related to the atomic orbitals from which they are built. If the original atomic orbitals were of type s, this band is called an s-band. Similarly if p-orbitals are considered, it will be a p-band. If the atomic p-orbital was less bound than the s one in the atom, the p-band will lie higher in energy than the s-band, but because both bands now have a finite energy width one may wonder whether the two bands will overlap or not. In bulk metals such two bands typically overlap, allowing electronic transport at no cost. If the two bands are separated one will speak of a band gap to label the energy difference between the lowest energy level of the high-energy band and the highest energy level of the low-energy band. The band gap is the generalization to bulk of the HOMO-LUMO concept introduced in finite systems in Sect. 1.3.2. It basically contains the same information. For example, in a metal atom like sodium the valence shell is partially occupied/unoccupied (see Fig. 1.19) and the corresponding bulk material is a conducting metal with conduction band intersecting valence band. In turn in a rare gas atom like neon (see again Fig. 1.19) the large HOMO–LUMO atomic gap leads to an insulating bulk material with a sizable band gap.

4.4 From Model Spaces to Realistic Calculations

The methods introduced in previous sections (VB, LCAO, Hückel) have turned extremely interesting for computing basic properties of simple molecules. The simplified Hückel approach even allows, with dedicated approximations, to treat relatively large systems at a reasonable expense and essentially in an analytic way. This technical simplicity allows to keep key aspects rather transparent and is thus extremely interesting for understanding basic effects in molecular binding. It should nevertheless be kept in mind that these approaches suffer from a certain lack of flexibility. Indeed, the many-electron problem in a molecule is a true many-body problem involving subtle correlation effects which cannot be accounted for in simplified models. The basic defect of the methods we have discussed above (which *a contrario* makes their interest due to the simplicity it implies) lies in the fact that many-electron wave functions are built from a rather limited set of atomic wave functions. This causes a possibly strong reduction of the Hilbert space of accessible many-electron wave functions, and it was exactly this deficiency of the VB method which led us to consider the larger Hilbert space provided by LCAO. Still, even LCAO in general leads to a too restricted space. One can simply illustrate the point by considering the Hamiltonian (4.4) for the simple two electrons dimer molecules. It is clear that the electron–electron interaction term may lead to sizable effects. Remember that this interaction was not accounted for by the atomic wave functions in the case of one-electron atoms. One is thus bound to consider more complicated approaches than the mere LCAO method, as soon as dealing with complex many-electron systems. In order to illustrate such approaches we first consider an extension of the LCAO approach by using tunable basis wave functions, then switch to more general cases and briefly discuss standard computational methods in quantum chemistry.

4.4.1 Example: H_2^+ Dimer in LCAO with Gaussians

A large manifold of quantum chemistry calculations uses LCAO with Gaussian basis functions. We exemplify this approach using the simple H_2^+ dimer as example for which all calculations can be carried through analytically. The Gaussian ansatz furthermore allows a simple enlargement of the Hilbert space by letting the Gaussian width be a free (optimizable) parameter, as shall be seen below. The H_2^+ dimer is a one-electron system with the Hamiltonian

$$\hat{H} = \frac{\hat{p}^2}{2m} + V + \frac{e^2}{|\mathbf{d}|}, \quad V = -\frac{e^2}{|\mathbf{r} - \mathbf{d}/2|} - \frac{e^2}{|\mathbf{r} + \mathbf{d}/2|}. \tag{4.25}$$

The two protons are placed at $\pm\mathbf{d}/2$ and $\mathbf{d}$ is the distance vector between them. The last term accounts for the ionic Coulomb repulsion and has no effect on the electrons.

The ansatz for the electronic wave functions is a superposition of two Gaussians $\mathcal{G}$ centered at the two proton sites

$$\varphi_\nu = c_\nu \left[\mathcal{G}(\mathbf{r}-\mathbf{d}/2;\sigma) + \nu\mathcal{G}(\mathbf{r}+\mathbf{d}/2;\sigma)\right], \quad \nu \in \{+,-\}, \tag{4.26a}$$

$$\mathcal{G}(\mathbf{x};\sigma) = \pi^{-3/4}\sigma^{-3/2}\exp\left(-\frac{\mathbf{x}^2}{2\sigma^2}\right), \tag{4.26b}$$

$$c_\nu = \left[\int d^3r \left(\mathcal{G}(\mathbf{r}-\mathbf{d}/2;\sigma) + \nu\mathcal{G}(\mathbf{r}+\mathbf{d}/2;\sigma)\right)^2\right]^{-1/2}. \tag{4.26c}$$

The pre-factor c_ν serves for proper normalization. The symmetric case $\nu = +$ will be the ground-state wave function and $\nu = -$ the anti-bonding excited state.

The free-model parameter is here the Gaussian width σ. We plan to optimize it by minimizing the total energy, which needs to be computed first. One key piece is the overlap of two Gaussians:

$$\begin{aligned}\langle\mathbf{d}_1|\mathbf{d}_2\rangle &= \int d^3r\, \mathcal{G}(\mathbf{r}-\mathbf{d}_1;\sigma)\mathcal{G}(\mathbf{r}-\mathbf{d}_2;\sigma) \\ &= \int \frac{d^3x}{\pi^{3/2}\sigma^3} \exp\left(-\frac{(\mathbf{r}-\mathbf{d}_1)^2 + (\mathbf{r}-\mathbf{d}_2)^2}{2\sigma^2}\right) = \exp\left(-\frac{(\mathbf{d}_1-\mathbf{d}_2)^2}{4\sigma^2}\right).\end{aligned}$$

This immediately allows to compute the normalization coefficients

$$\begin{aligned}c_\nu &= \left[\langle\mathbf{d}/2|\mathbf{d}/2\rangle + \nu\langle\mathbf{d}/2|-\mathbf{d}/2\rangle + \nu\langle-\mathbf{d}/2|\mathbf{d}/2\rangle + \langle-\mathbf{d}/2|-\mathbf{d}/2\rangle\right]^{-1/2} \\ &= \left\{2 + 2\nu\exp\left(-\frac{\mathbf{d}^2}{4\sigma^2}\right)\right\}^{-1/2}.\end{aligned}$$

The kinetic energy can be computed by parametric integration using $\nabla_{\mathbf{r}}\mathcal{G}(\mathbf{r}-\mathbf{d}_i;\sigma) = -\nabla_{\mathbf{d}_i}\mathcal{G}(\mathbf{r}-\mathbf{d}_i;\sigma)$ yielding

$$(\mathbf{d}_1|\overleftarrow{\nabla}\cdot\overrightarrow{\nabla}|\mathbf{d}_2) = \nabla_{\mathbf{d}_1}\cdot\nabla_{\mathbf{d}_2}\langle\mathbf{d}_1|\mathbf{d}_2\rangle = \left[\frac{3}{2\sigma^2} - \frac{(\mathbf{d}_1-\mathbf{d}_2)^2}{4\sigma^4}\right]\exp\left(-\frac{(\mathbf{d}_1-\mathbf{d}_2)^2}{4\sigma^2}\right) .$$

The expectation value of the kinetic energy thus becomes

$$\begin{aligned} E_{\mathrm{kin},\nu}(\mathbf{d}) &= \frac{\hbar^2}{2m}c_\nu^2\,\{(\mathbf{d}/2|\nabla\cdot\nabla|\mathbf{d}/2) + (-\mathbf{d}/2|\nabla\cdot\nabla|-\mathbf{d}/2) \\ &\qquad +\nu\,[(-\mathbf{d}/2|\nabla\cdot\nabla|\mathbf{d}/2) + (\mathbf{d}/2|\nabla\cdot\nabla|-\mathbf{d}/2)]\} \\ &= \frac{\hbar^2}{2m}\frac{3}{2\sigma^2}\frac{1+\nu\left[1-\frac{\mathbf{d}^2}{6\sigma^2}\right]\exp\left(-\frac{\mathbf{d}^2}{4\sigma^2}\right)}{1+\nu\exp\left(-\frac{\mathbf{d}^2}{4\sigma^2}\right)}. \end{aligned}$$

A bit more involved is the potential energy. First define the density associated with $\mathcal{G}$

$$\rho_{\mathcal{G}}(\mathbf{r};\sigma) = \mathcal{G}^2(\mathbf{r};\sigma)$$

and with it the basic Coulomb integral

$$\mathcal{I}_{\mathrm{C}}(a) = \int \mathrm{d}^3r'\frac{e^2}{|\mathbf{r}'-\mathbf{a}|}\rho_{\mathcal{G}}(\mathbf{r}';\sigma) \quad\longleftrightarrow\quad \Delta_a\mathcal{I}_{\mathrm{C}}(a) = -4\pi e^2\rho_{\mathcal{G}}(\mathbf{a};\sigma),$$

which is the electrostatic potential of a normalized Gaussian density at position $\mathbf{a}$ and which depends only on $a = |\mathbf{a}|$. Exploiting spherical symmetry, we have only to solve a radial integration

$$\begin{aligned} \mathcal{I}_{\mathrm{C}}(\mathbf{a}_1-\mathbf{a}_2) &= \int \mathrm{d}^3r\rho_{\mathcal{G}}(\mathbf{r}-\mathbf{a}_1;\sigma)\frac{e^2}{|\mathbf{r}-\mathbf{a}_2|} = \int \mathrm{d}^3r'\frac{e^2}{|\mathbf{r}'-\mathbf{a}_2+\mathbf{a}_1|}\rho_{\mathcal{G}}(\mathbf{r}';\sigma), \\ \frac{1}{a}\partial_a^2 a\mathcal{I}_{\mathrm{C}}(a) &= -4\pi e^2\rho_{\mathcal{G}}(a;\sigma), \\ \partial_a a\mathcal{I}_{\mathrm{C}}(a) &= -\frac{4\pi e^2}{\pi^{3/2}\sigma^3}\int_a^\infty \mathrm{d}r\,r\,\exp\left(-\frac{r^2}{\sigma^2}\right) = -\frac{2e^2}{\pi^{1/2}\sigma}\exp\left(-\frac{a^2}{\sigma^2}\right), \\ \mathcal{I}_{\mathrm{C}}(a) &= -\frac{1}{a}\frac{2e^2}{\pi^{1/2}\sigma}\int_0^a \mathrm{d}r\,y\exp\left(-\frac{r^2}{\sigma^2}\right) = -\frac{e^2}{a}\mathrm{erf}\left(\frac{a}{\sigma}\right), \end{aligned}$$

where the error function is defined as $\mathrm{erf}(x) = 2\pi^{-1/2}\int_0^r dx\exp(-x^2)$. This allows to evaluate the various Coulomb contributions

$$
\begin{aligned}
\langle \mathbf{d}/2|V|\mathbf{d}/2\rangle &= -\int \mathrm{d}^3 r \rho_{\mathcal{G}}(\mathbf{r}-\mathbf{d}/2;\sigma)\frac{e^2}{|\mathbf{r}-\mathbf{d}/2|} \\
&\quad -\int \mathrm{d}^3 r \rho_{\mathcal{G}}(\mathbf{r}-\mathbf{d}/2;\sigma)\frac{e^2}{|\mathbf{r}+\mathbf{d}/2|} \\
&= -\mathcal{I}_{\mathrm{C}}(0) - \mathcal{I}_{\mathrm{C}}(\mathbf{d}) = -\frac{2e^2}{\pi^{1/2}\sigma} - \frac{e^2}{d}\operatorname{erf}\left(\frac{d}{\sigma}\right) \\
&= \langle -\mathbf{d}/2|V|-\mathbf{d}/2\rangle, \\
\mathcal{G}(\mathbf{r}-\mathbf{d}/2;\sigma)\mathcal{G}(\mathbf{r}+\mathbf{d}/2;\sigma) &= \frac{1}{\pi^{3/2}\sigma^3}\exp\left(-\frac{(\mathbf{r}-\mathbf{d}/2)^2}{2\sigma^2} - \frac{(\mathbf{r}+\mathbf{d}/2)^2}{2\sigma^2}\right) \\
&= \rho_{\mathcal{G}}(\mathbf{r};\sigma)\exp\left(-\frac{\mathbf{d}^2}{4\sigma^2}\right), \\
\langle \mathbf{d}/2|V|-\mathbf{d}/2\rangle &= -\exp\left(-\frac{\mathbf{d}^2}{4\sigma^2}\right)\int \mathrm{d}^3 r \rho_{\mathcal{G}}(\mathbf{r};\sigma)\left[\frac{e^2}{|\mathbf{r}-\mathbf{d}/2|} + \frac{e^2}{|\mathbf{r}+\mathbf{d}/2|}\right] \\
&= -2\exp\left(-\frac{\mathbf{d}^2}{4\sigma^2}\right)\mathcal{I}_{\mathrm{C}}(d/2) \\
&= -2e^2\exp\left(-\frac{\mathbf{d}^2}{4\sigma^2}\right)\frac{2}{d}\operatorname{erf}\left(\frac{d}{2\sigma}\right)
\end{aligned}
$$

and to compute the electronic potential energy

$$
\begin{aligned}
\langle \varphi_\nu|V|\varphi_\nu\rangle &= 2c_\nu^2\left\{\langle \mathbf{d}/2|V|\mathbf{d}/2\rangle + \nu\langle -\mathbf{d}/2|V|\mathbf{d}/2\rangle\right\} \\
&= -e^2\frac{\dfrac{2}{\pi^{1/2}\sigma} + \dfrac{1}{d}\operatorname{erf}\left(\dfrac{d}{\sigma}\right) + 2\nu e^2\exp\left(-\dfrac{d^2}{4\sigma^2}\right)\dfrac{2}{d}\operatorname{erf}\left(\dfrac{d}{2\sigma}\right)}{1+\nu\exp\left(-\dfrac{d^2}{4\sigma^2}\right)}.
\end{aligned}
$$

Finally, we have the total energy

$$
\begin{aligned}
E_\nu(d,\sigma) &= \frac{\hbar^2}{2m}\frac{3}{2\sigma^2}\frac{1+\nu\left[1-\frac{d^2}{6\sigma^2}\right]\exp\left(-\frac{d^2}{4\sigma^2}\right)}{1+\nu\exp\left(-\frac{d^2}{4\sigma^2}\right)} \\
&\quad -e^2\frac{\dfrac{2}{\pi^{1/2}\sigma} + \dfrac{1}{d}\operatorname{erf}\left(\dfrac{d}{\sigma}\right) + 2\nu e^2\exp\left(-\dfrac{d^2}{4\sigma^2}\right)\dfrac{2}{d}\operatorname{erf}\left(\dfrac{d}{2\sigma}\right)}{1+\nu\exp\left(-\dfrac{d^2}{4\sigma^2}\right)} + \frac{e^2}{d}.
\end{aligned}
$$

There is one free parameter, the Gaussian width σ. The optimal choice is the one for which the energy becomes minimal

$$E_\nu(d) = \min_\sigma\{E_\nu(d,\sigma)\}.$$

This final step of minimization yields a transcendental equation which can only be solved by a numerical root-finding procedure. A small Fortran program `lcgo.f` which performs the minimization is given on the CD appended to this book. The resulting energy curves are shown and compared with an exact calculation in Fig. 4.8. The upper panel shows the total energies. The Gaussian approximation stays well above the exact energy with an error of about 7%. This is a satisfying result in view of the fact that the Gaussians are extremely simple wave functions which allow to easily evaluate all necessary integrals. Note that the error varies only little with bond length d. As a consequence, the dimer binding energy (shifted to asymptotic energy zero) performs much better. This can be seen from comparing the approximate energy E_+ with the exact one. The lower panel also shows the energy E_- of the anti-bonding state. The pattern of E_+ and E_- are precisely those previewed in the simple tight-binding picture in Fig. 4.3.

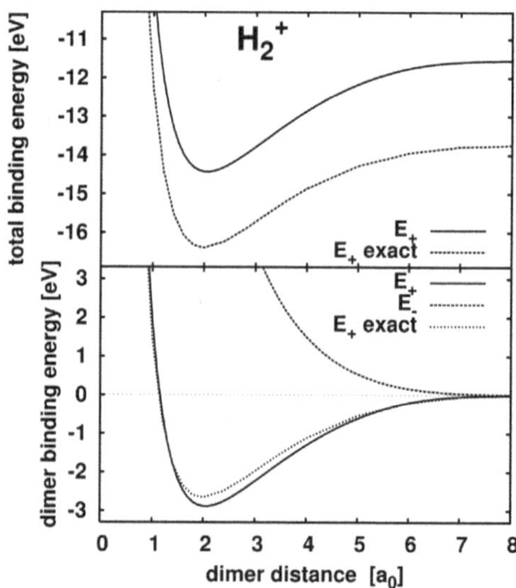

Fig. 4.8 The binding energy for H_2^+ as function of proton distance d. Compared are the results of the present LCAO model with an exact quantum-mechanical calculation. *Lower*: Dimer binding, i.e. energy relative to the asymptotic state of one H atom and one H^+ ion. *Upper*: Total binding energy on an absolute scale

4.4.2 Toward Fully Realistic Quantum Chemistry Calculations

The basic starting point of elaborate quantum chemistry calculations is the Hartree–Fock (mean-field, HF) approach which we shall detail in Chap. 5. This approach consists in treating the effect of electron–electron interactions by performing an educated average over them, while properly accounting for fermionic effects. This amounts to forcing the many-electron wave function to be a Slater determinant built out of single-electron orbitals. This theory will be discussed in many places in this book. Still, one should mention here two interesting aspects in relation to this mean-field approach. The first one concerns the nature of the HF equation and the way to solve it. The second one concerns the methods to go beyond such an approach.

In the HF approach, the average treatment of electron–electron interactions leads to a potential well (mean field) depending on the electronic wave functions themselves (self-consistency of the problem). The problem thus becomes nonlinear and one is bound to solve it in an iterative manner. Iteration methods can be accelerated by using educated guesses of the solutions. And LCAO approaches can again be very helpful to provide a starting point for iterative calculations via an acceptable "zeroth-order approximation" to the true solution. In any case the LCAO wave functions, and more generally speaking symmetry considerations, are extremely valuable in molecular species to construct initial guesses of electronic wave functions with the correct symmetries and node sequence. And this by itself constitutes a valuable guide toward the solution.

The HF approach is, nevertheless, known to provide only an approximate solution to the true many-electron problem in a molecule and quantum chemistry has for a long time developed methods to go beyond HF. These methods are usually called ab initio in quantum chemistry and consist, again, in allowing an enlarged Hilbert space. As we will see in Chap. 5, the HF approximation implies severe restrictions on the accessible Hilbert space for the many-electron problem (even if these restrictions are much less severe than in the VB, LCAO, or Hückel approaches). Ab initio methods propose a systematic way to construct extended Hilbert spaces on the basis of the HF Hilbert space. By applying particle–hole excitations on the HF ground-state wave function one can construct more and more complex wave functions and thus extend the Hilbert space in a progressive way. These methods are know as configuration interaction (CI) methods. They are based on a linear superposition (4.1) of Slater states. For a given set of Slater states the expansion coefficients c_α are determined by the Ritz variational principle [24], which again yields a secular equation like in the LCAO calculations, but this time built on self-consistent Slater wave functions. The quality of description depends on the choice of the expansion basis. Although the CI approaches are in principle extremely powerful, it turns out that high accuracy requires huge basis sets and the computations soon reach limits of todays computers even for moderately large systems of a few tens of electrons. One is then again bound to develop dedicated approximations and here the experience gathered from the simple, intuitive pictures like VB, LCAO, or Hückel is of invaluable help.

4.5 Concluding Remarks

This chapter has illustrated the capabilities of approaches based on model spaces. Such approaches amount to describing the many-fermion problems by considering a limited Hilbert space for single-particle wave functions. They have been largely developed in chemistry and molecular physics. The restriction of the Hilbert space allows a rather simple picture of the involved many-electron problem, at the price of reduced realism. There is a thus a balance to find between simplicity and degree of realism. All in all we have seen a few general trends emerging from the few examples treated. A first aspect is the fact that these methods offer a rather transparent description of the system under consideration. This is especially true in the case of covalent binding in which electron wave functions remain, up to simple linear combinations, rather close to atomic wave functions. This is exactly the reason why LCAO methods work so well, but the simplified picture can in fact be applied to a rather large set of systems and several examples thereof were presented, especially in the simplified Hückel approximation, which allows treating even extended systems and which has indeed been used in many systems from molecules to bulk solid. Finally one should remember that approaches based on models also provide powerful starting points for more elaborate methods. Most of these elaborate methods rely on self-consistent approaches, in which the effective Schrödinger equations to be solved effectively depend on their own solutions, calling for iterative solution methods. In such cases a good starting point is always a key ingredient and methods based on model spaces offer an efficient solution for obtaining it.

Chapter 5
Hartree–Fock

There are numerous situations of highly correlated many-body systems or field theories where a full treatment is too cumbersome, if not impossible. At the same time such systems display very simple properties if one looks at structure or low-energy excitations which can be described by one-body potentials and weak residual interactions; think, e.g., of atomic nuclei [17, 92, 42]. The nuclear interaction combines strong attraction and a huge short-range repulsion. These large and counteracting influences make an ab initio treatment extremely involved, up to these days not fully under control. Nonetheless, a simple nuclear shell model allows to sort all structural properties and basic low-energy excitations in the range up to giant resonances (around 20–30 MeV) [48]. Simple one-body models of that sort are ubiquitous in all areas of physics, see Chap. 3. They embody, however, a disquieting amount of arbitrariness as these shell-model potentials usually emerge from an educated guess.

The Hartree–Fock approximation developed in this chapter provides a *self-consistent* method to construct a mean field on the basis of the elementary interaction between constituents. The method is well known in many areas of physics with strongholds in nuclear physics and electronic systems. It often serves as a starting point for the construction of more elaborate descriptions. The latter can be attained either by using energy-density functional (Chap. 6) or by the inclusion of correlations (Chap. 9) on top of the mean field itself.

5.1 The Hartree–Fock Setup

5.1.1 Foundation

In the following, we assume a Hamiltonian containing only a one-body part $\hat{T}$ and a two-body interaction $\hat{V}$. In coordinate space this could look like

$$\hat{H} = \underbrace{\sum_{k=1}^{N}\left[\frac{-\hbar^2}{2m}\nabla_k^2 + U_{\text{ext}}(x_k)\right]}_{\hat{T}} + \underbrace{\frac{1}{2}\sum_{k,l=1}^{N} v(x_k, x_l, \ldots)}_{\hat{V}}, \tag{5.1}$$

J.A. Maruhn et al., *Simple Models of Many-Fermion Systems*,
DOI 10.1007/978-3-642-03839-6_5, © Springer-Verlag Berlin Heidelberg 2010

where x as usual stands for both spatial coordinates and spin and where the dots indicate that the potential might depend on additional properties of the particles like their momenta and possibly isospins. The one-body part of the Hamiltonian consists of the kinetic energy and a possible external one-body potential. Nuclei and drops of liquid ^{3}He have no external potential. Electronic systems (atoms, molecules, clusters, quantum dots, solids) are usually bound by the external Coulomb field of the ions. Atoms in a trap are, by definition, kept together in the (harmonic) potential of the trap (see Sect. 1.1).

In terms of Fermion operators (see Sect. A.4), the Hamiltonian can be expressed as

$$\hat{H} = \sum_{\alpha\beta} t_{\alpha\beta}\, \hat{a}_\alpha^\dagger \hat{a}_\beta + \tfrac{1}{2} \sum_{\alpha\beta\gamma\delta} v_{\alpha\beta\gamma\delta}\, \hat{a}_\alpha^\dagger \hat{a}_\beta^\dagger \hat{a}_\delta \hat{a}_\gamma \,, \tag{5.2}$$

where the indices α, β, γ, and δ label the single-particle states in some complete orthonormal basis and the $v_{\alpha\beta\gamma\delta}$ are the matrix elements of the two-body interaction. The eigenstates of this Hamiltonian are determined by the stationary Schrödinger equation

$$\hat{H}|\Psi\rangle_{\text{exact}} = E_{\text{exact}}|\Psi\rangle_{\text{exact}}. \tag{5.3}$$

Each such eigenstate can be expanded as a sum over N-particle Slater determinants corresponding to all possible occupations of the single-particle states. If $\alpha_1, \alpha_2, \ldots, \alpha_N$ indicate the states occupied in each term, the many-body wave function can be written as

$$|\Psi\rangle = \sum_{\alpha_1 < \alpha_2 < \cdots < \alpha_N} c_{\alpha_1,\alpha_2,\ldots,\alpha_N}\, \hat{a}_{\alpha_1}^\dagger \hat{a}_{\alpha_2}^\dagger \cdots \hat{a}_{\alpha_N}^\dagger |0\rangle. \tag{5.4}$$

Note that because of the Pauli principle the occupied states α_i must all be different and that their ordering is unimportant; this was indicated by choosing a special order in the sum.

5.1.2 The Hartree–Fock Approximation

We are mainly interested in the ground-state solution. Even with that restriction, the exact problem is exceedingly demanding. If we limit the basis to a finite number of $N_a \geq N$ single-particle states, the number of Slater determinants in the sum (5.4) is given by the number of combinations of N distinct indices drawn from a total of N_a available ones, i.e., $\binom{N_a}{N}$. Obviously the problem cannot be handled in practice for larger values of N_a which are, on the other hand, necessary to achieve a reasonably good solution. There are two ways out of this dilemma: either restricting the number of particles and states or, alternatively, replacing the general wave function (5.4) by an approximation. The first option often is too limiting, and the most useful case of the second approach is the *Hartree–Fock approximation*, which makes the quite radical assumption that the sum can be restricted to one term, i.e., a *single Slater determinant*

$$|\Psi\rangle \quad \longrightarrow \quad |\Phi\rangle = \hat{a}_1^\dagger \hat{a}_2^\dagger \cdots \hat{a}_N^\dagger |0\rangle, \tag{5.5}$$

which is composed of the energetically lowest single-particle states $\alpha = 1, \ldots, N$. Since a single Slater determinant corresponds to non-interacting particles, this approximation is also called the *independent-particle model*. If the wave functions in the ansatz (5.5) were just chosen from an arbitrary basis, however, this approximation would be uncontrollably bad. The quality of the approximation comes about by adjusting the wave functions themselves to produce an optimal result.

In principle thus the stationary Schrödinger equation (5.3) is to be solved restricting the many-body wave function to a Slater determinant $|\Phi\rangle$, see (5.5). A Slater determinant is of course not a solution to (5.3), so that the best we can hope is to make it an optimal approximation to the exact wave function. A method familiar from elementary quantum mechanics is the *Ritz variational principle* [24] (which was also used, e.g., in LCAO, see Sect. 4.2). It states that a best approximation to the *ground state* of $\hat{H}$ can be found by varying $|\Phi\rangle$ in such a way that the energy $\langle\Phi|\hat{H}|\Phi\rangle$ is minimized. In essence the variational principle is based on the idea that if we expand the approximate solution into all the exact eigenstates, any contribution from excited states raises the expectation value of $\hat{H}$, so that minimizing the energy expectation value will also minimize those contributions. The energy minimization is to be augmented with the additional constraint that the state remain normalized to $\langle\Phi|\Phi\rangle = 1$ during variation. This is taken care of by implementing the constraint together with a Lagrangian multiplier, yielding the variational equation

$$\delta\big(\langle\Phi|\hat{H}|\Phi\rangle - E\langle\Phi|\Phi\rangle\big) = 0\,. \tag{5.6}$$

It will become clear in a moment that the Lagrange multiplier is indeed just the energy expectation value.

The variation in (5.6) can be carried out with respect to either $|\Phi\rangle$ or $\langle\Phi|$. Varying $\langle\Phi|$ and $|\Phi\rangle$ separately yields the equivalent equations

$$\langle\delta\Phi|\hat{H}|\Phi\rangle - E\langle\delta\Phi|\Phi\rangle = 0\,, \langle\Phi|\hat{H}|\delta\Phi\rangle - E\langle\Phi|\delta\Phi\rangle = 0\,. \tag{5.7}$$

If $|\delta\Phi\rangle$ were an arbitrary Hilbert space vector it would follow that $\hat{H}|\Phi\rangle - E|\Phi\rangle = 0$, leading back to the Schrödinger equation. In the case of interest here, of course, $|\delta\Phi\rangle$ is restricted and $|\Phi\rangle$ is not an eigenfunction of $\hat{H}$.

5.1.3 The Space of Variations of a Slater State

The approximate wave function (5.5) is a pure Slater determinant whose building blocks are the occupied single-particle states $\varphi_j(x)$. The variation $|\delta\Phi\rangle$ comes from the variation of the single-particle states which can be varied via an admixture of other single-particle wave functions from the complete set $\{\varphi_\alpha(x)\}$, i.e.,

$$\delta\varphi_j(x) = \sum_{\alpha\neq j} \varphi_\alpha(x)\delta c_{\alpha j} \, .$$

This is equivalent to varying the creation operators

$$\delta\hat{a}_j^\dagger = \sum_{\alpha\neq j} \hat{a}_\alpha^\dagger \delta c_{\alpha j} \, .$$

Inserting the varied operator into the wave function (5.5) yields

$$|\Phi + \delta\Phi_j\rangle = |\Phi\rangle + \sum_{\alpha\neq j} \delta c_{\alpha j}\, \hat{a}_1^\dagger \hat{a}_2^\dagger \cdots \hat{a}_{j-1}^\dagger \hat{a}_\alpha^\dagger \hat{a}_{j+1}^\dagger \cdots \hat{a}_N^\dagger |0\rangle \, .$$

The variation vanishes if the index α of the newly mixed-in wave function is among the indices of occupied states, i.e., if $\alpha \in \{1, \ldots, N\}$, because of the Pauli principle [24]. This leads to two important consequences:

- Any replacement of the occupied wave functions by linear combinations among themselves leaves the total many-particle wave function invariant (this is also familiar from the fact that replacing a column in a determinant by a linear superposition with other columns does not change the value of the determinant).
- The only non-vanishing variations of a single Slater determinant are admixtures of unoccupied wave functions to any of the occupied ones.

In consequence, the variational principle can only define an optimal separation between occupied and unoccupied states. There is space, or need, for further conditions if one wants to specify the single-particle states uniquely.

The above formulation of the variation is still a bit cumbersome. It can be expressed more concisely by using an annihilation operator to depopulate the wave function j before populating the new state n which, as pointed out, is restricted to unoccupied states only. For the variation of $|\Phi\rangle$, if one occupied state j is varied, this yields

$$|\delta\Phi_j\rangle = \sum_{n>N} c_{nj}\, \hat{a}_n^\dagger \hat{a}_j |\Phi\rangle \, ,$$

where any sign arising from a reordering of the operators has been absorbed into the definition of the c_{nj}. The most general variation possibly affects every occupied state and is

$$|\delta\Phi\rangle = \sum_{n>N, j\leq N} c_{nj}\, \hat{a}_n^\dagger \hat{a}_j |\Phi\rangle\,. \tag{5.8}$$

Any choice of the (small) expansion coefficients c_{kj} provides one valid variation. Thus it suffices for the variational principle to consider *all* variations of the form

$$|\delta\Phi\rangle = \eta \hat{a}_n^\dagger \hat{a}_j |\Phi\rangle\,, \quad n > N\,, \quad j \leq N\,, \tag{5.9a}$$

or equivalently

$$\langle\delta\Phi| = \eta^* \langle\Phi|\hat{a}_j^\dagger \hat{a}_n\,. \tag{5.9b}$$

The parameter η is an arbitrary parameter, sufficiently small to consider $|\delta\Phi\rangle$ or $\langle\delta\Phi|$ a small variation. The state $\hat{a}_n^\dagger \hat{a}_j |\Phi\rangle$ is called a *one-particle–one-hole excitation* ($1ph$) because one occupied state j is emptied and replaced by one hitherto unoccupied state n (see Sect. 1.3.2 and Appendix A.4).

Notation: Throughout this chapter it will be necessary to distinguish three types of indices: those referring to occupied or unoccupied states only and others that are unrestricted. To facilitate the notation we use the following convention:

- The indices h, h' and their subscripted forms h_1, h_2, etc., refer to occupied states only, i.e., they take values from 1 through N exclusively, where N is the number of occupied states: $h, h' = 1, \ldots, N$.
- The indices p, p' and their subscripted forms refer to unoccupied single-particle states only: $p, p' = N+1, \ldots, \infty$.
- Greek letters like α are reserved for unrestricted indices: $\alpha = 1, \ldots, \infty$.

5.2 The Hartree–Fock Equations

5.2.1 In Terms of Fermion Operators

It is now straightforward to evaluate the variational equations (5.7) formally. The $1ph$ states are orthogonal to the underlying Slater state, i.e., $\langle\Phi|\hat{a}_h^\dagger \hat{a}_p|\Phi\rangle = 0$ and correspondingly $\langle\Phi|\hat{a}_p^\dagger \hat{a}_h|\Phi\rangle = \langle\Phi|\hat{a}_h^\dagger \hat{a}_p|\Phi\rangle^* = 0$, leaving

$$\langle\Phi|\hat{a}_h^\dagger \hat{a}_p \hat{H}|\Phi\rangle = 0\,, \quad \langle\Phi|\hat{H}\hat{a}_p^\dagger \hat{a}_h|\Phi\rangle = 0\,. \tag{5.10}$$

The Hartree–Fock optimization consists in decoupling the ground state from the $1ph$ space. So far so simple. Further handling and practical applications require to reshape these equations in different forms. In a first step, we will derive an effective single-particle equation and, in a second step, translate it into the coordinate-space picture.

5.2.2 In Configuration Space (Matrix Elements)

The evaluation of the matrix element of $\hat{H}$ is straightforward. Nevertheless we present the steps in detail as they constitute a useful exercise. Using the fermion algebra (A.12) the order of two fermion operators may be permuted as

$$\hat{a}^\dagger_\alpha \hat{a}^\dagger_\beta = -\hat{a}^\dagger_\beta \hat{a}^\dagger_\alpha\,, \quad \hat{a}_\alpha \hat{a}_\beta = -\hat{a}_\beta \hat{a}_\alpha\,, \quad \hat{a}^\dagger_\alpha \hat{a}_\beta = \delta_{\alpha\beta} - \hat{a}_\beta \hat{a}^\dagger_\alpha\,.$$

The creation and annihilation operator properties with respect to the Slater state are given by

$$\hat{a}_p|\Phi\rangle = 0\,,\ \hat{a}^\dagger_h|\Phi\rangle = 0\,,\ \langle\Phi|\hat{a}_h = 0\,,\ \langle\Phi|\hat{a}^\dagger_p = 0\,,\ p > N\,,\ h \leq N\,,$$

and both together produce a basic expectation value

$$\begin{aligned}\langle\Phi|\hat{a}^\dagger_\alpha \hat{a}_\beta|\Phi\rangle &= \langle\Phi|\hat{a}^\dagger_\alpha \hat{a}_\beta|\Phi\rangle\vartheta(\beta \leq N)\vartheta(\alpha \leq N)\\ &= \delta_{\alpha\beta}\vartheta(\beta \leq N) - \langle\Phi|\hat{a}_\beta \underbrace{\hat{a}^\dagger_\alpha|\Phi\rangle}_{=0}\,\vartheta(\beta \leq N)\vartheta(\alpha \leq N)\,,\end{aligned}$$

in short

$$\langle\Phi|\hat{a}^\dagger_\alpha \hat{a}_\beta|\Phi\rangle = \delta_{\alpha\beta}\vartheta(\beta \leq N)\,, \quad \vartheta(\beta \leq N) = \begin{cases} 1 & \text{for } \beta \leq N \\ 0 & \text{for } \beta > N \end{cases}\,. \tag{5.11}$$

Note that the ϑ-function notation was adapted a bit here to make things more legible.

The steps for the one-body operator of the kinetic energy then are in detail

$$\begin{aligned}\langle\Phi|\hat{a}^\dagger_h \hat{a}_p \hat{T}|\Phi\rangle &= \sum_{\alpha\beta} t_{\alpha\beta}\langle\Phi|\hat{a}^\dagger_h \hat{a}_p \hat{a}^\dagger_\alpha \hat{a}_\beta|\Phi\rangle\\ &= \sum_{\alpha\beta} t_{\alpha\beta}\delta_{p\alpha}\langle\Phi|\hat{a}^\dagger_h \hat{a}_\beta|\Phi\rangle - \sum_{\alpha\beta} t_{\alpha\beta}\langle\Phi|\hat{a}^\dagger_h \hat{a}^\dagger_\alpha \hat{a}_p \hat{a}_\beta|\Phi\rangle\\ &= \sum_\beta t_{p\beta}\delta_{h\beta} - \sum_\beta t_{p\beta}\langle\Phi|\hat{a}_\beta|\underbrace{\hat{a}^\dagger_h\Phi\rangle}_{=0} + \sum_{\alpha\beta} t_{\alpha\beta}\langle\Phi|\hat{a}^\dagger_h \hat{a}^\dagger_\alpha \hat{a}_\beta|\underbrace{\hat{a}_p\Phi\rangle}_{=0}\\ &= t_{ph}\,.\end{aligned}$$

After this detailed exercise, we can step more quickly through the evaluation of the two-body operator for the potential:

$$
\begin{aligned}
2\langle\Phi|\hat{a}_h^\dagger\hat{a}_p\hat{V}|\Phi\rangle &= \sum_{\alpha\beta\gamma\delta} v_{\alpha\beta\gamma\delta}\langle\Phi|\hat{a}_h^\dagger\hat{a}_p\hat{a}_\alpha^\dagger\hat{a}_\beta^\dagger\hat{a}_\delta\hat{a}_\gamma|\Phi\rangle \\
&= \sum_{\beta\gamma\delta} v_{p\beta\gamma\delta}\langle\Phi|\hat{a}_h^\dagger\hat{a}_\beta^\dagger\hat{a}_\delta\hat{a}_\gamma|\Phi\rangle - \sum_{\alpha\beta\gamma\delta} v_{\alpha\beta\gamma\delta}\langle\Phi|\hat{a}_h^\dagger\hat{a}_\alpha^\dagger\hat{a}_p\hat{a}_\beta^\dagger\hat{a}_\delta\hat{a}_\gamma|\Phi\rangle \\
&= -\sum_{\beta\gamma\delta} v_{p\beta\gamma\delta}\langle\Phi|\hat{a}_\beta^\dagger\hat{a}_h^\dagger\hat{a}_\delta\hat{a}_\gamma|\Phi\rangle - \sum_{\alpha\gamma\delta} v_{\alpha p\gamma\delta}\langle\Phi|\hat{a}_h^\dagger\hat{a}_\alpha^\dagger\hat{a}_\delta\hat{a}_\gamma|\Phi\rangle \\
&= -\sum_{\beta\gamma} v_{p\beta\gamma h}\langle\Phi|\hat{a}_\beta^\dagger\hat{a}_\gamma|\Phi\rangle + \sum_{\beta\gamma\delta} v_{p\beta\gamma\delta}\langle\Phi|\hat{a}_\beta^\dagger\hat{a}_\delta\hat{a}_h^\dagger\hat{a}_\gamma|\Phi\rangle \\
&\quad + \sum_{\alpha\gamma\delta} v_{\alpha p\gamma\delta}\langle\Phi|\hat{a}_\alpha^\dagger\hat{a}_h^\dagger\hat{a}_\delta\hat{a}_\gamma|\Phi\rangle \\
&= -\sum_{\beta\gamma} v_{p\beta\gamma h}\underbrace{\langle\Phi|\hat{a}_\beta^\dagger\hat{a}_\gamma|\Phi\rangle}_{\delta_{\beta\gamma}\theta(\beta\le N)} + \sum_{\beta\delta} v_{p\beta h\delta}\underbrace{\langle\Phi|\hat{a}_\beta^\dagger\hat{a}_\delta|\Phi\rangle}_{\delta_{\beta\delta}\theta(\beta\le N)} \\
&\quad + \sum_{\alpha\gamma} v_{\alpha p\gamma h}\underbrace{\langle\Phi|\hat{a}_\alpha^\dagger\hat{a}_\gamma|\Phi\rangle}_{\delta_{\alpha\gamma}\theta(\alpha\le N)} - \sum_{\alpha\delta} v_{\alpha p h\delta}\underbrace{\langle\Phi|\hat{a}_\alpha^\dagger\hat{a}_\delta|\Phi\rangle}_{\delta_{\alpha\delta}\theta(\alpha\le N)} \\
&= \sum_{h'\le N}\left[-v_{ph'h'h} + v_{ph'hh'} + v_{h'ph'h} - v_{h'phh'}\right] \\
&= 2\sum_{h'\le N}\left[v_{ph'hh'} - v_{ph'h'h}\right]
\end{aligned}
$$

where the symmetry $v_{\alpha\beta\gamma\delta} = v_{\beta\alpha\delta\gamma}$ of the two-body matrix element was exploited. Summarizing the results, we have

$$
\begin{aligned}
\langle\Phi|\hat{a}_h^\dagger\hat{a}_p\hat{H}|\Phi\rangle &= t_{ph} + \sum_{h'}\bar{v}_{ph'hh'}, \quad & \bar{v}_{ph'jh'} &= v_{ph'hh'} - v_{ph'h'h}, \\
\langle\Phi|\hat{H}\hat{a}_p^\dagger\hat{a}_h|\Phi\rangle &= t_{hp} + \sum_{h'}\bar{v}_{hh'ph'}, \quad & \bar{v}_{hh'ph'} &= v_{hh'ph'} - v_{hh'h'p},
\end{aligned}
$$

and so get the *Hartree–Fock equations* in terms of matrix elements as

$$
h_{ph} = 0, \quad h_{hp} = 0, \tag{5.12a}
$$

$$
h_{\alpha\beta} = t_{\alpha\beta} + \sum_{i=1}^{N}\bar{v}_{\alpha i\beta i} = t_{\alpha\beta} + \sum_{i=1}^{N}\left[v_{\alpha i\beta i} - v_{\alpha ii\beta}\right]. \tag{5.12b}
$$

To understand the physics contained in this equation, remember the range the individual indices vary over: h and h' denote occupied states, p an unoccupied one, while α and β are arbitrary. The equation demands that the single-particle states should be chosen such that the matrix elements $h_{\alpha\beta}$ vanish between occupied and unoccupied states. If we allow (α, β) to refer to arbitrary combinations of states, this defines a single-particle operator

$$\hat{h} = \sum_{\alpha\beta} h_{\alpha\beta} \hat{a}_\alpha^\dagger \hat{a}_\beta \,, \tag{5.13}$$

which is usually called the *mean-field* or *Hartree–Fock Hamiltonian*. As expected, the condition (5.12a) characterizes the set of occupied states and not those states individually.

The definition (5.12b) has already introduced the mean-field Hamiltonian for all pairs of single-particle states, although the Hartree–Fock conditions fix it only in $1ph$ space. The parts of the matrix purely in occupied space or purely in unoccupied space are yet undetermined. We can exploit the freedom and ask additionally that it become diagonal. This yields the Hartree–Fock equations in the form of a one-body Schrödinger equation in configuration space, i.e.,

$$\begin{aligned} h_{\alpha\beta} &= \varepsilon_\alpha \delta_{\alpha\beta} \\ \varepsilon_\alpha &= t_{\alpha\alpha} + \sum_{h=1}^{N} \bar{v}_{\alpha h \alpha h} \,. \end{aligned} \tag{5.14}$$

It is, however, more involved than a simple one-body matrix equation because the mean-field Hamiltonian $\hat{h}$ is an effective one-body operator which itself depends on the (occupied) single-particle states on which it acts, see (5.12b). Thus (5.14) can in general only be solved iteratively.

5.2.3 In Coordinate Space

It is instructive from a formal side and necessary for many practical applications to reformulate the Hartree–Fock equation (5.14) in coordinate space. To that end, we assume a purely local two-body interaction $V(\mathbf{r} - \mathbf{r}')$. The matrix elements of $\hat{h}$ thus read

$$\begin{aligned} h_{\alpha\beta} = &\int \mathrm{d}x\, \varphi_\alpha^\dagger(x) \left[-\frac{\hbar^2}{2m} \nabla^2 + U_{\mathrm{ext}}(x) \right] \varphi_\beta(x) \\ &+ \sum_i \int \mathrm{d}x\, \mathrm{d}x'\, \varphi_\alpha^\dagger(x) \varphi_i^\dagger(x') V(x, x') \varphi_\beta(x) \varphi_i(x') \\ &- \sum_i \int \mathrm{d}x\, \mathrm{d}x'\, \varphi_\alpha^\dagger(x) \varphi_i^\dagger(x') V(x, x') \varphi_i(x) \varphi_\beta(x') \,. \end{aligned}$$

From this result the Hartree–Fock equation in coordinate space can be deduced simply by removing the $\int dx$ and $\varphi_\alpha^\dagger(x)$. This yields

$$\hat{h}\varphi_\alpha = \varepsilon_\alpha \varphi_\alpha, \tag{5.15a}$$

$$\begin{aligned}(\hat{h}\varphi_\alpha)(x) = &-\frac{\hbar^2}{2m}\nabla^2\varphi_\alpha(x) + U_{\text{ext}}(x)\varphi_\alpha(x) + U_{\text{dir}}(x)\varphi_\alpha(x)\\ &-\int dx'\, U_{\text{ex}}(x,x')\varphi_\alpha(x'),\end{aligned} \tag{5.15b}$$

$$U_{\text{dir}}(x) = \int dx'\, v(x,x')\sum_{i=1}^{N}|\varphi_i(x')|^2, \tag{5.15c}$$

$$U_{\text{ex}}(x,x') = v(x,x')\sum_{i=1}^{N}\varphi_i(x)\varphi_i^\dagger(x'). \tag{5.15d}$$

The equations are quite similar in form to a Schrödinger equation for single-particle states. The second term on the right-hand side, U_{dir}, is the *average potential* which has the simple interpretation of the potential generated by the density distribution of the particles. It is called the *direct* term. The third term carrying U_{ex} is the *exchange term*. Note that the exchange kernel $U_{\text{ex}}(x, x')$ is still an operator in spin space. Exchange together with the average potential defines the self-consistent *mean field*. The Hartree–Fock approximation is called a *mean-field approximation*.

The exchange term of course makes the problem quite a bit more complicated than a simple one-particle Schrödinger equation. Since it changes the Schrödinger equation into an integral equation, it is much harder to deal with in practice, and various approximations are employed. One popular approach is that of using effective zero-range interactions like — in nuclear physics — the Skyrme forces (see Sect. 6.4) for which the exchange term can be combined with the direct one. For forces with a finite range – unavoidable in the case of the Coulomb interaction – one often employs dedicated approximations, see Sect. 6.1 on density functional theory.

Another simplification is the *Hartree approximation*, in which the exchange term is simply neglected. Since the exchange term also subtracts the interaction of a particle with itself, the sum over single-particle states in the direct term has then to be modified to exclude the wave function being calculated; there is then a different single-particle Hamiltonian for each particle and the states are no longer orthogonal. This aspect is further explored in Sect. 6.3.

The mean field embodies a large part of the two-body interaction, but not all. The full two-body interaction, to the extent that it is not included in the mean field, is called the *residual interaction* which builds correlations of various sorts, see Chap. 9. It is also a crucial ingredient for computing the excitation spectrum, see Chap. 8. The Hartree–Fock equations form a *self-consistent* problem in the sense that the wave functions determine the mean field, while the mean field, in turn, determines the wave functions. In practice this leads to iterative solutions in which one starts from an initial guess for the wave functions and determines the mean field from them. Solving the Schrödinger equations then yields a new set of wave functions, and this process is repeated until convergence is achieved.

5.3 The Hartree–Fock Energy

5.3.1 *The Total Energy*

Let us now investigate the properties of the states of the many-body system in the Hartree–Fock approximation. First, we will have a look at the energy of the Hartree–Fock state $|\Phi\rangle$ which is the expectation value of $\hat{H}$.

The expectation values for $\hat{T}$ becomes

$$\langle\Phi|\hat{T}|\Phi\rangle = \sum_{\alpha\beta} t_{\alpha\beta} \underbrace{\langle\Phi|\hat{a}^\dagger_\alpha \hat{a}_\beta|\Phi\rangle}_{\delta_{\alpha\beta}\theta(\beta\leq N)} = \sum_{i=1}^{N} t_{ii},$$

and for the two-body interaction

$$\begin{aligned}
\langle\Phi|\hat{V}|\Phi\rangle &= \frac{1}{2}\sum_{\alpha\beta\gamma\delta} v_{\alpha\beta\gamma\delta}\langle\Phi|\hat{a}^\dagger_\alpha\hat{a}^\dagger_\beta\hat{a}_\delta\hat{a}_\gamma|\Phi\rangle \\
&= \frac{1}{2}\sum_{\alpha\beta=1}^{N}\sum_{\gamma\delta} v_{\alpha\beta\gamma\delta}\langle\Phi|\hat{a}^\dagger_\alpha\hat{a}^\dagger_\beta\hat{a}_\delta\hat{a}_\gamma|\Phi\rangle \\
&= \frac{1}{2}\sum_{\alpha\beta=1}^{N}\sum_{\gamma\delta} v_{\alpha\beta\gamma\delta}\delta_{\beta\delta}\langle\Phi|\hat{a}^\dagger_\alpha\hat{a}_\gamma|\Phi\rangle - \frac{1}{2}\sum_{\alpha\beta=1}^{N}\sum_{\gamma\delta} v_{\alpha\beta\gamma\delta}\langle\Phi|\hat{a}^\dagger_\alpha\hat{a}_\delta\hat{a}^\dagger_\beta\hat{a}_\gamma|\Phi\rangle \\
&= \frac{1}{2}\sum_{\alpha\beta=1}^{N} v_{\alpha\beta\alpha\beta} - \frac{1}{2}\sum_{\alpha\beta=1}^{N} v_{\alpha\beta\gamma\beta}\delta_{\beta\gamma}\langle\Phi|\hat{a}^\dagger_\alpha\hat{a}_\delta|\Phi\rangle \\
&= \frac{1}{2}\sum_{hh'=1}^{N}\left[v_{hh'hh'} - v_{hh'h'h}\right] = \frac{1}{2}\sum_{hh'=1}^{N}\bar{v}_{hh'hh'}.
\end{aligned}$$

The total energy thus becomes

$$\begin{aligned} E_{\mathrm{HF}} = \langle \Phi|\hat{H}|\Phi\rangle &= \sum_{h=1}^{N} t_{hh} + \tfrac{1}{2}\sum_{hh'=1}^{N} \bar{v}_{hh'hh'} \\ &= \sum_{h=1}^{N} h_{hh'} - \tfrac{1}{2}\sum_{hh'=1}^{N} \bar{v}_{hh'hh'} = \sum_{h=1}^{N} \varepsilon_h - \tfrac{1}{2}\sum_{hh'=1}^{N} \bar{v}_{hh'hh'} . \end{aligned} \quad (5.16)$$

It is important to realize that the energy of the Hartree–Fock ground state is not simply the sum of the individual single-particle energies, but has an additional contribution from the potential interactions. The mathematical reason is that the Hamiltonian of the many-particle system is not the sum of the single-particle Hamiltonians, but contains the interactions as two-body matrix elements.

5.3.2 *The Interpretation of Single-Particle Energies: Koopman's Theorem*

A similar analysis also sheds light on the physical meaning of the single-particle energies ε_α. Compare, for example, the energy of the system with N particles to that with $N-1$ particles with one particle removed from the occupied state h. The latter is described by the wave function

$$|h\rangle = \hat{a}_h|\Phi\rangle ,$$

where it was neglected that a change in occupation numbers will also change all the single-particle states because of self-consistency. Its energy is given by

$$E_h = \sum_{h'\neq h} t_{h'h'} + \tfrac{1}{2}\sum_{h_1,h_2\neq h} \bar{v}_{h_1h_2h_1h_2} ,$$

and the difference from the ground-state energy becomes

$$\begin{aligned} E_h - E_{\mathrm{HF}} &= -t_{hh} - \tfrac{1}{2}\sum_{h'} \bar{v}_{h'hh'h} - \tfrac{1}{2}\sum_{h'} \bar{v}_{hh'hh'} = -t_{hh} - \sum_{h'} \bar{v}_{h'hh'h} \\ &= -\varepsilon_h . \end{aligned} \quad (5.17)$$

Here the symmetry of the matrix elements $\bar{v}_{h'hh'h} = \bar{v}_{hh'hh'}$ was used. Thus *the single-particle energy ε_h corresponds (approximately, namely when overlooking impact of rearrangement) to the energy required to remove one particle from state h of the system.* This is the content of *Koopman's theorem*.

The results of Hartree–Fock calculations can thus be used not only to predict the bulk properties of the ground state of the many-body system, such as the binding energy, mean square radius, surface thickness, but also, within the further approximations discussed above, for the description of the ionization potential (as it is called in atomic physics) or separation energy (as it is called in nuclear physics), respectively.

Note that Koopman's theorem is approximate: it cannot be used recursively: if it were reapplied to remove another particle, the potential interactions between the two particles (rearrangement energies) accumulate and thus would be treated incorrectly (otherwise the total energy of the ground state should be given by $\sum_i \varepsilon_i$, which was seen to be wrong).

5.3.3 Excitations and Excitation Energies

It appears a simple matter now to construct excited states of an N-body system based on the Hartree–Fock ground state: these should simply be given by the particle–hole excitations of various orders. In principle this is not quite true, because the mean field also depends on the states actually occupied. The density changes with occupation, leading to changes in the mean potential and in turn to the single-particle states themselves. So in principle all the single-particle wave functions are affected by changing the state of a single particle. In practice, however, this problem is often ignored for larger particle number with the argument that the change of state of one single particle will change the mean field only negligibly, and the associated change of the single-particle states is even smaller. We thus construct excited states as *one-particle–one-hole* ($1ph$) excitations

$$|ph\rangle = \hat{a}_p^\dagger \hat{a}_h |\Phi\rangle ,$$

two-particle–two-hole ($2ph$) excitations

$$|p'phh'\rangle = \hat{a}_{p'}^\dagger \hat{a}_p^\dagger \hat{a}_{h'} \hat{a}_h |\Phi\rangle ,$$

and so on, using the single-particle states of the Hartree–Fock ground state. The expectation value of the energy of such states can easily be calculated. For the one-particle–one-hole excitations, for example, one obtains the excitation energy

$$\begin{aligned} E_{ph} - E_{\mathrm{HF}} &= \langle ph|\hat{H}|ph\rangle - E_{\mathrm{HF}} \\ &= E_{\mathrm{HF}} + t_{pp} - t_{hh} + \sum_{\substack{h'=1 \\ h'\neq h}}^{N} (\bar{v}_{ph'ph'} - \bar{v}_{hh'hh'}) - E_{\mathrm{HF}} \\ &= \varepsilon_p - \varepsilon_h - \bar{v}_{phph} . \end{aligned}$$

This is predominantly the difference of the single-particle energies of the two states which are connected by the transition plus a contribution arising from the additional change in potential interaction. It is to be noted, however, that this is only the expectation value of the residual interaction over the $1ph$ state. There are also non-vanishing off-diagonal matrix elements

$$\langle ph|\hat{H}|p'h'\rangle = -\bar{v}_{ph'p'h}$$

which require a diagonalization by considering a coherent superposition of $1ph$ states. In many cases, though, the potential contribution can be neglected and the particle–hole states can be treated as approximate eigenstates of the problem. This is, e.g., often a reasonable approach for atomic excitation. If the simple $1ph$ picture turns out to be insufficient, one needs to proceed to a more complete calculation of excitation properties like the *random-phase approximation* (RPA), see Chap. 8. That is typically necessary for nuclei and the electron cloud in metals or quantum dots.

5.4 The Density Matrix Formulation

The Hartree–Fock equations take a particularly simple form when expressed in terms of the one-particle density matrix $\hat{\varrho}$. As is discussed in Appendix A.5, the one-particle density matrix corresponding to a single Slater-determinant state fulfills $\hat{\varrho}^2 = \hat{\varrho}$, so that it is idempotent. Mathematically, an idempotent operator describes a projection, since acting once with the operator produces a state that is not changed any further by repeated application. From the special form it takes if expressed in the single-particle states contained in $|\Phi\rangle$,

$$\varrho_{kl} = \begin{cases} \delta_{kl} & \text{for } k \text{ and } l \text{ occupied in } |\Phi\rangle \\ 0 & \text{otherwise,} \end{cases}$$

it is clear that it is the identity operator for occupied states and produces zero for unoccupied ones: thus it projects onto the space of occupied single-particle states.

This last property of the one-particle density matrix allows a simple formulation of the decomposition of a matrix into particle–particle, particle–hole, etc., contributions (related to a given Slater determinant). Returning to the Hartree–Fock case, the hole–hole part $\hat{h}_{\mathrm{hh}}$ of the single-particle Hamiltonian $\hat{h}$ can immediately be written as

$$\hat{h}_{\mathrm{hh}} = \hat{\varrho}\hat{h}\hat{\varrho}\,.$$

To find similar expressions for the other parts of $\hat{h}$, note that the matrix $\hat{\sigma} = 1 - \hat{\varrho}$ projects onto the space of empty single-particle states. It is easy to show that $\hat{\sigma}$ is also a projector, i.e., $\hat{\sigma}^2 = \hat{\sigma}$. It is, in fact, the complementing projector to $\hat{\varrho}$. This immediately yields

$$\hat{h}_{\text{hp}} = \hat{\varrho}\hat{h}\hat{\sigma}\,, \quad \hat{h}_{\text{ph}} = \hat{\sigma}\hat{h}\hat{\varrho}\,, \quad \hat{h}_{\text{pp}} = \hat{\sigma}\hat{h}\hat{\sigma}\,.$$

The Hartree–Fock conditions (5.10) or (5.12a), respectively, were that $\hat{h}_{\text{hp}} = \hat{h}_{\text{ph}} = 0$. They can be rewritten now as

$$\hat{\sigma}\hat{h}\hat{\varrho} = 0\,, \quad \hat{\varrho}\hat{h}\hat{\sigma} = 0\,,$$

and, upon inserting the definition of $\hat{\sigma}$,

$$\hat{h}\hat{\varrho} - \hat{\varrho}\hat{h}\hat{\varrho} = 0\,, \quad \hat{\varrho}\hat{h}\hat{\varrho} - \hat{\varrho}\hat{h} = 0\,,$$

from which we conclude that $\hat{\varrho}\hat{h} = \hat{h}\hat{\varrho}$ or in the most elegant formulation of the Hartree–Fock conditions

$$[\hat{\varrho}, \hat{h}] = 0\,. \tag{5.18}$$

It is clear why formulations using the density matrix are often more useful for formal derivations: there is no need to separate occupied and empty single-particle states in this approach; this distinction is handled by the density matrix. Simple matrix manipulations can then replace cumbersome sum expressions that have the additional problem of having to distinguish these different index ranges.

5.5 A Simple Test Case: Two Particles in 1D

5.5.1 The Model

An instructive test case for self-consistent models is the Negele–Yoon model [115, 74]. We consider a system of two fermions in one spatial dimension coupled via a zero-range two-body interaction. Its Hamiltonian is

$$\hat{H} = \frac{\hat{p}_1^2 + \hat{p}_2^2}{2m} - \frac{c}{2}\delta(x_1 - x_2)\,. \tag{5.19}$$

Note that, unlike the usage elsewhere in this book, the x_i here stands just for the spatial x-coordinate and spin is handled in terms of Pauli spinors $\chi_{+\frac{1}{2}}$ and $\chi_{-\frac{1}{2}}$. Note, furthermore, that this two-body Hamiltonian is translationally invariant. Consider a translation $x_i \longrightarrow x_i' = x_i + \delta$ where the shift δ is the same for both coordinates. The derivative is invariant, i.e., $\partial_{x_i'} = \partial_{x_i}$ and so is the whole kinetic energy operator (first term in the Hamiltonian). The difference $x_1 - x_2$ is also invariant because the constant shift δ cancels. One would expect a solution which in some way reproduces this invariance. This surely holds for the exact solution, but it has yet to be seen how it is sustained through approximations.

The aim of the Hartree–Fock approach is to find the energetically optimal solution in independent-particle form, i.e., as a Slater determinant of single-particle wave functions

$$\Phi = \frac{1}{\sqrt{2}}\mathcal{A}\left\{\varphi_0(x_1)\chi^{(1)}_{+\frac{1}{2}}\,\varphi_0(x_2)\chi^{(2)}_{-\frac{1}{2}}\right\} \tag{5.20a}$$

$$= \frac{\varphi_0(x_1)\varphi_0(x_2)}{\sqrt{2}}\left[\chi^{(1)}_{+\frac{1}{2}}\chi^{(2)}_{-\frac{1}{2}} - \chi^{(1)}_{-\frac{1}{2}}\chi^{(2)}_{+\frac{1}{2}}\right],$$

$$\int_{-\infty}^{\infty} \mathrm{d}x\, |\varphi_0(x)|^2 = 1\,. \tag{5.20b}$$

The total wave function is normalized as $\langle\Phi|\Phi\rangle = 1$ due to the normalization (5.20b) and the orthonormality of the Pauli spinors $\chi^\dagger_{\sigma'}\chi_\sigma = \delta_{\sigma'\sigma}$. In fact, we have two spinor spaces, one for particle 1 and another for particle 2. The spinors are augmented by an upper index which indicates for which of these two spaces it applies. It is self-explaining that spinor overlaps can only be taken with spinors belonging to the same space. The spatial wave function φ_0 is the same for both spins, making the total wave function spatially symmetric under exchange. Antisymmetrization is taken up fully by the spin part. Other choices, e.g., antisymmetric spatial part and symmetric spins, are also conceivable, but correspond to excited states which are not of interest here. The disentangling of spatial and spinor parts is a crucial issue in atoms and molecules and requires wise bookkeeping, the more elaborate the more electrons are involved. The present case remains simple with the ansatz (5.20) for the ground-state wave function. The spin part is fixed by the requirement of antisymmetry. It remains to determine the spatial wave function $\varphi_0(x)$. Considering the ground state, which is a stationary state, $\varphi_0(x)$ can be assumed to be a real function.

5.5.2 The Total Energy

The total energy (5.16) for the ground state (5.20) becomes

$$\langle\Phi|\hat{H}|\Phi\rangle = \frac{1}{2m}\sum_{\sigma\in\{+\frac{1}{2},-\frac{1}{2}\}}\int \mathrm{d}x\,\varphi_0(x)\chi^\dagger_\sigma\hat{p}^2\varphi_0(x)\chi_\sigma$$

$$-\frac{c}{4}\int \mathrm{d}x_1\,\mathrm{d}x_2\,\varphi_0^*(x_1)\varphi_0^*(x_2)\delta(x_1-x_2)\varphi_0(x_1)\varphi_0(x_2)$$

$$\left[\chi^{(1)}_{+\frac{1}{2}}\chi^{(2)}_{-\frac{1}{2}} - \chi^{(1)}_{-\frac{1}{2}}\chi^{(2)}_{+\frac{1}{2}}\right]^+\left[\chi^{(1)}_{+\frac{1}{2}}\chi^{(2)}_{-\frac{1}{2}} - \chi^{(1)}_{-\frac{1}{2}}\chi^{(2)}_{+\frac{1}{2}}\right].$$

As the ground state is spin-saturated, one can easily evaluate the spin summations. The spinor in the kinetic term just yields a factor of two, while the spinor part in the potential term looks more involved.

They become

$$\begin{aligned}&\left[\chi^{(1)}_{+\frac{1}{2}}\chi^{(2)}_{-\frac{1}{2}}-\chi^{(1)}_{-\frac{1}{2}}\chi^{(2)}_{+\frac{1}{2}}\right]^{+}\left[\chi^{(1)}_{+\frac{1}{2}}\chi^{(2)}_{-\frac{1}{2}}-\chi^{(1)}_{-\frac{1}{2}}\chi^{(2)}_{+\frac{1}{2}}\right]\\&=\underbrace{\chi^{(1)\,\dagger}_{+\frac{1}{2}}\chi^{(1)}_{+\frac{1}{2}}}_{=1}\chi^{(2)\,\dagger}_{-\frac{1}{2}}\chi^{(2)}_{-\frac{1}{2}}-\underbrace{\chi^{(1)\,\dagger}_{+\frac{1}{2}}\chi^{(1)}_{-\frac{1}{2}}}_{=0}\chi^{(2)\,\dagger}_{-\frac{1}{2}}\chi^{(2)}_{+\frac{1}{2}}\\&\quad-\chi^{(1)\,\dagger}_{-\frac{1}{2}}\chi^{(1)}_{+\frac{1}{2}}\chi^{(2)\,\dagger}_{+\frac{1}{2}}\chi^{(2)}_{-\frac{1}{2}}+\chi^{(1)\,\dagger}_{-\frac{1}{2}}\chi^{(1)}_{-\frac{1}{2}}\chi^{(2)\,\dagger}_{+\frac{1}{2}}\chi^{(2)}_{+\frac{1}{2}}\\&=2\,.\end{aligned}$$

There remain the spatial parts. The double integration in the potential term shrinks to a single integration by virtue of the Dirac δ-distribution, reducing $x_1 = x_2 \to x$. The total energy then is

$$E[\varphi_0]=\langle\Phi|\hat{H}|\Phi\rangle=\underbrace{2\frac{\hbar^2}{2m}\int_{-\infty}^{\infty}\mathrm{d}x\,\varphi_0^*(x)(-\partial_x^2)\varphi_0(x)}_{E_{\mathrm{kin}}}\underbrace{-\frac{c}{2}\int_{-\infty}^{\infty}\mathrm{d}x\,|\varphi_0(x)|^4}_{E_{\mathrm{pot}}}\,. \tag{5.21}$$

The energy is a functional of the spatial single-particle wave function φ_0, a fact which is expressed by the square brackets.

5.5.3 The Hartree–Fock Equation

We start from the Hartree–Fock equations (5.15) formulated in coordinate space. The local density and consequently the direct mean field (5.15c) in the present case become

$$\rho(x)=\sum_{\sigma}|\varphi_0(x)|^2\chi_\sigma^\dagger\chi_\sigma=2|\varphi_0(x)|^2\quad\Longrightarrow\quad U_{\mathrm{dir}}=-c|\varphi_0(x)|^2\,. \tag{5.22}$$

For the exchange contribution, we remember the completeness relation in spinor space $\sum_\sigma \chi_\sigma\chi_\sigma^\dagger = \hat{1}$ where $\hat{1}$ stands for the unity operator in 2×2 spinor space. Using (5.15d) and remembering that integration has to be performed in both real space (x) and spinor space we get

$$\begin{aligned}U_{\mathrm{ex}}\varphi_0(x)\chi_\sigma&=\frac{c}{2}\int\mathrm{d}x'\sum_{\sigma'}\delta(x-x')\sum_{\sigma}\varphi_0(x)\varphi_0(x')\chi_{\sigma'}\chi_{\sigma'}^\dagger\varphi_0(x')\chi_\sigma\\&=\frac{c}{2}|\varphi_0(x)|^2\varphi_0(x)\chi_\sigma\,,\end{aligned}$$

which up to a factor $\frac{1}{2}$ is the same as the direct term. This finally yields the mean-field equation as

$$\left[-\frac{\hbar^2}{2m}\partial_x^2 + U_{\rm mf}(x)\right]\varphi_0 = \epsilon\varphi_0 \quad , \quad U_{\rm mf}(x) = -\frac{c}{2}\varphi_0^2(x)\,. \tag{5.23}$$

This is a nonlinear equation. The mean-field potential $U_{\rm mf}$ depends on the wave function φ_0, which is a solution of the Schrödinger equation with $U_{\rm mf}$. That poses a self-consistency problem which can in general only be attacked by iterative methods.

5.5.4 A Closed Solution

The present model is just simple enough that it allows a closed solution. The ansatz is the function $1/\cosh(ax)$. To avoid a second round through formal steps, start with the normalized ansatz:

$$\varphi_0 = \sqrt{\frac{a}{2}}\frac{1}{\cosh(ax)}\,.$$

Using the well-known property $\partial_x \tanh(x) = 1/\cosh^2(x)$ normalization is proved via

$$\begin{aligned}\int_{-\infty}^{\infty} \mathrm{d}x\,\varphi_0^2 &= \int_{-\infty}^{\infty} \mathrm{d}x\,\frac{a}{2}\frac{1}{\cosh^2(ax)} = \frac{1}{2}\int_{-\infty}^{\infty} d(ax)\,\frac{1}{\cosh^2(ax)} \\ &= \frac{1}{2}\tanh(ax)\Big|_{-\infty}^{\infty} = 1\,.\end{aligned}$$

Next, evaluate the second derivative which is needed for the kinetic energy:

$$\begin{aligned}\partial_x\varphi_0 &= \sqrt{\frac{a}{2}}\,\frac{-a\sinh(ax)}{\cosh^2(ax)}\,, \\ \partial_x^2\varphi_0 &= \sqrt{\frac{a}{2}}\left[\frac{-a^2}{\cosh(ax)} + \frac{2a^2\sinh^2(ax)}{\cosh^3(ax)}\right] = \sqrt{\frac{a}{2}}\left[\frac{a^2}{\cosh(ax)} - \frac{2a^2}{\cosh^3(ax)}\right] \\ &= \left[a^2 - \frac{2a^2}{\cosh^2(ax)}\right]\varphi_0\,.\end{aligned}$$

Inserting this result into the mean-field equation (5.23) yields

$$\left[-\frac{\hbar^2a^2}{2m} + \frac{\hbar^2a^2}{m\cosh(ax)^2} - \frac{ca}{4\cosh(ax)^2}\right]\varphi_0 = \varepsilon\varphi_0\,,$$

from which the final solution with its self-consistent mean field can be read off as

$$\varphi_0 = \sqrt{\frac{a}{2}}\frac{1}{\cosh(ax)} , \quad a = \frac{cm}{4\hbar^2} , \quad \epsilon = -\frac{\hbar^2 a^2}{2m} = -\frac{c^2 m}{32\hbar^2} , \tag{5.24a}$$

$$U_{\rm mf} = -\frac{c^2 m}{2\hbar}\frac{1}{\cosh^2(ax)} . \tag{5.24b}$$

Herein ϵ is the single-particle energy. The total energy is obtained from inserting the solution (5.24a) into (5.21), resulting in

$$\begin{aligned} E &= 2\frac{\hbar^2}{2m}\frac{a}{2}\int_{-\infty}^{\infty} \mathrm{d}x \frac{1}{\cosh(ax)}(-\partial_x^2)\frac{1}{\cosh(ax)} - \frac{c}{2}\frac{a^2}{4}\int_{-\infty}^{\infty} \mathrm{d}x \frac{1}{\cosh^4(ax)} \\ &= 2\frac{\hbar^2}{2m}\frac{a^3}{2}\int_{-\infty}^{\infty} \mathrm{d}x \left[-\frac{1}{\cosh^2(ax)} + \frac{2}{\cosh^4(ax)}\right] - \frac{c}{2}\frac{a^2}{4}\int_{-\infty}^{\infty} \mathrm{d}x \frac{1}{\cosh^4(ax)} . \end{aligned}$$

We use the normalization integral and $\int_{-\infty}^{\infty} \mathrm{d}y \cosh^{-4} = 4/3$ to obtain

$$E = \frac{2}{3}\frac{\hbar^2 a^2}{2m} - \frac{ca}{6} = -\frac{c^2 m}{48\hbar^2} . \tag{5.25}$$

In the present model, it has the same trend as the single-particle energy ϵ in (5.24a), but differs in the numerical factor.

5.5.5 *Symmetry Breaking*

The most interesting aspect of this solution is the self-stabilization of a finite wave packet (5.24a) together with a binding potential well (5.24b). The original many-body Hamiltonian (5.19) was translational invariant. It does not show any seed of a localization, but the Hartree–Fock solution is localized, in the case given in (5.24), around $x = 0$. The restriction to antisymmetrized product wave functions leads to a nonlinear equation which, in turn, here shows the feature of self-focusing. Translational invariance is manifestly broken by the Hartree–Fock solution.

Translational invariance, however, shows up in a different manner. The solution (5.24) can be shifted by any distance x_0 and the shifted wave function remains a solution. It is straightforward to check that

$$\varphi_0 = \sqrt{\frac{a}{2}}\frac{1}{\cosh(a(x - x_0))} , \quad \epsilon(x_0) = -\frac{m}{32\hbar^2}c^2 = \epsilon(0) \quad , \tag{5.26a}$$

$$U_{\rm mf} = -\frac{c^2 m}{2\hbar}\frac{1}{\cosh^2(a(x - x_0))} , \tag{5.26b}$$

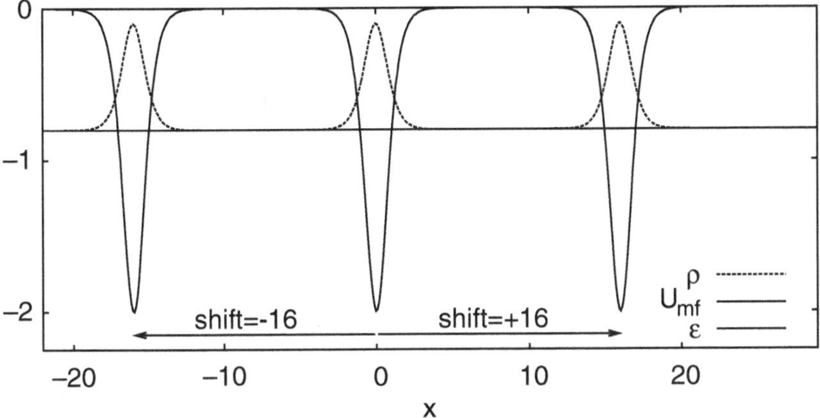

Fig. 5.1 Illustration of the solution of the 1D Hartree–Fock model. It shows the mean-field potential U_{mf} (*heavy line*) localized at $x = 0$ with the corresponding density ρ (*dashed line*). The latter is drawn with respect to the single-particle energy ϵ (*faint horizontal line*). Furthermore, equivalent solutions with shifts to $x = -16$ and +16 are shown

also satisfies the Hartree–Fock equation (5.23). Note that all solutions have the same single-particle energy ϵ and the same total energy E, i.e., $E(x_0) = E(0) = \text{constant}$. Thus not only one solution, but a multitude of degenerate solutions are obtained, which are generated from one special solution by applying the symmetry operation, here translation by x_0. Figure 5.1 serves to illustrate the solution. It also indicates the variety of equivalent solutions with a few examples. The whole multitude is given by a continuum of shifts x_0, impossible to represent graphically. All cases shown have the same total energy (5.25). The degeneracy suggests to consider an improved solution as a coherent superposition of these Slater states. This yields a particular form of a correlated state which restores the translational symmetry.

The above case illustrates the generic mechanism of symmetry breaking which occurs in many places in physics: A mean-field treatment of a complex system produces nonlinear equations and it can happen under certain conditions that the stability point lies at a configuration which breaks a symmetry originally given in the Hamiltonian.

5.6 An Illustrative Example: The He Atom

We consider the He atom as a test case to exemplify the performance of HF in a realistic case. This two-electron system does not have a closed analytical solution and thus provides an opportunity to test variational approaches as such. The benchmark is the experimental electronic binding energy of the He atom, $E_{\mathrm{B}}(He) = -5.80$ Ry $= -78.9$ eV. We thus compute the HF solution and compare it directly to the experimental energy.

5.6.1 Hamiltonian and HF Equations

The Hamiltonian for the electron cloud (2 electrons, 1 and 2) of the He atom is composed as in (6.1) with the particular potential operator

$$\hat{U}_{\text{ext}} = -\frac{2e^2}{|\mathbf{r}_1|} - \frac{2e^2}{|\mathbf{r}_2|} . \tag{5.27}$$

The full ground-state wave function is, of course, highly correlated and far beyond analytical treatment. We consider the independent-particle state $\Phi \propto \mathcal{A}\{\phi_1\phi_2\}$ composed from the two single-particle states

$$\phi_1 = \varphi(\mathbf{r}_1)\chi_{+\frac{1}{2}} , \quad \phi_2 = \varphi(\mathbf{r}_1)\chi_{-\frac{1}{2}} , \tag{5.28}$$

where φ is a purely spatial wave function and χ_σ a Pauli 2-spinor. The two fermions are distinguished by spin only while occupying the same spatial wave function. This will be the wave function with lowest energy in the emerging mean field corresponding to the lowest possible two-particle state. The spatial density becomes, similar to (5.22),

$$\rho(\mathbf{r}) = 2\varphi^*(\mathbf{r})\varphi(\mathbf{r}) . \tag{5.29}$$

The Hamiltonian (5.27) does not depend on spin at all. Thus in a first step all energy expectation values can be reduced to purely spatial integrals. These are

$$E = E_{\text{kin}} + E_{\text{ext}} + E_{\text{coul,dir}} + E_{\text{coul,ex}} , \tag{5.30a}$$

$$E_{\text{kin}} = \sum_\sigma \frac{\hbar^2}{2m_{\text{e}}} \int \mathrm{d}^3r \, |\nabla\varphi|^2 \chi_\sigma^\dagger \chi_\sigma = 2\frac{\hbar^2}{2m_{\text{e}}} \int \mathrm{d}^3r \, |\nabla\varphi|^2 , \tag{5.30b}$$

$$E_{\text{ext}} = -\sum_\sigma \int \mathrm{d}^3r \, \varphi^*(\mathbf{r}) \frac{2e^2}{|\mathbf{r}|} \varphi(\mathbf{r}) \chi_\sigma^\dagger \chi_\sigma = -\int \mathrm{d}^3r \, \frac{2e^2 \rho(\mathbf{r})}{|\mathbf{r}|} , \tag{5.30c}$$

$$\begin{aligned} E_{\text{coul,dir}} &= \int \mathrm{d}^3r_1 \mathrm{d}^3r_2 \, \varphi^*(\mathbf{r}_1)\varphi^*(\mathbf{r}_2) \frac{e^2}{|\mathbf{r}_1 - \mathbf{r}_2|} \varphi(\mathbf{r}_1)\varphi(\mathbf{r}_2) \\ &\quad \underbrace{\sum_{\sigma_1\sigma_2} \chi_{\sigma_1}^\dagger \chi_{\sigma_1} \chi_{\sigma_2}^\dagger \chi_{\sigma_2}}_{=4} \\ &= \int \mathrm{d}^3r_1 \mathrm{d}^3r_2 \, \rho(\mathbf{r}_1) \frac{e^2}{|\mathbf{r}_1 - \mathbf{r}_2|} \rho(\mathbf{r}_2) , \end{aligned} \tag{5.30d}$$

$$E_{\text{coul,ex}} = -\int \mathrm{d}^3r_1 \mathrm{d}^3r_2\, \varphi^*(\mathbf{r}_1)\varphi^*(\mathbf{r}_2)\frac{e^2}{|\mathbf{r}_1-\mathbf{r}_2|}\varphi(\mathbf{r}_1)\varphi(\mathbf{r}_2)$$
$$\underbrace{\sum_{\sigma_1\sigma_2} \chi^\dagger_{\sigma_1}\chi_{\sigma_2}\chi^\dagger_{\sigma_2}\chi_{\sigma_1}}_{=2}$$
$$= -\frac{1}{2}\int \mathrm{d}^3r_1 \mathrm{d}^3r_2\, \rho(\mathbf{r}_1)\frac{e^2}{|\mathbf{r}_1-\mathbf{r}_2|}\rho(\mathbf{r}_2) = -\frac{1}{2}E_{\text{coul,dir}}\,. \tag{5.30e}$$

All potential energies can be expressed through the local density, even the exchange term. This is a special property of this simple two-electron system where both electrons still can share the same spatial wave function. It also produces an exchange term which is simply proportional to the direct term with a factor minus one half. The Hartree–Fock equation is obtained by variation with respect to φ^* yielding

$$\left[-\frac{\hbar^2}{2m_\text{e}}\Delta - \frac{2e^2}{|\mathbf{r}|} + \left(1-\frac{1}{2}\right)\int \mathrm{d}^3r'\, \frac{e^2}{|\mathbf{r}-\mathbf{r}'|}\rho(\mathbf{r}')\right]\varphi(\mathbf{r}) = \varepsilon\varphi(\mathbf{r})\,, \tag{5.31}$$

where the single-particle energy ε is the same for ϕ_1 and ϕ_2. The third term in the mean-field Hamiltonian contains the direct term with the factor $+1$ and the exchange term with factor $-1/2$. The Hartree approximation would consist in dropping the $-1/2$ for the exchange term.

5.6.2 *Variational Ansatz*

Although seemingly simple, the mean-field equation (5.31) does not have a closed analytical solution. We take that as a chance to exemplify a further approximation, namely variational optimization of a chosen ansatz for a searched wave function. We briefly recall the general scheme: the goal is a solution of the Schrödinger equation $\hat{H}|\Psi\rangle = E|\Psi\rangle$. By experience, intuition or whatever means, an educated guess for the solution is made, which has a few free parameters $\alpha_1 \ldots \alpha_P$, i.e., $|\Psi\rangle = |\Psi(\alpha_1 \ldots \alpha_P)\rangle$. This defines a sub-manifold of the Hilbert space for the system. The Ritz variational principle as discussed in Sect. 5.1.2 now states that the best possible approximation within that sub-manifold is the state with the lowest energy. The necessary condition for a minimum is

$$\partial_{\alpha_i}\frac{\langle\Psi|\hat{H}|\Psi\rangle}{\langle\Psi|\Psi\rangle} = 0\,, \quad i = 1 \ldots P\,. \tag{5.32}$$

For the present example of a He atom, the single-particle wave functions may be assumed to approximately maintain the shape of the Coulomb solutions, i.e.,

$$\varphi = \sqrt{\frac{\alpha^3}{\pi}}\exp(-\alpha r)\,, \quad r = |\mathbf{r}|\,, \tag{5.33}$$

where the width parameter α remains open. It will become the one free parameter for variation.

To compute the energy, insert the ansatz (5.33) into the energy (5.30), noting that the density is $\rho = 2(\alpha^3/\pi)\exp(-2\alpha r)$, and recall the following relations:

$$\nabla \exp(-\lambda r) = \mathbf{e}_r \partial_r \exp(-\lambda r) = -\lambda \exp(-\lambda r)\mathbf{e}_r\,,$$

$$\int_0^\infty \mathrm{d}x\, x^n \exp(-\lambda x) = \frac{n!}{\lambda^{n+1}}\,,$$

$$\int \mathrm{d}x\, x \exp(-\lambda x) = \frac{1-\lambda x}{\lambda^2}\exp(-\lambda x)\,,$$

$$\int_{-1}^1 \mathrm{d}y \frac{1}{\sqrt{a^2+b^2-2aby}} = -\frac{\sqrt{a^2+b^2-2aby}}{ab}\Big|_{-1}^1 = 2\frac{1}{\max(a,b)}\,,$$

$$\int \mathrm{d}^3r \cdots = \int_0^\infty \mathrm{d}r\, r^2 \int_{-1}^1 \mathrm{d}(\cos\theta) \int_0^{2\pi} \mathrm{d}\phi \cdots$$
$$\longrightarrow 4\pi \int_0^\infty \mathrm{d}r\, r^2 \cdots\,,$$

where the last step (the "$\longrightarrow$") applies only to radially symmetric integrands. For the one-body terms this leads to

$$E_{\mathrm{kin}} = = 2\frac{\hbar^2}{2m_\mathrm{e}}\frac{\alpha^5}{\pi}4\pi \int_0^\infty \mathrm{d}r\, r^2\, \mathrm{e}^{-2\alpha r} = 2\frac{\hbar^2}{2m_\mathrm{e}}\alpha^2\,, \tag{5.34a}$$

$$E_{\mathrm{ext}} = -8\pi\frac{\alpha^3}{\pi}\int_0^\infty \mathrm{d}r\, r^2\, \mathrm{e}^{-2\alpha r}\frac{2e^2}{r} = -4\alpha e^2\,. \tag{5.34b}$$

The sixfold integration in the two-body terms requires careful choice of the reference frame for the spherical coordinates. The same reasoning as already used for the computation of the exchange energy in the electron gas (see Sect. 2.5) can be applied here. The only angular dependence occurs in the denominator $|\mathbf{r}_1 - \mathbf{r}_2| = \sqrt{r_1^2 + r_2^2 - 2r_1 r_2\cos(\theta)}$ where θ is the angle between $\mathbf{r}_1$ and $\mathbf{r}_2$. Let us consider d^3r_1 as the outer integration. The direction of $\mathbf{r}_1$ is then defined before the d^3r_2 integration starts. We chose for $\mathbf{r}_2$ the frame where the direction of $\mathbf{r}_1$ is the z-axis. Thus the angle in the $\mathbf{r}_2$ frame becomes identical with the angle between $\mathbf{r}_1$ and $\mathbf{r}_2$, i.e., $\theta_2 = \theta$. There is no dependence on the two azimuthal angles (ordinarily denoted as ϕ) whose integration then produces a factor $(2\pi)^2$. After these preliminaries, the electron–electron Coulomb energy can be evaluated as

$$E_{\mathrm{coul,dir}} = (2\pi)^2 \frac{\alpha^6}{\pi^2} \int_0^\infty \mathrm{d}r_1\, r_1^2 \underbrace{\int_{-1}^{1} \mathrm{d}(\cos(\theta_1))}_{=2} \int_0^\infty \mathrm{d}r_2\, r_2^2 \int_{-1}^{1} \mathrm{d}(\cos(\theta_2))$$

$$\mathrm{e}^{-2\alpha r_1} \frac{e^2}{\sqrt{r_1^2 + r_2^2 - 2r_1 r_2 \cos(\theta_2)}} \mathrm{e}^{-2\alpha r_2}\,,$$

$$= 64\alpha^6 \int_0^\infty \mathrm{d}r_1\, r_1^2\, \mathrm{e}^{-2\alpha r_1} \int_{r_1}^\infty \mathrm{d}r_2\, r_2^2 \frac{1}{r_2} \mathrm{e}^{-2\alpha r_2}$$

$$= 16\alpha^4 e^2 \int_0^\infty \mathrm{d}r_1\, r_1^2\, \mathrm{e}^{-4\alpha r_1} (2\alpha r_1 - 1) = \frac{5}{4}\alpha e^2, \tag{5.34c}$$

$$E_{\mathrm{coul,ex}} = -\frac{5}{8}\alpha e^2\,. \tag{5.34d}$$

The total HF energy for the ansatz (5.33) then becomes

$$E_{\mathrm{HF}}(\alpha) = \frac{\hbar^2}{m}\alpha^2 - \frac{27e^2}{8}\alpha\,. \tag{5.35}$$

The variational equation $\partial_\alpha E_{\mathrm{HF}}(\alpha) = 0$ yields, after some simple algebraic manipulations,

$$\alpha_{\mathrm{HF}} = \frac{27}{16}\frac{me^2}{\hbar^2} = \frac{27}{16}\frac{1}{\mathrm{a}_0} \quad \Longrightarrow$$

$$E_{\mathrm{HF,opt}} = \left(\frac{27}{16}\right)^2 \left(\frac{\hbar^2}{m} - 2e^2\right) = -5.69\,\mathrm{Ry}\,. \tag{5.36}$$

This value is to be compared with the measured exact energy $E_{\mathrm{exp}} = -5.80\,\mathrm{Ry}$. The deviation is less than 2%, which is a very good result in view of the simplicity of the calculation.

5.6.3 Full HF Solution

The comparison between the variational approach with ansatz (5.33) and a full numerical solution of the mean-field equations is also enlightening. In the case of a full numerical solution one obtains a binding energy $E_{\mathrm{HF}} = -5.72$ Ry, a bit closer to the experimental value, but remember that the difference to the variational ansatz is only about 0.5%. The quality of the simplest variational ansatz here is thus already quite good, about the level of the quality of HF as such. An obvious further improvement would be to extend (5.33) to a linear superposition with a wave function in form of a $2s$ state. That could still be worked out analytically with a bit more patience and this strategy would immediately lead to LCAO as discussed in Sect. 4.2, but it could only lead to an improvement less than the full numerical HF

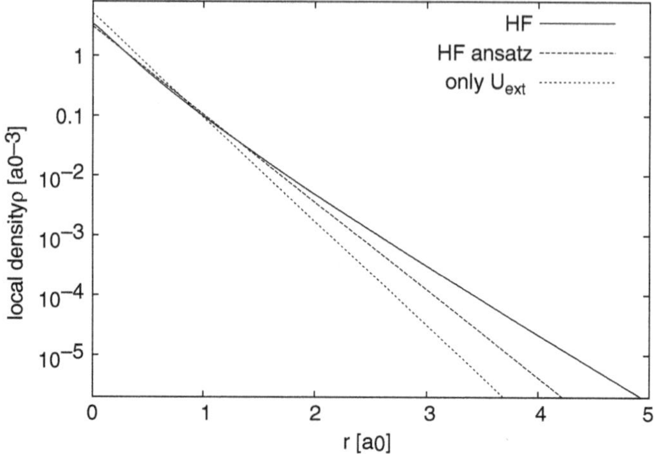

Fig. 5.2 The local density $\rho(r)$ for the electron cloud of the He atom, computed in three different approaches as indicated

solution (because of Hilbert space restriction) which will correspond to a fraction of a percent improvement on the total energy.

Figure 5.2 compares the electron densities for various approaches. It is drawn on a logarithmic scale to emphasize the crucial asymptotic behavior. The difference between purely external field and the Hartree approach illustrates the orders of magnitude and associated variations involved. The HF result lies in between these two extremes as expected. Most interesting is the difference between the two HF results, one with ansatz (5.33) and one from the numerical solution. The inner region, where the bulk of the density is concentrated, shows a very nice agreement corroborating the favorable situation already seen for the energies. Systematic differences develop concerning the exponential decay at large distances. This illustrates that the performance of an approximation depends not only on the forces in the system but also on the observable which one is interested in. Those who are particularly interested in the outer tail of the electron density will not be satisfied with the simple variational approach on the basis of ansatz (5.33). One may even suspect that correlations beyond HF could grow more important for such subtle observables. The dotted line, finally, shows the result from using the unmodified external Coulomb field. The difference to the other results indicates the importance of the electron–electron interaction to the binding properties. They are sizable already for this small two-electron system.

5.7 Concluding Remarks

The Hartree–Fock approximation constitutes a simple and well-founded method to construct a mean-field description of a many-fermion system. We have shown in this chapter how it can be constructed and what are its inherent limitations. We have also seen that it can perform very well even in comparison to experimental results.

The Hartree–Fock approximation furthermore provides a sound basis for the construction of more elaborate approaches in which interactions would be taken into account in a more detailed way. The general procedure here is known as the inclusion of correlations and can be achieved in different ways. We illustrate two strategies along that line in the following chapters. In Chap. 6 correlations are introduced in an effective way through a modified interaction on which a Hartree mean field is constructed. A more explicit account of correlations is presented in Chap. 9 in some specific cases.

Chapter 6
Density Functional Theory

The Hartree–Fock (HF) method provides a straightforward approach to construct mean fields in many-fermion systems. We have exemplified its capabilities in a few examples (Chap. 5). The HF approach, however, suffers from some limitations, on the formal side due to the average treatment of interactions and on the practical side due to the involved exchange term. To go beyond HF strictly speaking implies including correlations, which will be discussed in Chap. 9. Still, there is an alternative "effective" path which consists in building a simple HF approximation (or even Hartree approximation, namely without including exchange), on top of *effective interactions*. The construction of such effective interactions of course requires some elaborate calculations but it allows to include correlation effects in the interaction itself, which allows a mean-field description of the given system, mostly at the technically simple Hartree level. Such effective approaches usually constitute a large step forward to more self-consistent modeling. The formally sound way to formulate such effective theories proceeds through energy-density functionals as a starting point. Examples for effective self-consistent models are the Skyrme–Hartree–Fock method in nuclear physics [92, 42], chiral nonlinear spinor modeling (e.g., Nambu–Jona–Lasinio theory) in field theory [33], and, of course, the well-known *density functional theory* (DFT) for electronic systems [56].

In the following, we will briefly recapitulate the basic steps of electronic DFT finalizing with the He atom as example. Although we will comment on the formal basis, the focus is on the practitioners "bottom line," the compact recipes which emerge at the end and make DFT so useful in many areas of physics. For those who are interested in a more detailed understanding, we refer to books and articles specialized to electronic DFT, e.g., [80, 54, 27]. It is also interesting to note that DFT was used in practical applications first, e.g., in terms of *Thomas–Fermi theory* [34] or the *Slater approximation* to exchange [99]; the theoretical foundation as outlined below came later. The associated deeper understanding, however, has given direction to further improvements which allowed the steady development up to the very powerful method of today. We shall discuss some of these aspects at the end of the chapter together with an excursion into the case of nuclear energy-density functionals.

J.A. Maruhn et al., *Simple Models of Many-Fermion Systems*,
DOI 10.1007/978-3-642-03839-6_6, © Springer-Verlag Berlin Heidelberg 2010

6.1 Basics of Electronic Density Functional Theory (DFT)

6.1.1 The Hohenberg–Kohn Theorem

The exact Schrödinger equation for N-electron systems is given as

$$\hat{H}|\Psi\rangle = E|\Psi\rangle\,, \tag{6.1a}$$

$$\hat{H} = \hat{T} + \hat{U}_{\text{ext}} + \hat{V}_{\text{coul}}\,, \tag{6.1b}$$

$$\hat{T} = \sum_{n=1}^{N} \frac{\hat{p}_p^2}{2m}\,, \tag{6.1c}$$

$$\hat{U}_{\text{ext}} = \sum_{n=1}^{N} U_{\text{ext}}(\mathbf{r}_p)\,, \tag{6.1d}$$

$$\hat{V}_{\text{coul}} = \frac{1}{2}\sum_{n\neq m=1}^{N} \frac{e^2}{|\mathbf{r}_p - \mathbf{r}_m|}\,, \tag{6.1e}$$

where $|\Psi\rangle$ is the fully correlated N-body state, $\hat{T}$ the kinetic energy, $\hat{U}_{\text{ext}}$ an external potential coming, e.g., from the ionic cores of an atom or molecule, and $\hat{V}_{\text{coul}}$ the Coulomb interaction between the electrons. Kinetic and Coulomb energy are inevitable features of electrons, while the external potentials $U_{\text{ext}}(\mathbf{r})$ change and with them the actual quantum state of the electrons. It is obvious that a given $U_{\text{ext}}(\mathbf{r})$ uniquely leads to a many-body wave function as solution of the Schrödinger equation, and this in turn determines a certain local density $\rho(\mathbf{r})$. This suggests that one may equally well characterize the state of the system through the density $\rho(\mathbf{r})$. The advantage is that we so get rid of the external potential and can formulate everything in terms of purely electronic quantities, kinetic energy, Coulomb energy, and density. The ultimate dream is a formulation of all electronic correlations in terms of the simple one-body density which is universal in the sense that, once established, it can be applied to varying situations, i.e., external potentials $\hat{U}_{\text{ext}}$.

We jump over a series of motivating steps and come directly to the famous Hohenberg–Kohn theorem [50] which, to a large extent, is self-explanatory. From the total energy $E = \langle\Psi|\hat{H}|\Psi\rangle$, only the electronic part, i.e.,

$$E_{\text{HK}} = \langle\Psi|\hat{T} + \hat{V}_{\text{coul}}|\Psi\rangle \tag{6.2}$$

is considered. The Hohenberg–Kohn theorem now consists of two statements:

1. E_{HK} is a unique and universal functional of the local density $\rho(\mathbf{r})$

$$E_{\text{HK}} = E_{\text{HK}}[\rho(\mathbf{r})]\,. \tag{6.3a}$$

2. The correct ground-state density for a given external field $\hat{U}_{\text{ext}}$ is obtained from variation of the total energy, i.e.,

$$E[\rho(\mathbf{r})] = E_{\mathrm{HK}}[\rho(\mathbf{r})] + \int \mathrm{d}^3r\,\rho(\mathbf{r})U_{\mathrm{ext}}(\mathbf{r}) \tag{6.3b}$$

with respect to the density $\rho(\mathbf{r})$.

Note that the functional is tied to a given electron number N. It is universal in the sense that it applies to any local external potential. The simplification of the description is dramatic. Assume that we are given a manageable energy functional $E_{\mathrm{HK}}[\rho(\mathbf{r})]$: then a large number of changing situations can be evaluated by a simple local equation emerging from variation.

The optimism should be mellowed by mentioning two open problems. The development started from the observation that each given local potential $U_{\mathrm{ext}}(\mathbf{r})$ uniquely leads to one certain density. The reverse, however, is not true. Not every conceivable local density can be produced by a local potential. So caution has to be applied when shuffling pieces around. One remains on the safe side as long as one proceeds forward from U_{ext}. The other problem is more severe. The Hohenberg–Kohn theorem proves the existence of such a functional but does not provide any means for its construction. That step requires further approximations, one of which will be presented in the next section.

6.1.2 The Local Density Approximation

As pointed out above, a universal density functional is proven to exist, in principle. In practice, one has to approach it in modest steps. The most widely used construction is provided by the local density approximation (LDA), which we will now briefly discuss. The starting point is the homogeneous gas, i.e., a continuum of electrons in a spatially homogeneous external field (see Chap. 2). The electron density then also becomes homogeneous, $\rho(\mathbf{r}) = \rho_0 = \text{constant}$, and the energy is a simple function of the density ρ_0. We write it in a suggestive fashion as

$$E_{\mathrm{HK}}(\rho_0) = \int \mathrm{d}^3r\,\frac{E_{\mathrm{elgas}}}{V}(\rho_0) = \int \mathrm{d}^3r\,\rho_0\frac{E_{\mathrm{elgas}}}{N}(\rho_0),$$

where E_{elgas}/N is the energy per particle in the electron gas covering kinetic, exchange, and correlation energy. The direct term, the Coulomb Hartree energy, is compensated by the positive jellium background approximation. We have deliberately written the energy as a spatial integral, which is an overkill for a homogeneous system because the integrand would be constant anyway. Now, though, we release that restriction and allow for slight inhomogeneities. The spatial changes should remain very gentle, so that we can still assume a state of piecewise infinite homogeneous matter. Thus we daringly generalize to

$$E_{\mathrm{HK}}[\rho(\mathbf{r})] \approx E_{\mathrm{HK-LDA}}[\rho(\mathbf{r})] = \int \mathrm{d}^3r\,\rho(\mathbf{r})\frac{E_{\mathrm{elgas}}}{N}(\rho(\mathbf{r})) \tag{6.4}$$

by replacing ρ_0 by $\rho(\mathbf{r})$. This is a functional of $\rho(\mathbf{r})$, even a particularly simple one because it is a local functional. The energy is associated point by point with a locally given density. Variation of this expression is straightforward.

The full correlation energy can only be computed numerically [21]. We exemplify the procedure for kinetic and exchange energy, whose extremely simple expressions were given in Sect. 2.5. In addition, it is advantageous for further extension to write the terms separately. We start from the kinetic energy (2.24) and the exchange energy (2.48), identify $k_F = \left(3\pi^2\rho_0\right)^{1/3}$ according to (2.18), and let $\rho_0 \longrightarrow \rho(\mathbf{r})$. This leads to the LDA functional as

$$E_{\text{HK-LDA}} = E_{\text{kin-LDA}} + E_{\text{ext}} + E_{\text{H}} + E_{\text{x-LDA}}\,, \tag{6.5a}$$

$$E_{\text{ext}} = \int \mathrm{d}^3\, \rho(\mathbf{r}) U_{\text{ext}}(\mathbf{r})\,, \tag{6.5b}$$

$$E_{\text{H}} = \int \mathrm{d}^3 r_1 \mathrm{d}^3 r_2\, \rho(\mathbf{r}_1) \frac{e^2}{|\mathbf{r}_1 - \mathbf{r}_2|} \rho(\mathbf{r}_2)\,, \tag{6.5c}$$

$$E_{\text{kin-LDA}} = \int \mathrm{d}^3 r\, \beta_{\text{kin}} \rho^{5/3}(\mathbf{r})\,, \tag{6.5d}$$

$$E_{\text{x-LDA}} = -\int \mathrm{d}^3 r\, \beta_{\text{x}} \rho^{4/3}(\mathbf{r})\,, \tag{6.5e}$$

$$\beta_{\text{kin}} = \frac{3}{5} \frac{(3\pi^2)^{2/3}}{2m_e} \approx 5.74\,\text{Ry a}_0^2\,, \tag{6.5f}$$

$$\beta_{\text{x}} = \frac{3e^2(3\pi^2)^{1/3}}{4\pi} \approx 1.48\,\text{Ry a}_0\,, \tag{6.5g}$$

where the energy from the external field E_{ext} and the Hartree energy E_{H} (which is the direct part of the Coulomb interaction energy between the electrons) are density functionals from the outset and need no further approximation. The LDA applies to the two remaining terms. The LDA expression (6.5d) for the kinetic energy is also known as the Thomas–Fermi approximation [34, 20] and the exchange energy (6.5e) as the Slater approximation [99]. In fact, we have here derived the basics of the semi-classical Thomas–Fermi approximation for electronic systems. This will be discussed in more detailed applications in Sect. 6.5.

One piece is missing in this very simple motivation of LDA. There is no criterion what "slightly inhomogeneous" means quantitatively. To that end, look at the LDA for the one-body density matrix $\varrho(x, x')$. Starting from the expression (2.20) for the density matrix in homogeneous matter, remember $k_F = k_F(\rho_0)$ and again replace $\rho_0 \longrightarrow \rho(\mathbf{r})$. This yields

$$\varrho(\mathbf{r}\nu,\mathbf{r}'\nu') = \frac{\delta_{\nu\nu'}}{2}\rho(\overline{\mathbf{r}})\mathcal{J}\left((3\pi^2\rho(\overline{\mathbf{r}}))^{1/3}|\mathbf{r}-\mathbf{r}'|\right) \; , \; \overline{\mathbf{r}} = \frac{1}{2}\left(\rho(\mathbf{r})+\rho(\mathbf{r}')\right), \tag{6.6}$$

with $\mathcal{J}(x)$ as given in (2.20). We recall Fig. 2.3 which shows that $\mathcal{J}$ is practically confined to a region $|\mathbf{r}-\mathbf{r}'| \leq 4/k_{\mathrm{F}}$. The LDA expression (6.6) suggests that $\rho(\overline{\mathbf{r}})$ should not vary too much over a spatial region of $2/k_{\mathrm{F}} \approx r_s$. This is a very conservative estimate. Many applications in atomic and molecular physics deal with stronger inhomogeneities and, nevertheless, work surprisingly well. There seems to be a friendly cancellation of LDA errors between exchange and correlation term which improves the quality beyond the rough estimate [27].

LDA, although an extremely clever compromise between expense and return, is often not precise enough for high demands like the computation of subtle bonding properties in molecules. An obvious step further is to consider effects from first-order inhomogeneities characterized in terms of the gradient $\nabla\rho$. Such a gradient expansion for the kinetic energy leads to the extended Thomas–Fermi approximation [20]. Application to exchange and correlation energy yields the generalized gradient approximation (GGA) [81] which may be a key ingredient in the success of applied DFT, see, e.g., [27, 56].

6.1.3 Kohn–Sham Approach

The DFT presented above at the level of the Hohenberg–Kohn theorem and LDA for all parts of the energy leads to the semi-classical Thomas–Fermi theory. It describes the average trends of binding energies and other bulk observables well, but overrides any quantum-mechanical shell effects. That is a major drawback because shell effects are crucial ingredients in many systems and processes, e.g., nuclear fission [18], the Jahn–Teller effect in molecules and solids [32], or the variations of chemical bonding, see Chap. 4. Perfect density functionals would, of course, reproduce these quantum effects, but the LDA with its inherent smoothness and regularity wipes them out. An extension of the density functional by imprinting quantum effects has not been successfully achieved hitherto. The solution is to involve quantum-mechanical single-particle wave functions besides the local density. That is done in the Kohn–Sham (KS) scheme [57], which we are going to briefly review here.

The aim is to develop a description at the level of simplicity and complexity of the Hartree theory with a mean-field equation

$$\left[-\frac{\hbar^2}{2m_{\mathrm{e}}}\Delta + U_{\mathrm{ext}}(\mathbf{r}) + U_{\mathrm{KS}}(\mathbf{r})\right]\varphi_\alpha(\mathbf{r}) = \varepsilon_\alpha\varphi_\alpha(\mathbf{r}), \tag{6.7}$$

for the occupied single-particle wave functions φ_α and employing a local mean-field potential, the KS potential $U_{\mathrm{KS}}(\mathbf{r})$, besides the external field, and an explicit operator for the kinetic energy. The KS potential should again be a functional of the local density $\rho(\mathbf{r})$, deduced from a corresponding energy-density functional. The proof of existence, universality, and uniqueness is very similar to the case of the Hohenberg–Kohn theorem, but now using the kinetic energy from the occupied single-particle states, $\propto \int d^3r|\nabla\varphi_\alpha|^2|$, explicitly. It remains to construct the energy-density functional for the potential energy from electron–electron Coulomb interaction. This again requires further approximations and LDA is the most widely used method. The recipe then becomes very simple: we start from the LDA expression (6.5) for the Hohenberg–Kohn energy functional and replace the LDA for the kinetic energy by the single-particle kinetic energy. This yields the KS energy

$$E_{\mathrm{KS-LDA}} = \sum_{\alpha=1}^{N} \int d^3r|\nabla\varphi_\alpha|^2| + E_{\mathrm{ext}}[\rho] + E_{\mathrm{H}}[\rho] + E_{\mathrm{x-LDA}}[\rho], \tag{6.8}$$

with the remaining contributions still as energy-density functionals in precisely the same form as given in (6.5).

The local potential in the KS equation (6.7) is to be obtained by the KS variational principle

$$\delta\varphi_{\alpha^\dagger}\left\{E_{\mathrm{KS-LDA}} - \sum_{\alpha=1}^{N}\varepsilon_\alpha \int d^3r|\varphi_\alpha|^2|\right\} = 0, \tag{6.9}$$

where the single-particle energies ε_α appear as Lagrange multipliers for the constraint on normalization of the single-particle wave functions $\int d^3r|\varphi_\alpha|^2| = 1$ (see also the derivation of the HF equations in Chap. 5). Note that the ε_α are theoretical tools; a sensible physical interpretation need not be guaranteed, and in fact there are intriguing problems with deducing band gaps or ionization potentials from them [27]. Variation of the kinetic energy term produces the standard kinetic energy in (6.7), while variation of E_{ext} naturally yields the external potential U_{ext}. Finally variation of the Hartree energy and LDA exchange term results in

$$U_{\mathrm{KS}}(\mathbf{r}) = \frac{\delta(E_{\mathrm{H}} + E_{\mathrm{x-LDA}})}{\delta\rho(\mathbf{r})} = \int d^3r_2 \frac{e^2}{|\mathbf{r}-\mathbf{r}_2|}\rho(\mathbf{r}_2) - \frac{4\beta_{\mathrm{x}}}{3}\rho^{1/3}(\mathbf{r}). \tag{6.10}$$

Nearly all practical applications of the KS scheme also employ an energy-density functional for correlations, which is usually combined with the exchange functional $E_{\mathrm{xc}}[\rho]$, see, e.g., [82]. These functionals look more involved than the bare exchange functional in LDA, but their application is as straightforward.

The KS equations provide an enormous simplification because every ingredient is handled through the local density and local potentials. The technically most expensive part for large system size is the Hartree potential due to the Coulomb force involved. It cannot be reduced because it carries the crucial information on the long range of the Coulomb interaction. The next expensive one is the kinetic energy operator, which cannot be omitted without losing the quantum structure. The remaining pieces, Coulomb exchange and correlations, would in principle be considerably more cumbersome to evaluate, but in practice these have been dramatically simplified to an almost trivial local variation.

6.2 Back to the Example of the He Atom

In Sect. 5.6 we considered the simple He atom as a realistic test case for exploring the capabilities of HF and of a variational ansatz for it. We now want to use these results as benchmark for testing DFT. We shall consider a simple model neglecting correlations, for then the HF approach becomes the theoretical benchmark and one can test DFT (exchange only) with respect to the HF result. The He atom case is a very critical test because the external potential, the Coulomb field of the He nucleus, is far from being "slowly varying," and so is the density, not to mention the technical difficulties outlined in Sect. 5.6.

Our starting point is thus exactly the model Hamiltonian equation (5.27) and we compute average values of the various terms exactly the same way, in the Kohn–Sham approach, but for the exchange term which is treated at LDA level. This amounts to replacing $E_{\mathrm{coul,ex}}$ in (5.30) by the exchange energy in LDA as given in (6.5e). Evaluating this for the ansatz (5.33) as

$$\begin{aligned} E_{\mathrm{ex-LDA}} &= -\beta_{\mathrm{x}} \int \mathrm{d}^3r\, \rho^{4/3}(\mathbf{r}) = -4\pi\beta_{\mathrm{x}} \left(\frac{2}{\pi}\right)^{4/3} \alpha^4 \int_0^\infty \mathrm{d}r\, r^2 \exp\left(-\frac{8\alpha}{3}r\right) \\ &= -4\pi \frac{3e^2(3\pi^2)^{1/3}}{4\pi} \left(\frac{2}{\pi}\right)^{4/3} \alpha \frac{27\times 6}{8^3} = -e^2\alpha\frac{3^5}{2^7\pi} \end{aligned} \tag{6.11}$$

produces the energy at the level of LDA as

$$\begin{aligned} E_{\mathrm{LDA}}(\alpha) &= E_{\mathrm{kin}} + E_{\mathrm{ext}} + E_{\mathrm{coul,dir}} + E_{\mathrm{x-LDA}} \\ &= \frac{\hbar^2}{m}\alpha^2 + e^2\alpha\left(-4 + \frac{5}{4} - \frac{3^5}{2^7\pi}\right) 2\,\mathrm{Ry}\left(\alpha^2 - 3.35\,\alpha\right). \end{aligned} \tag{6.12}$$

Minimization of this energy yields

$$\alpha_{\mathrm{LDA}} = 1.68\,\mathrm{a}_0^{-1} \quad \Longrightarrow \quad E_{\mathrm{LDA,opt}} = -5.61\,\mathrm{Ry}. \tag{6.13}$$

This value reproduces the HF energy, at which it aims, within 1.4% — which is a very satisfying result in view of the simplicity of the LDA functional (6.5e) for exchange. This is even more true as the He atom is far from a homogeneous density distribution. One should, however, remain cautious. One satisfying result is not a proof of generally good performance; it is, at best, an encouragement.

Table 6.1 summarizes the findings in connection with the test case He atom. We also show the energies from the pure external potential without electron–electron

Table 6.1 Binding energies of the electron cloud of the He atom obtained with various approaches as indicated. Benchmark is the experimental value given in the first line. The second block shows results obtained analytically with a variational ansatz as outlined in this section, while the third block shows results from a numerical calculation

	E_B [Ry]	Deviation %
Experiment	−5.81	
Pure U_{ext}	−8.00	−38.0
Hartree $e^{-\alpha r}$-ansatz	−3.80	35.0
HF $e^{-\alpha r}$-ansatz	−5.69	2.1
KS x-LDA $e^{-\alpha r}$-ansatz	−5.61	3.4
Hartree	−3.90	33.0
HF	−5.72	1.5
KS x-LDA	−5.71	1.7

interaction and the Hartree result, where exchange is ignored. This serves to illustrate the balance of energies. The energy in line two from pure external field shows dramatic over-binding. The Hartree approach (line 3 or 6) overcompensates and pushes the energy far above the wanted value. Including exchange corrects that and shifts the result pretty close to the goal. A deviation of 1.5% may be insufficient for many practical applications, but one cannot overemphasize the huge energies involved, which so nicely cancel. The missing correlation effects are a perturbation, crucial for quantitative success, but small. This shows that successive corrections with perturbation theory are a promising strategy in atomic physics and this approach is, indeed, widely used [100]. The LDA for exchange also works surprisingly well, as discussed above.

Figure 6.1 complements Fig. 5.2 on the density distribution in the He atom. We compare the LDA results with HF. The differences in ρ as such are found to be so small that they can hardly be distinguished graphically. Therefore the difference of the LDA densities in relation to the HF density is shown. The kink in the absolute value indicates a change of sign in the deviation. What concerns the sign, the LDA results fall off slightly more slowly, so that the deviation is positive for large r. This complies with the fact that the asymptotic exponential decrease is determined by the ionization potential (last bound single-particle state) and this is somewhat

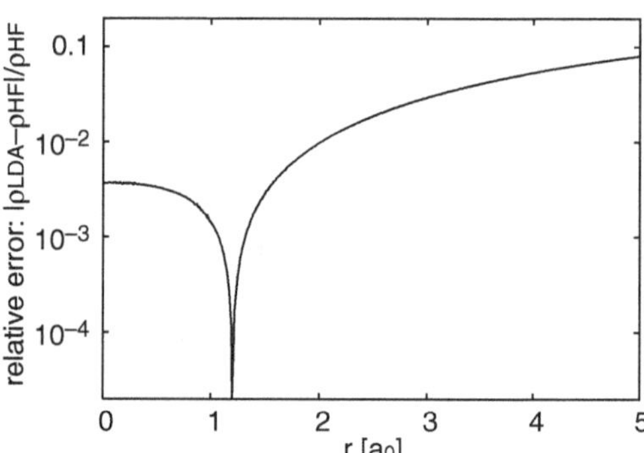

Fig. 6.1 The relative error of local densities, DFT-LDA as compared to HF for the electron cloud of the He atom: $|\rho_{\text{LDA}} - \rho_{\text{HF}}|/\rho_{\text{HF}}$

underestimated in LDA (for an improvement see the next section). Nonetheless, the agreement between the two methods is remarkable. It degrades, of course, for the far asymptotics, a region which is extremely sensitive to details of the model.

6.3 Self-Interaction Correction (SIC)

Let us consider the Kohn–Sham equations (6.7 and 6.10) for the case of one particle only. There exists only one occupied state $\varphi_0(\mathbf{r})\chi_{+\frac{1}{2}}$ and the density simply becomes $\rho(\mathbf{r}) = |\varphi_0(\mathbf{r})|^2$. We insert that into the KS equations and obtain

$$\begin{aligned}\varepsilon_0\varphi_0(\mathbf{r}) = \Big[&-\frac{\hbar^2}{2m_\mathrm{e}}\Delta + U_\mathrm{ext}(\mathbf{r}) \\ &\underbrace{+\int \mathrm{d}^3r_2 \frac{e^2}{|\mathbf{r}-\mathbf{r}_2|}\,|\varphi_0(\mathbf{r}_2)|^2 - \frac{4\beta_\mathrm{x}}{3}\,|\varphi_0(\mathbf{r})|^{2/3}}_{\neq 0 \quad \text{lightning}} \Big]\varphi_0(\mathbf{r})\,.\end{aligned}$$

The interaction obviously does not vanish, implying that the particle experiences a self-interaction which appears unphysical. We countercheck with the Hartree–Fock equations (5.15) for the case of one particle. The self-consistent field from the interaction becomes

$$\begin{aligned}\left(U_\mathrm{dir} - \hat{U}_\mathrm{ex}\right)\phi_\alpha &= \int \mathrm{d}^3r'\; v(\mathbf{r}-\mathbf{r}')\,|\varphi_0(\mathbf{r})|^2\,\phi_0(\mathbf{r}) - \int \mathrm{d}^3r'\; v(\mathbf{r}-\mathbf{r}')\varphi_0(\mathbf{r})\varphi_0^*(\mathbf{r}')\varphi_0(\mathbf{r}') \\ &= 0\,.\end{aligned}$$

It vanishes as it should be. The spurious self-interaction has sneaked in somehow on the way to LDA. One could argue that LDA is an approximation for *many*-fermion systems and that the self-interaction error becomes relatively unimportant for large N. Moreover, it turns out that the LDA description of the total energy and bulk density is surprisingly reliable. And yet, there remain problems in detail. Consider, e.g., the asymptotic behavior of the mean field at large distances for a neutral atom. The exchange potential in LDA falls off very quickly. The direct Coulomb potential stems from the total density of N electrons, thus fully compensating the attractive nuclear Coulomb potential (from N positive charges). Thus the mean-field potential falls off exponentially, whereas the correct behavior when removing one electron to large distances r would be that only the remaining $N - 1$ electrons contribute to the total Coulomb potential, leaving an asymptotic trend $\propto -e^2/r$. It is plausible that this self-interaction error may concern only details and that the global properties remain basically correct. Experience confirms that in most cases. The "details" which are corrupted are, e.g., the tail of the charge density, Koopman's theorem (see Sect. 5.3.2), and with it all single-particle energies.

There are various solutions to this problem. The simplest, and most widely followed, strategy is to refrain from looking at single-particle energies, claiming that these do not belong to the safe observables of DFT. There exist several recent developments which try to unify exact exchange with DFT by developing better manageable schemes in the spirit of the optimized effective potential approach, see, e.g., [61]. A conceptually simple and intuitive remedy is the self-interaction correction (SIC) initiated in [83]: the spurious self-interaction is simply subtracted from the energy functional for each particle. Let us assume that we are given an LDA functional $E_{\rm LDA}[\rho]$. The SIC functional is then composed as

$$E_{\rm SIC} = E_{\rm LDA}[\rho] - \sum_{i=1}^{N} E_{\rm LDA}[\rho_i] \quad , \quad \rho_i(x) = |\varphi_i(x)|^2 \; . \tag{6.14a}$$

Variation with respect to φ_j^* then yields the SIC mean-field Hamiltonian as

$$\hat{h}_j^{\rm (SIC)}\varphi_j = \left[-\frac{\hbar^2}{2m_{\rm e}}\Delta + U_{\rm ext}(\mathbf{r}) + \frac{\delta E_{\rm LDA}}{\delta\rho}[\rho] - \frac{\delta E_{\rm LDA}}{\delta\rho}[\rho_j]\right]\varphi_j \, . \tag{6.14b}$$

The LDA mean-field $\delta E_{\rm LDA}/\delta\rho$ appears with the total density ρ, which is the standard DFT-LDA term, but it reappears in a next term with the single-particle density ρ_j and that term correctly removes the self-interaction, as can quickly be seen from the limit $N = 1$. Now, however, the effective Hamiltonian depends on the state j on which it acts. Extra measures have to be taken to guarantee orthonormality of the set of occupied single-particle states $\{\varphi_i\}$. These technical details go beyond the scope of this book. We merely want to illustrate the effect for a realistic example. Figure 6.2 shows (numerical) results for the Ne atom using the elaborate LDA functional from [82]. The effect is obvious. LDA alone produces too weak binding for the electrons. The SIC produces the correct (attractive) asymptotics, thus enhancing binding and shifting at once the single-particle energies into the correct range. We

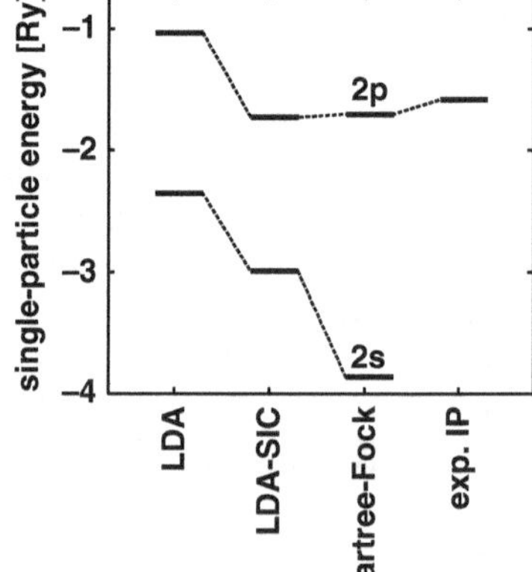

Fig. 6.2 The single-electron energy in the $2s$–$2p$ shell of the Ne atom described at various levels of approximation

mention in passing that the SIC also restores Koopman's theorem which is grossly violated for pure LDA, but which worked fine for HF.

6.4 The Skyrme Energy Functional in Nuclear Physics

From a slightly different perspective, namely that of an effective interaction, Skyrme proposed a simple ansatz for the self-consistent description of nuclei [98]. The model has developed into a widely used tool in nuclear structure physics with the appearance of a first quality parametrization [109] (for a recent review see [8]). It can be considered as an effective energy-density functional along the lines of DFT as outlined above. As nuclei have a large spin–orbit coupling, however, the functional includes other densities besides the local density ρ, namely at least a spin–orbit density and often also a kinetic density (to model the effective nucleon mass in the nuclear medium). The full functional is thus very involved. We refer the reader to more specialized presentations for details [92, 42, 8]. Here as a practical example we will discuss a highly simplified version for a zeroth-order description of nuclei.

The key building block of the potential energy functional of the Skyrme force consists of a zero-range attractive two-body term $\propto \rho^2$ and a counterweighting repulsive one with a higher density dependence $\propto \rho^{2+\alpha}$. We follow the lines of [58] and discuss a simple model where the spin–orbit contributions and the kinetic corrections are ignored. This is acceptable for small nuclei. Furthermore, the complications of introducing isovector terms (i.e., terms which are sensitive to the difference between proton and neutron densities) are circumvented by confining the discussion to nuclei with $N = Z$ (N being the neutron number and Z the proton number). The reduced Skyrme functional then becomes

$$E_{\text{Sk}} = \frac{\hbar^2}{2m}\sum_{\alpha=1}^{A}\int \mathrm{d}^3r|\nabla\varphi_\alpha|^2| + \frac{1}{2}\int \mathrm{d}^3r\,\left[b_0\rho^2 + b_3\rho^{2+\alpha}\right] + E_{\text{Coul}}[\rho_p], \quad (6.15)$$

where $A = N + Z$, ρ denotes the total density, and ρ_p the proton density. The term E_{Coul} stands for the Coulomb energy as given in (6.5c). In the spirit of LDA, the model is calibrated in homogeneous nuclear matter. The ansatz (6.15) then yields for the energy per nucleon in symmetric matter

$$\frac{E}{N} = \frac{\hbar^2}{2m}\frac{6}{5}\left(\frac{3\pi^2}{2}\right)^{2/3}\rho^{2/3} + \frac{1}{2}\left[b_0\rho + b_3\rho^{1+\alpha}\right]. \quad (6.16)$$

In the spirit of LDA, we compare that with results of microscopic calculations for symmetric nuclear matter [5], as shown in Fig. 6.3 (solid line). The most important part of that curve resides around the minimum characterized by the equilibrium density $\rho_0 = 0.155\,\text{fm}^{-3}$, the corresponding energy $(E/N)_{\text{eq}} = -15.6\,\text{MeV}$, and the curvature at the minimum, the incompressibility $K = 9\rho^2\partial_\rho^2(E/N) = 285\,\text{MeV}$. The ansatz (6.16) has three free parameters: b_0, b_3, and α. They are adjusted to these

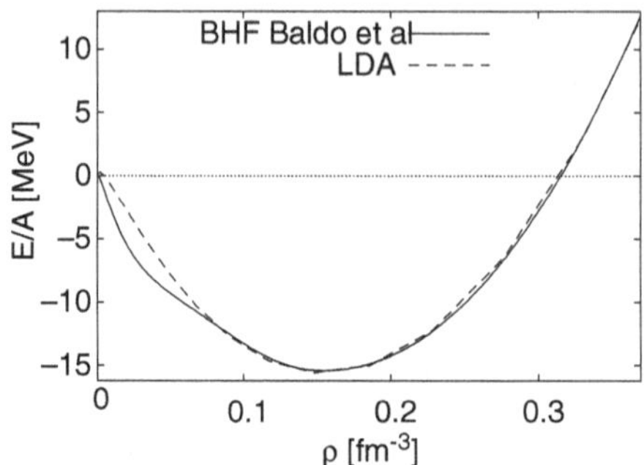

Fig. 6.3 Binding energy per nucleon of symmetric nuclear-matter versus total density. The solid line shows a result from "ab initio" calculations taking care of two-nucleon correlations [5]. The *dashed line* is a fit to the form (6.16) adjusting the free parameters b_0, b_3, and α

three basic equilibrium properties. The resulting binding-energy curve is shown as a dashed line in Fig. 6.3. It is obvious that the very simple ansatz provides a surprisingly good description of the nuclear-matter binding-energy curve. Larger deviations occur only at low densities, a region which is known to be extremely hard to describe. We now take the energy functional (6.15) with the parameters fitted to the nuclear-matter curve and solve the corresponding Kohn–Sham equations numerically for the finite (rather small) nucleus ^{16}O. The result is a binding energy of 163 MeV (as compared to the experimental value of 126 MeV), and an r.m.s. radius of 2.50 fm as compared to 2.68 fm. That is a very satisfying result in view of the extreme simplicity of modeling and adjustment. At the same time, the still sizeable discrepancies indicate that the nuclear functional has to be considerably more complicated than the ansatz (6.15). The most important next ingredient is a gradient correction $\propto (\nabla\rho)^2$, which is important for describing surface tension, and, of course, the spin–orbit density. The modeling of the (volume) density dependence can remain as naive as conjectured in ansatz (6.15). The hierarchy of importance differs from electronic LDA.

6.5 The Thomas–Fermi Approach

The Thomas–Fermi approximation [105, 34] is a DFT at the level of the Hohenberg–Kohn theorem, where the kinetic energy is not treated in detail like in the Kohn–Sham scheme, but also approximated as a functional of density using the LDA for the kinetic energy (see also (2.24))

$$\epsilon_{\text{kin}}[\rho] = \frac{E_{\text{kin}}}{N} = \frac{3}{5}\frac{\hbar^2}{2m}k_{\text{F}}^2 = (3\pi^2)^{2/3}\frac{3}{5}\frac{\hbar^2}{2m}\rho^{2/3} \tag{6.17}$$

This yields a total energy for spin 1/2 particles.

$$E_{\text{tot}} = \int \mathrm{d}^3r\rho(\mathbf{r})\left(\epsilon_{\text{kin}}[\rho(\mathbf{r})] + \epsilon_{\text{pot}}[\rho(\mathbf{r})] + U_{\text{ext}}(\mathbf{r})\right), \tag{6.18}$$

where the potential energy functional covers all interaction effects (direct term, exchange, correlations) and U_{ext} is a possible external field. This energy is a functional of the local density $\rho(\mathbf{r})$ throughout and so provides a particularly simple description lacking, of course, any shell effects because the LDA functional has wiped them out. The variation of the energy to reach the ground state is here done with respect to the density. There is also the constraint to a prescribed particle number $N = \int d^3r\, \rho$. This yields the variational condition

$$\delta_{\rho(\mathbf{r})}\Big[E_{\text{tot}} - \mu_0 \int d^3r\, \rho\Big] = 0,$$

and consequently the Thomas–Fermi equation

$$(3\pi^2)^{2/3}\frac{\hbar^2}{2m}(\rho(\mathbf{r}))^{2/3} + U_{\text{LDA}}[\rho(\mathbf{r})] = \mu_0 - U_{\text{ext}}(\mathbf{r}), \tag{6.19a}$$

$$U_{\text{LDA}} = \frac{\partial(\rho\epsilon_{\text{pot}})}{\partial\rho}, \tag{6.19b}$$

$$\mu_0 \quad \text{fixed by condition} \quad \int d^3r\rho(\mathbf{r}) = N, \tag{6.19c}$$

where the Lagrangian parameter μ_0 is the Fermi energy, equivalent to the chemical potential μ at temperature $T = 0$. Note that U_{LDA} here is formally the same as U_{KS} in (6.10). Equation (6.19a) determines the spatial density distribution $\rho(\mathbf{r})$ as the solution of a transcendental equation, which may be expressed in closed form for sufficiently simple U_{LDA}. The solution then provides the total energy as well as local density and any observable which can be directly computed from these two quantities. More involved observables may also be computed but require to go back to the Fermi gas and to perform the steps of LDA for that particular observable (an example is the discussion of the one-body density matrix in relation to Fig. 2.4).

In the following, we will present two examples for the Thomas–Fermi method, namely the electron cloud of atoms and fermionic atoms in a harmonic trap. The first test case is the oldest example [105, 34], worked out long before DFT was established, while the second one is a quite modern application.

6.5.1 The Thomas–Fermi Model for Atoms

The atomic Thomas–Fermi model is one of the first self-consistent approaches to the many-fermion problem, providing a simple approximation to the electronic density in atoms. While approximating the electron density by its "Fermi-gas" value it fully accounts for the Coulomb interaction, at least its direct part. It furthermore turns out that with a clever change of variables the solution of the problem becomes a universal function and thus needs to be solved for only once. It then provides the electronic density $\rho(\mathbf{r})$ and the associated Coulomb field $V(\mathbf{r})$.

The nucleus (charge Z) is placed at $\mathbf{r} = 0$ and outside that point the electronic density $\rho(\mathbf{r})$ and the Coulomb potential $V(\mathbf{r})$ are linked together by the Poisson equation

$$\Delta V = 4\pi e\rho(\mathbf{r}), \tag{6.20}$$

where we use the electronic charge density $-e\rho(\mathbf{r})$. Note that the obvious spherical symmetry of the problem allows to reduce $V(\mathbf{r})$ to $V(r)$. Two boundary conditions complement this equation. For $r \to 0$, V should reduce to the nuclear contribution:

$$V \to \frac{Ze}{r} \quad \text{for} \quad r \to 0, \tag{6.21}$$

and outside the atomic radius a net charge z associated to the corresponding ion is assumed:

$$V \to \frac{ze}{r} \quad \text{for} \quad r \geq R, \tag{6.22}$$

where R is the atomic radius. A neutral atom is associated with the choice $z = 0$. These boundary conditions are complemented by a continuity condition at $r = R$, namely

$$\frac{\mathrm{d}V}{\mathrm{d}r}_{|r=R} = -\frac{ze}{R^2}. \tag{6.23}$$

The second component of the model concerns the treatment of the electron density. One first introduces the energy E of a single electron which can be written as

$$E = \frac{\hbar^2\mathbf{k}^2}{2m_e} - eV(r). \tag{6.24}$$

Demanding that the electron should be bound leads to the condition $E \leq -eV(R)$ which can be rewritten as

$$\mathbf{k}^2 \leq k_{\mathrm{F}}^2(r) = \frac{2m_e}{\hbar^2} e\big(V(r) - V(R)\big) \tag{6.25}$$

and defines an r-dependent maximum value of the momentum k_{max}. Assuming a local Fermi-gas description of the electron gas one can then express the electron density $\rho(\mathbf{r})$ as a function of $k_{\mathrm{max}} = k_F(r)$ following the Fermi-gas expression (2.17) as

$$\rho(\mathbf{r}) = \frac{k_F^3}{3\pi^2} = \frac{1}{3\pi^2}\Big[\frac{2m_e e}{\hbar^2}(V(r) - V(R))\Big]^{3/2}. \tag{6.26}$$

Note that the density is defined only in the classically allowed region, namely for $r \leq R$.

Putting all the pieces together finally allows writing down a self-contained equation for the potential $V(r)$. Inserting the expression for $\rho(\mathbf{r}) = \rho(r)$ into the Poisson equation and using the Laplacian in spherical symmetry as $\Delta = r^{-1}\partial_r^2 r$ leads to

$$\frac{\mathrm{d}^2}{\mathrm{d}r^2}(rV) = \frac{4e^2 r}{3\pi}\Big[\frac{2m_e}{\hbar^2}(V(r) - V(R))\Big]^{3/2}. \tag{6.27}$$

This equation can be further simplified by introducing the intermediate quantity

$$v(r) = \frac{r}{Ze^2}\big(V(r) - V(R)\big) \tag{6.28}$$

and the reduced variable

$$x = \frac{r}{a} \qquad \text{with} \qquad a = \left(\frac{9\pi^2}{128Z}\right)^{1/3}\frac{\hbar^2}{m_e e} \simeq 0.88534Z^{-1/3}\mathrm{a}_0. \tag{6.29}$$

With this variable and the intermediate function v a dimensionless equation for v is finally obtained:

$$\frac{\mathrm{d}^2 v}{\mathrm{d}x^2} = \frac{v(x)^{3/2}}{\sqrt{x}} \tag{6.30}$$

together with the boundary conditions (introducing $X = R/a$)

$$v(0) = 1 \qquad v(X) = 0 \qquad X\left.\frac{\mathrm{d}v}{\mathrm{d}x}\right|_{x=X} = -\frac{z}{Z}. \tag{6.31}$$

The first two conditions are boundary conditions which specify the solution of the second-order differential equation (6.30) uniquely. The third condition describes the connection between the (scaled) radius X and the relative ionization z/Z. The appropriate X for a desired z has to be found iteratively. Once the solution for the dimensionless potential $v(x)$ is found, one can easily recover the associated Thomas–Fermi density as

$$\rho(r) = \frac{1}{3\pi^2}\left[\frac{2Z}{\mathrm{a}_0}\right]^{3/2}\left(\frac{v(x)}{xa}\right)^{3/2} \approx 0.0796\frac{Z}{a^3}\left(\frac{v(x)}{x}\right)^{3/2}. \tag{6.32}$$

The solutions of (6.30) are universal for any Z but require a numerical calculation. Results for v are plotted in Fig. 6.4 and also reflect the behavior of the density

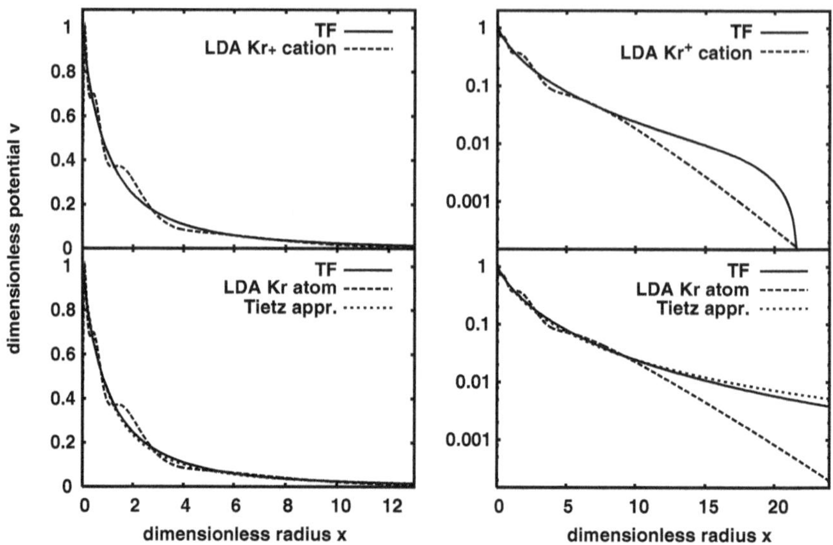

Fig. 6.4 The universal potential $v(x)$ from (6.30) for two effective ionizations z/Z. *Lower*: Neutral atom $z/Z = 0$. *Upper*: $z/Z = 1/36$ which is a singly charged cation for Kr. The *left panels* show the results in the relevant region of substantial densities. The *right panels* show the asymptotics by using a logarithmic scale and larger range of x. For comparison, the results of a fully quantum-mechanical LDA-SIC calculation (see Sect. 6.1) for the Kr atom (*lower*) and Kr^+ cation (*upper*) are also shown. The "potential" is here a way to plot the density as $v = x(\rho\, a/(0.0796\, Z))^{2/3}$. In case of the neutral atom, the Tietz approximation [106] (6.33) is also given

according to (6.32). Figure 6.4 displays two cases: an ion (solution with $z = 1$ for $Z = 36$) and a neutral atom (solution with $z = 0$ for $Z = 36$). The left panels show the region where most of the density is concentrated. The TF results have a smooth trend in contrast to the fully quantum-mechanical result for Kr, which shows the typical shell fluctuations of the density (see also Figs. 1.3 and 2.1), but it is gratifying to see how nicely the TF trends average through the quantal results. For the neutral atom, one should note that there exists an analytic approximation to the exact solution $v(x)$. It is the Tietz approximation [106]

$$v(x) \simeq \tilde{v}(x) = \frac{1}{(1+\alpha x)^2}, \quad \alpha = 0.53625 \tag{6.33}$$

with the constant α ensuring normalization. It is also shown in Fig. 6.4 and it reproduces the bulk density fairly well (lower left panel). The right panels show the asymptotic behavior for large x. Here we see large differences, if not pathologies. The neutral atom in TF approximation has an infinite extension, contrary to reality. The TF density of the cation terminates at a finite X, see (6.31), again at variance with the realistic case which falls off exponentially. Still, the atomic Thomas–Fermi model provides an interesting, close to analytical, solution to the many-electron problem in atoms. It describes the average trends and it may also serve as a starting point in iterative processes for more involved theories like HF or Kohn–Sham (see above sections).

6.5.2 Energy Functional for Atoms in a Trap

Fermionic atoms in a trap span an enormous range of physical conditions from a weakly perturbed Fermi gas to a highly correlated system with possibly different outcome: as a Bose–Einstein condensate (of dimers) or as a well-paired BCS phase (for a brief introduction see Chap. 1, for a review on the experimental status [14], and for a theoretical survey [38]). A key parameter deciding on the regime is the scattering length a which is deduced from the total cross section at $k \longrightarrow 0$ as $\sigma(k = 0) = \pi a^2$ [62]. A highly interesting regime is reached in what is called the *unitary limit* $k_F a \longrightarrow \infty$. This allows, e.g., a system which is dilute and at the same time highly correlated. Again, it is possible to incorporate the correlations into an energy-density functional and to obtain a surprisingly simple description at the level of the Thomas–Fermi approach. This is what we are going to briefly present in this section. A very broad and detailed discussion of fermions in a trap is found in [38].

The unitary limit establishes simple scaling behavior because one length scale, the scattering length, disappears. On general grounds, one can then assume that the total energy at temperature $T = 0$ becomes

$$\varepsilon(\rho) = \frac{E}{N} = \frac{3}{5}\frac{\hbar^2 k_F^2}{2m}(1 + \beta), \tag{6.34}$$

where more elaborate calculations predict $\beta = -0.58$ (see Sect. IV.F. in [19]). Traps usually provide a harmonic external potential which we assume here as being spherical to simplify notations. The corresponding Thomas–Fermi equation then reads

$$\mu_0 = U_{TF}[\rho(\mathbf{r})] + V_{ho}(\mathbf{r}), \tag{6.35a}$$

$$U_{TF}[\rho] = \frac{\partial(\rho\epsilon)}{\partial\rho}[\rho] = (3\pi^2)^{2/3}\frac{\hbar^2}{2m}\rho^{2/3}(1 + \beta), \tag{6.35b}$$

$$V_{ho} = \frac{m}{2}\omega^2 r^2. \tag{6.35c}$$

The solution is

$$\rho(\mathbf{r}) = \frac{(1+\beta)^{-3/2}}{3\pi^2}\left(\frac{2m}{\hbar^2}\right)^{3/2}\left(\mu_0 - \frac{m}{2}\omega^2 r^2\right)^{3/2}\theta\left(\sqrt{\frac{2\mu_0}{m\omega^2}} - r\right),$$

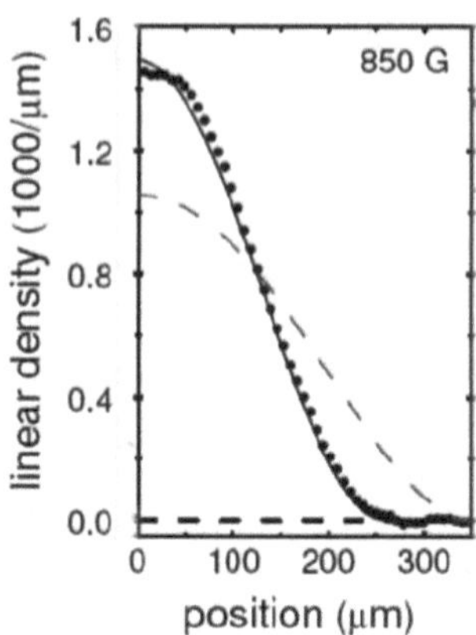

Fig. 6.5 The x–y-integrated density profile $n^{(1)}(z)$ for a trapped ensemble of ^{6}Li atoms under the conditions close to the unitary limit. The *dots* are experimental results. The *dashed line* corresponds to the prediction of a free Fermi gas. The *solid line* represents the Thomas–Fermi approach (6.37) with a freely fitted $\beta = -0.73$, adapted from [38]

with μ_0 to be fixed using (6.19c). This can be converted into one closed expression

$$\rho(\mathbf{r}) = \frac{8N}{\pi^2 R^3}\left(1 - \frac{r^2}{R^2}\right)^{3/2} \theta(R - r), \quad R = (1+\beta)^{1/4}\sqrt{\frac{\hbar}{m\omega}}(24N)^{1/6}. \quad (6.36)$$

The experimental observation refers to the integrated density

$$n^{(1)}(z) = \int \mathrm{d}x\,\mathrm{d}y\, n(\mathbf{r}) = \frac{N}{R_z}\frac{16}{5\pi}\left(1 - \frac{z^2}{R_z^2}\right)^{5/2}. \quad (6.37)$$

Figure 6.5 compares the result of this simple Thomas–Fermi model with an experimental density distribution for a gas of ^{6}Li atoms driven in the regime of the unitary limit. The agreement is very satisfying. Comparison with the free gas ($\beta = 0$, dashed line) shows the impact of interaction effects, but these amount to a simple rescaling of the "kinetic energy" (6.34).

6.6 Concluding Remarks

Density functionals provide a remarkably efficient tool for the description of many-fermion systems. They are especially used in electronic systems and nuclear physics, although they may differ in detail from one field to the other. The interest of such density functional approaches is that they allow to treat the problems at the formally simple level of the Hartree approximation by using an effective interaction which practically includes correlation effects in a simple way.

Although introduced from a practical point of view in the 1930s in atoms, density-functional theories were only much later founded on a strong formal basis which then motivated numerous formal and practical developments especially in electronic systems. The nuclear case progressed in a somewhat parallel way. The situation today is the one of a very active field in which various facets of the theory and possible applications are explored in parallel. There obviously remains in this domain several aspects to be worked out in deeper detail and this constitutes a further motivation for the topic which joins formal simplicity to efficiency.

Chapter 7
Quasispin Models

In elementary quantum mechanics, the spin-$\frac{1}{2}$ system is often taken as the simplest, still non-trivial, example of a two-state system. It allows to point out major quantum properties in the purely abstract space of spin degrees of freedom. Assemblies of spins provide simple models for interacting systems. They are a key model system in solid-state theory, e.g., for spin glasses or strongly interacting electrons (Hubbard model). Here we want to exploit the simplicity provided by a two-level system to understand how elaborate techniques of the many-fermion problem perform in practice. The advantage of such model systems, even if not fully realistic, lies in the fact that most calculations can be performed analytically thanks to straightforward algebraic techniques of angular-momentum algebra. We shall thus use such methods here, in particular to analyze mean-field approaches in some detail.

Although somewhat inspired by spin systems, the models we want to consider in this chapter are more general and can be derived as a simplification of realistic situations leading to a simplified two-level Hamiltonian, hence bearing some flavor of a spin-$\frac{1}{2}$ system, especially from the formal point. In order to keep in mind the formal similarities, but to exemplify the generality of such approaches, they are usually called quasispin models. We will investigate their formal properties and see how the mean-field approximation performs in comparison to exact solutions. We will also consider some realistic applications, especially in the case of deformation effects in molecules and nuclei.

7.1 Construction of Quasispin Models

7.1.1 Examples and Motivation

Probably the simplest quantum-mechanical example for a quasispin model is a spin-$\frac{1}{2}$ system. It allows a fully analytic treatment using algebraic methods and is a generic system which turns out to be applicable as a simple model for a broad variety of physical systems which are seemingly very different, but share a common formal structure. In this section, we will discuss quasispin models in the spirit of the

J.A. Maruhn et al., *Simple Models of Many-Fermion Systems*,
DOI 10.1007/978-3-642-03839-6_7, © Springer-Verlag Berlin Heidelberg 2010

Lipkin–Meshkov–Glick (LMG) model [64] as a particular case which we develop in depth. Other applications are then discussed in brief.

Let us consider many-fermion compounds with a closed shell like, e.g., rare-gas atoms, magic nuclei, metal clusters, ^{3}He clusters, or quantum dots. The typical spectral situation is sketched in Fig. 7.1. The occupied and unoccupied levels next to the Fermi energy are the ones reacting most sensitively to any perturbation. The $1ph$ energies are lowest and the coupling matrix elements are generally largest for these transitions. For highly symmetric systems (as closed-shell systems usually are), the levels are bunched in shells such that there emerges a more or less clear distinction between the last occupied shell and those farther below and between the first unoccupied one and others farther up, see, e.g., Fig. 3.3. The most active zone is given by the two shells just below and just above the Fermi energy. It can be described schematically by two highly degenerate levels as indicated in the step from the left part to the right part in Fig. 7.1, further simplified by assuming the same degree of degeneracy in both shells and by considering only vertical transitions. The occupied states share one level, denoted $m = -1$, and the relevant unoccupied ones the other level, $m = +1$. The energy difference ε between these two levels represents the shell gap. The states within each level are labeled through the secondary quantum number $\alpha = 1 \ldots N$. The zeroth-order Hamiltonian then defines these single-particle energies $\varepsilon_{m\alpha}$ and the transitions go from the lower to the upper level. For sake of simplicity we consider only transitions which preserve the α-quantum number, as indicated.

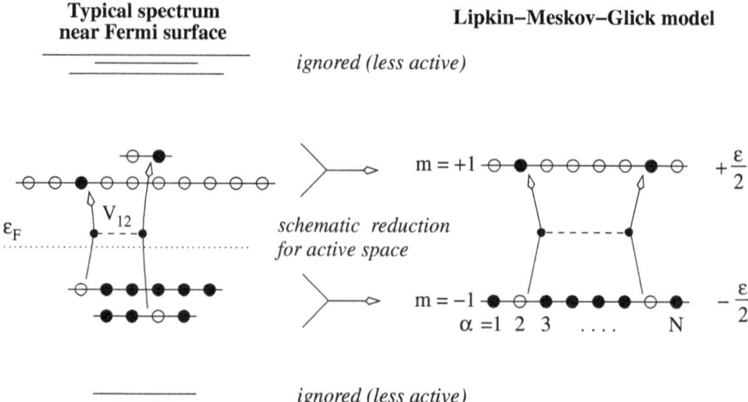

Fig. 7.1 Schematic view of the LMG model. *Left panel*: A typical spectrum for a small spherical system (atom, nucleus, metal cluster, quantum dots) with a closed shell near the Fermi surface. Occupied states are indicated by a *filled circle*, unoccupied ones by an *empty circle*, and levels farther away from the Fermi energy only by a *line*. Transitions are indicated by *arrows*. Interactions induce two simultaneous, coupled transitions. That is indicated by a *horizontal dashed line* (labeled by V_{12}). *Right panel*: The same for the LMG model. Labels and notations for the model states are also indicated

7.1.2 The Model Hamiltonian

It is now crucial to realize that an isolated two-level system is isomorphous to a spin-$\frac{1}{2}$ system. This is visualized in Fig. 7.2. The three operators for excitation $\hat{a}^\dagger_{+1}\hat{a}_{-1}$, de-excitation $\hat{a}^\dagger_{-1}\hat{a}_{+1}$, and measurement of status $\hat{a}^\dagger_{+1}\hat{a}_{+1} - \hat{a}^\dagger_{-1}\hat{a}_{-1}$ obey the same commutator algebra as the three spin-$\frac{1}{2}$ operators $\hat{s}_+$, $\hat{s}_-$, and $\hat{s}_z$, as will be shown in Sect. 7.2. Many-particle systems translate to systems of many coupled spins, and this is the point where the powerful techniques from angular-momentum algebra can be applied.

Having sorted out this analogy to spin, we continue with the construction of the Hamiltonian for the quasispin model of a many-fermion system near the Fermi energy. The intended $\alpha \leftrightarrow \alpha'$ symmetry of the model suggests a two-body interaction which has the same strength for all $2ph$ excitations, i.e., for all $\hat{a}^\dagger_{+1,\alpha}\hat{a}^\dagger_{+1,\alpha'}\hat{a}_{-1,\alpha'}\hat{a}_{-1,\alpha}$, and similarly for the other ups and downs. All this together yields the model Hamiltonian

$$\hat{H} = \hat{H}_0 + \hat{V}, \quad \hat{H}_0 = \varepsilon \hat{J}_0, \quad \hat{V} = \frac{G}{2}\left(\hat{J}_+ + \hat{J}_-\right)^2, \tag{7.1a}$$

$$\hat{J}_0 = \sum_\alpha \frac{1}{2}[\hat{a}^\dagger_{1,\alpha}\hat{a}_{1,\alpha} - \hat{a}^\dagger_{-1,\alpha}\hat{a}_{-1,\alpha}] \equiv \sum_\alpha \hat{s}_0^{(\alpha)}, \tag{7.1b}$$

$$\hat{J}_+ = \sum_\alpha \hat{a}^\dagger_{+1,\alpha}\hat{a}_{-1,\alpha} \equiv \sum_\alpha \hat{s}_+^{(\alpha)}, \tag{7.1c}$$

$$\hat{J}_- = \sum_\alpha \hat{a}^\dagger_{-1,\alpha}\hat{a}_{+1,\alpha} \equiv \sum_\alpha \hat{s}_-^{(\alpha)}, \quad \hat{J}_- = \hat{J}_+^\dagger, \tag{7.1d}$$

where the fermion operators obey, of course, the standard fermion algebra as summarized in Appendix A.4.

In spite of the dramatic simplifications, the setup still reflects many essential features of a realistic situation. In the stationary regime, it is a useful model to explore the restructuring of states in the Hartree–Fock ground state including the occurrence of spontaneous symmetry breaking (see Sect. 7.7) and related questions like restoration of symmetry and soft modes. In the dynamical regime, the LMG model is perfectly suited to explain the emergence of collectivity out of a coherent superposition of single-particle excitations like, e.g., the plasmons in metallic systems and quantum dots (see Sects. 1.1.5 and 8.4.2.2).

The quasispin algebra defines a quite generic scenario. There is a large number of similar models in many different fields. There are applications where one would not expect a relation to quasispin at first glance. In this book, we will have the example of pairing in the BCS approximation, as introduced in Sect. 9.3.1 and worked out in Sect. 9.4, for which we also find a close relation to quasispin. In addition there is, of

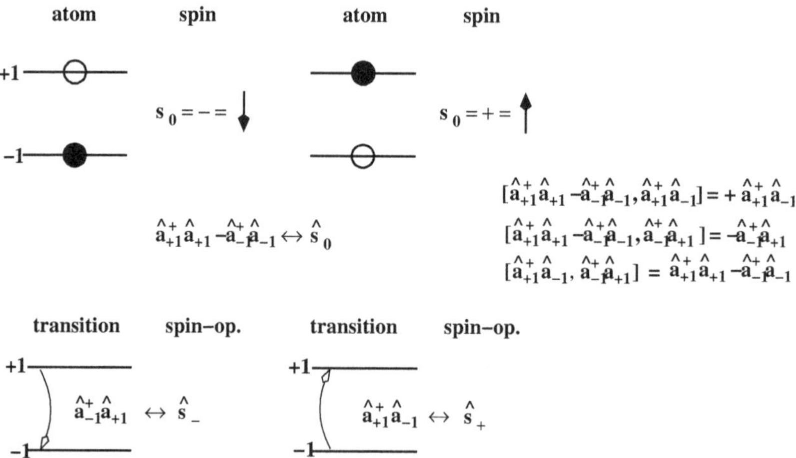

Fig. 7.2 Illustration of the equivalence of a two-level scenario with a spin-$\frac{1}{2}$ system

course, the huge family of genuine spin models going back to the celebrated Ising model [12], whose Hamiltonian reads in its most general form

$$\hat{H} = \epsilon \sum_{\alpha} \hat{s}_0^{(\alpha)} - \sum_{\alpha\beta} \sum_{i \in \{x,y,z\}} V_{\alpha\beta,i} \hat{s}_i^{(\alpha)} \hat{s}_i^{(\beta)} . \tag{7.2}$$

This covers a large variety of applications depending on the choice of the interaction matrix elements $V_{\alpha\beta,i}$. The above LMG model is recovered with $V_{\alpha\beta,i} = 2G\delta_{\alpha\beta}$. A linear chain with nearest-neighbor coupling is realized with $V_{\alpha\beta,i} = \delta_{\alpha,\beta+1}$ and it can be closed to a ring with $V_{\alpha\beta,i} = \delta_{\alpha,\text{mod}(\beta+1,N)}$ where $\alpha, \beta \in \{0 \ldots N-1\}$. In the case of magnetic systems, the first term in the Hamiltonian (7.2) stems from an external magnetic field B_0 producing a basic energy splitting $\epsilon = \mu B_0$, where μ is the magnetic moment of the basic constituents. The example of an anti-ferromagnetic ring makes it clear, though, that this modeling is not restricted to plain spins. The ring is also used to describe the electron cloud in a circular quantum dot [87], and "spin" can stand for any decision among two choices. Thus it is natural that coupled spin models play a role in quantum computing [51]. There is also a close relationship to lattice gas models like, e.g., the Hubbard model [94]. We cannot cover the full richness of quasispin models in this book, so that we confine the discussion to a few crucial examples which illustrate basic physical mechanisms (mean field, symmetry breaking, collective motion) and which also serve to develop some experience with the handling of quasispin models.

In the remainder of this chapter, we will work out the structure of the model and its operator algebra, derive the exact ground-state solution, develop and discuss the HF solutions, and finally remark on spontaneous symmetry breaking. Dynamical aspects will be taken up in Chap. 8 at the example of the random-phase approximation (RPA), the dynamic mean-field theory for small-amplitude oscillations.

7.2 The Quasispin Algebra

Figure 7.2 has suggested an identification of a two-level system with a spin-$\frac{1}{2}$ system. Many-fermion systems and other applications provide several similar two-level systems, or spins respectively, labeled here by the index α. We summarize the properties:

$$\begin{aligned} \hat{a}^\dagger_{+1,\alpha}\hat{a}_{-1,\alpha} &\equiv \hat{s}^{(\alpha)}_+ & \left[\hat{s}^{(\alpha)}_0, \hat{s}^{(\alpha')}_+\right] &= \delta_{\alpha\alpha'}\hat{s}^{(\alpha)}_+ \\ \hat{a}^\dagger_{-1,\alpha}\hat{a}_{+1,\alpha} &\equiv \hat{s}^{(\alpha)}_- & \left[\hat{s}^{(\alpha)}_0, \hat{s}^{(\alpha')}_-\right] &= -\delta_{\alpha\alpha'}\hat{s}^{(\alpha)}_- \\ \hat{a}^\dagger_{+1,\alpha}\hat{a}_{+1,\alpha} - \hat{a}^\dagger_{-1,\alpha}\hat{a}_{-1,\alpha} &\equiv 2\hat{s}^{(\alpha)}_0 & \left[\hat{s}^{(\alpha)}_-, \hat{s}^{(\alpha')}_+\right] &= 2\delta_{\alpha\alpha'}\hat{s}^{(\alpha)}_0 \end{aligned} \tag{7.3}$$

$$\hat{J}_\nu = \sum_{\alpha=1}^{N} \hat{s}^{(\alpha)}_\nu, \quad [\hat{J}_0, \hat{J}_\pm] = \pm\hat{J}_\pm, \quad [\hat{J}_+, \hat{J}_-] = 2\hat{J}_0 .$$

In the following, we will briefly provide the proof for the commutator algebra in the setup (7.3). The factors $\delta_{\alpha\alpha'}$ are trivial as different spins (levels) do not communicate. It remains to prove the basic spin commutators for one system α. We simplify the notations by dropping the counter α.

$$\begin{aligned} [\hat{s}_0, \hat{s}_+] &= \left[\frac{1}{2}(\hat{a}^\dagger_{+1}\hat{a}_{+1} - \hat{a}^\dagger_{-1}\hat{a}_{-1}), \hat{a}^\dagger_{+1}\hat{a}_{-1}\right] \\ &= \frac{1}{2}\Big\{ \underbrace{\left[\hat{a}^\dagger_{+1}\hat{a}_{+1}, \hat{a}^\dagger_{+1}\hat{a}_{-1}\right]}_{\hat{a}^\dagger_{+1}\hat{a}_{-1}} - \underbrace{\left[\hat{a}^\dagger_{-1}\hat{a}_{-1}, \hat{a}^\dagger_{+1}\hat{a}_{-1}\right]}_{-\hat{a}^\dagger_{+1}\hat{a}_{-1}} \Big\} = \frac{1}{2}2\hat{a}^\dagger_{+1}\hat{a}_{-1} = +\hat{s}_+. \end{aligned}$$

The proof for the commutator with $\hat{s}_-$ proceeds in the same fashion. We leave it as an exercise to the reader. The third commutator reads in detail

$$\begin{aligned} [\hat{s}_+, \hat{s}_-] &= \left[\hat{a}^\dagger_{+1}\hat{a}_{-1}, \hat{a}^\dagger_{-1}\hat{a}_{+1}\right] = \hat{a}^\dagger_{+1}\hat{a}_{-1}\hat{a}^\dagger_{-1}\hat{a}_{+1} - \hat{a}^\dagger_{-1}\hat{a}_{+1}\hat{a}^\dagger_{+1}\hat{a}_{-1} \\ &= \hat{a}^\dagger_{+1}\hat{a}_{+1} - \hat{a}^\dagger_{+1}\hat{a}^\dagger_{-1}\hat{a}_{-1}\hat{a}_{+1} - \hat{a}^\dagger_{-1}\hat{a}_{-1} + \underbrace{\hat{a}^\dagger_{-1}\hat{a}^\dagger_{+1}}_{\hat{a}^\dagger_{+1}\hat{a}^\dagger_{-1}}\underbrace{\hat{a}_{+1}\hat{a}_{-1}}_{\hat{a}_{-1}\hat{a}_{+1}} \\ &= \left\{\hat{a}^\dagger_{+1}\hat{a}_{+1} - \hat{a}^\dagger_{-1}\hat{a}_{-1}\right\} = 2\hat{s}_0. \end{aligned}$$

The algebra for the total quasispin $\hat{J}_\nu$ follows trivially from these basic commutators.

The analogy to angular momentum also suggests an alternative representation of the algebra. The operators $\hat{J}_\pm$ are extremely useful for stepping through the chains of excitations, as we will see, but they are not Hermitian and thus not suited as observables. We introduce instead the Hermitian combinations

$$\hat{J}_x = \frac{1}{2}\left(\hat{J}_+ + \hat{J}_-\right), \quad \hat{J}_y = -\frac{\mathrm{i}}{2}\left(\hat{J}_+ - \hat{J}_-\right), \quad \hat{J}_z \equiv \hat{J}_0, \tag{7.4}$$

where the third operator has just been renamed to fit better into the usual x–y–z notations. These three operators fulfill the angular momentum algebra in the better-known form

$$\left[\hat{J}_x, \hat{J}_y\right] = \mathrm{i}\hat{J}_z, \quad \left[\hat{J}_y, \hat{J}_z\right] = \mathrm{i}\hat{J}_x, \quad \left[\hat{J}_z, \hat{J}_x\right] = \mathrm{i}\hat{J}_y, \tag{7.5}$$

It is interesting to express the two-body interaction $\hat{V}$ in the model Hamiltonian (7.1) through these operators, which yields

$$\hat{V} = 2G\hat{J}_x^2. \tag{7.6}$$

7.3 The Unperturbed Ground State

The unperturbed system is the case when the two-body interaction $\hat{V}$ disappears, i.e., for $G = 0$ or $\hat{H} = \hat{H}_0$. The ground state of this unperturbed system is simply

$$|\Phi_0\rangle = \hat{a}^\dagger_{-1,N} \cdots \hat{a}^\dagger_{-1,1}|0\rangle \equiv \chi_\downarrow^{(1)} \cdots \chi_\downarrow^{(N)}, \tag{7.7a}$$

with the basic properties

$$\langle\Phi_0|\Phi_0\rangle = 1, \quad \hat{J}_-|\Phi_0\rangle = 0, \quad \langle\Phi_0|\hat{J}_+ = 0. \tag{7.7b}$$

Note that it is an independent-particle state, which is no surprise as $\hat{H}_0$ is a one-body Hamiltonian. It is an eigenstate of quasispin with the following properties:

$$\hat{\mathbf{J}}^2|\Phi_0\rangle = J(J+1)|\Phi_0\rangle, \quad \hat{J}_0|\Phi_0\rangle = -J|\Phi_0\rangle, \quad J = \tfrac{N}{2} \tag{7.8a}$$

$$\overline{J_x} = \langle\Phi_0|\hat{J}_x|\Phi_0\rangle = 0, \quad \overline{J_y} = \langle\Phi_0|\hat{J}_y|\Phi_0\rangle = 0, \tag{7.8b}$$

$$\overline{\Delta^2 J_x} = \langle\Phi_0|\hat{J}_x^2|\Phi_0\rangle = \tfrac{N}{4}\,, \quad \overline{\Delta^2 J_y} = \langle\Phi_0|\hat{J}_y^2|\Phi_0\rangle = \tfrac{N}{4}\,,$$
$$\overline{\Delta^2 J_z} = \langle\Phi_0|\hat{J}_z^2 - \overline{J_z}^2|\Phi_0\rangle = 0\,. \tag{7.8c}$$

This means that the unperturbed ground state has total angular momentum $J = N/2$ and its z component is $J_z = -N/2$, so that it is fully aligned in the negative z direction. This nicely corresponds to the view of a system composed of N times spin-$\frac{1}{2}$, initially all aligned with $-\frac{1}{2}$.

The expectation values (7.8b) are derived simply from the combinations (7.4) and

$$\langle\Phi_0|\underbrace{\hat{J}_-|\Phi_0\rangle}_{=0}=0,\quad \underbrace{\langle\Phi_0|\hat{J}_+}_{=0}|\Phi_0\rangle=0.$$

The eigenvalue relation (7.8a) is obvious for $\hat{J}_0$. For $\hat{\mathbf{J}}^2$ we first prove

$$\hat{J}_+\underbrace{\hat{J}_-|\Phi_0\rangle}_{=0}=0,\quad \hat{J}_-\hat{J}_+|\Phi_0\rangle=\underbrace{\left[\hat{J}_-,\hat{J}_+\right]}_{-2\hat{J}_0}|\Phi_0\rangle+\hat{J}_+\underbrace{\hat{J}_-|\Phi_0\rangle}_{=0}=N|\Phi_0\rangle,$$

and can then deduce

$$\begin{aligned}\hat{\mathbf{J}}^2|\Phi_0\rangle&=\left(\hat{J}_x^2+\hat{J}_y^2+\hat{J}_z^2\right)|\Phi_0\rangle=\frac{1}{2}\left(\hat{J}_-\hat{J}_++\hat{J}_+\hat{J}_-+2\hat{J}_0^2\right)|\Phi_0\rangle\\&=\frac{N}{2}\left(\frac{N}{2}+1\right)|\Phi_0\rangle.\end{aligned}$$

To evaluate the variances, heavy use is again made of $\hat{J}_-|\Phi_0\rangle=0$ and $\langle\Phi_0|\hat{J}_+=0$ to obtain

$$\begin{aligned}\overline{\Delta^2 J_x}&=\langle\Phi_0|\hat{J}_x^2|\Phi_0\rangle=\frac{1}{4}\langle\Phi_0|\hat{J}_+\hat{J}_++\hat{J}_+\hat{J}_-+\hat{J}_-\hat{J}_++\hat{J}_-\hat{J}_-|\Phi_0\rangle\\&=\frac{1}{4}\langle\Phi_0|\hat{J}_-\hat{J}_+|\Phi_0\rangle=\frac{N}{4},\\\overline{\Delta^2 J_y}&=\langle\Phi_0|\hat{J}_y^2|\Phi_0\rangle=\frac{1}{4}\langle\Phi_0|-\hat{J}_+\hat{J}_++\hat{J}_+\hat{J}_-+\hat{J}_-\hat{J}_+-\hat{J}_-\hat{J}_-|\Phi_0\rangle\\&=\frac{1}{4}\langle\Phi_0|\hat{J}_-\hat{J}_+|\Phi_0\rangle=\frac{N}{4},\\\overline{\Delta^2 J_0}&=\langle\Phi_0|\hat{J}_0^2-\overline{J_0}^2|\Phi_0\rangle=\left(\frac{N}{2}\right)^2-\left(\frac{N}{2}\right)^2=0,\end{aligned}$$

7.4 The Elementary Excitations

An elementary excitation of an N-spin system is $\hat{s}_+^{(\alpha)}|\Phi_0\rangle$, the lifting of one spinor α from $\downarrow$ to $\uparrow$. There are N different basic excitations. All α are equivalent in the LMG model. We have learned in the tight-binding model for benzene, see (4.22), and we will find for the dynamics of the LMG model (see Sect. 8.3.7) that the excitation eigenmodes of such symmetric systems are superpositions with cyclic symmetry, i.e., $\hat{A}_\nu^\dagger=\sum_{\alpha=1}^N e^{\mathrm{i}\nu 2\pi/N}\hat{s}_+^{(\alpha)}$ with $\nu=0,1,\dots N-1$. It is obvious that mode $\nu=0$ which excites all α-modes with equal phase and weight plays a preferred role. It is $A_\nu^\dagger=\hat{J}_+$, just the operator which appears in the interaction term in the Hamiltonian (7.1). Thus we can simplify the further considerations by dealing only with that coherent excitation

$$\hat{J}_+|\Phi_0\rangle = \sum_{\alpha=1}^{N} s_+^{(\alpha)}|\Phi_0\rangle = \sum_{\alpha=1}^{N} \hat{a}^\dagger_{+1,\alpha}\hat{a}_{-1,\alpha}|\Phi_0\rangle.$$

This corresponds to a collective mode where each particle contributes a small piece to a large deformation of the whole system. We will see that this collective deformation exclusively accumulates all interaction effects and thus plays the dominant role in finding the HF ground state as well as collective excitation modes (see Sect. 8.3.7). The same reasoning applies when stepping up to higher excitations. We will continue the restriction to the coherent mode and consider the family of n-particle–n-hole states

$$|\Phi_n\rangle = \hat{J}_+^n|\Phi_0\rangle\sqrt{\frac{(N-n)!}{N!n!}}, \tag{7.9a}$$

with the properties

$$\langle\Phi_n|\Phi_{n'}\rangle = \delta_{nn'}, \quad \hat{H}_0|\Phi_n\rangle = \left(\frac{n}{2} - \frac{N-n}{2}\right)\varepsilon|\Phi_n\rangle. \tag{7.9b}$$

The structure (7.9) is easily proven with the equivalence to the quasispin states and the known properties of the angular-momentum stepping operators

$$\begin{aligned}
|\Phi_0\rangle \equiv |J,-J\rangle, \quad |\Phi_n\rangle &\equiv |J,\underbrace{-J+n}_{M}\rangle,\\
|J\,M{+}1\rangle &= \hat{J}_+|J\,M\rangle\frac{1}{\sqrt{(M{+}1)(2J{-}M)}},\\
|J\,M{-}1\rangle &= \hat{J}_-|J\,M\rangle\frac{1}{\sqrt{M(2J{+}1{-}M)}}.
\end{aligned} \tag{7.10}$$

Orthonormality thus holds by construction and the normalization factor $\sqrt{(N-n)!/N!n!}$ is easily derived by iterating the factors in each step.

7.5 Exact Solution

The exact solution is a superposition of all conceivable excitations. We know that it will preserve the α-symmetry of the Hamiltonian (7.1), so that it can only be composed of the coherent excitations (7.9b). What remains is the rather simple expansion

$$|\Psi\rangle = c_0|\Phi_0\rangle + c_1|\Phi_1\rangle + c_2|\Phi_2\rangle + \cdots = \sum_{n=0}^{N} c_n|\Phi_n\rangle. \tag{7.11a}$$

The expansion coefficients are determined in standard manner [24] by the Schrödinger equation in matrix form

$$\sum_{n'} H_{nn'}c_{n'} = Ec_{n'}\ , \qquad H_{nn'} = \langle\Phi_n|\hat{H}|\Phi_{n'}\rangle. \tag{7.11b}$$

The matrix elements are trivial for the unperturbed Hamiltonian $\hat{H}_0$ because the $|\Phi_n\rangle$ are eigenstates, see (7.9b). For the two-body interaction we first derive, invoking (7.9a) and (7.10),

$$\begin{aligned}
\langle\Phi_n|\hat{J}_+^2|\Phi_{n'}\rangle &= \delta_{n,n'+2}\sqrt{n(n-1)(N+2-n)(N+1-n)},\\
\langle\Phi_n|\hat{J}_-^2|\Phi_{n'}\rangle &= \delta_{n,n'-2}\sqrt{n'(n'-1)(N+2-n')(N+1-n')}\,,\\
\langle\Phi_n|\hat{J}_-\hat{J}_+|\Phi_{n'}\rangle &= \delta_{n,n'}(n+1)(N-n)\,,
\end{aligned}$$

and decompose

$$\hat{V} = 2G\hat{J}_x^2 = \frac{G}{2}\left(\hat{J}_+^2 + \hat{J}_-^2 + 2\hat{J}_0 + 2\hat{J}_-\hat{J}_+\right).$$

Altogether this yields the Hamiltonian matrix elements

$$\begin{aligned}
\langle\Phi_n|\hat{H}|\Phi_{n'}\rangle = \delta_{n,n'}&\left[\left(-\frac{N}{2}+n\right)\varepsilon + G\left(-\frac{N}{2}+n+(n+1)(N-n)\right)\right]\\
&+\delta_{n,n'+2}\frac{G}{2}\sqrt{n(n-1)(N+2-n)(N+1-n)}\\
&+\delta_{n,n'-2}\frac{G}{2}\sqrt{n'(n'-1)(N+2-n')(N+1-n')}.
\end{aligned} \tag{7.12}$$

There is no simple closed form for the solution of (7.11b) with the matrix elements (7.12), but the resulting $N \times N$ eigenvalue problem can easily be solved on a computer, thus providing exact solutions as benchmarks for testing approximations.

7.6 Hartree–Fock Solution

7.6.1 The Space of Slater States in the LMG

The Hartree–Fock approximation seeks to find an optimum solution to the full Hamiltonian $\hat{H}$ in the space of independent particle states, called Slater states. The aim of this section is to develop a convenient parametrization of the space of Slater states in the LMG model. The most general Slater state in the LMG model is composed of transformed single-particle states as

$$|\Phi\rangle = \hat{b}^\dagger_{-1,1} \ldots \hat{b}^\dagger_{-1,N}|0\rangle, \quad \hat{b}^\dagger_{m\alpha} = \sum_{m'\alpha} u^{(\alpha)}_{m,m'} \hat{a}^\dagger_{m'\alpha}.$$

It is natural to confine the considerations to "relevant" Slater states, i.e., those states which preserve the α-symmetry of the LMG model. This means that the transformation becomes independent of α, i.e., $u^{(\alpha)}_{m,m'} \longrightarrow u_{m,m'}$, and $u_{m,m'}$ is the matrix of a 2×2 unitary transformation $\hat{U}$, which can be expressed through the generators of the Lie group SU(2) (i.e., the spin operators) as a rotation characterized by three Euler angles [30, 70] as

$$\hat{b}^\dagger_{m\alpha} = e^{\mathrm{i}\phi\hat{s}^{(\alpha)}_z} e^{\mathrm{i}\theta\hat{s}^{(\alpha)}_y} e^{\mathrm{i}\psi\hat{s}^{(\alpha)}_z} \hat{a}^\dagger_{m\alpha} e^{-\mathrm{i}\psi\hat{s}^{(\alpha)}_z} e^{-\mathrm{i}\theta\hat{s}^{(\alpha)}_y} e^{-\mathrm{i}\phi\hat{s}^{(\alpha)}_z}.$$

The fermion operator $\hat{a}^\dagger_{m\alpha}$ generates an eigenstate of $\hat{s}^{(\alpha)}_z$. Thus the innermost rotation about the angle ψ induces only an irrelevant phase factor. Moreover, we exploit the fact that all $\hat{s}^{(\alpha')}_\nu$ commute with $\hat{a}^\dagger_{m\alpha}$ and $\hat{s}^{(\alpha)}_\nu$ for $\alpha \neq \alpha'$ to define a coherent rotation for all α as

$$\hat{b}^\dagger_{m\alpha} = \hat{U}(\theta,\phi)\hat{a}^\dagger_{m\alpha}\hat{U}^\dagger(\theta,\phi), \tag{7.13a}$$

$$\hat{U}(\theta,\phi) = \exp(\mathrm{i}\phi\hat{J}_z)\exp(\mathrm{i}\theta\hat{J}_y), \; -\pi < \theta \leq \pi, \; -\frac{\pi}{2} < \phi \leq \frac{\pi}{2}. \tag{7.13b}$$

This allows a compact expression of the correspondingly transformed Slater state

$$\begin{aligned} |\Phi(\theta,\phi)\rangle &= \hat{b}^\dagger_{-1,1} \ldots \hat{b}^\dagger_{-1,N}|0\rangle \\ &= \hat{U}\hat{a}^\dagger_{-1,1} \underbrace{\hat{U}^\dagger \hat{U}}_{=\infty} \hat{a}^\dagger_{-1,2}\hat{U}^\dagger \ldots \hat{U}\hat{a}^\dagger_{-1,N} \underbrace{\hat{U}^\dagger|0\rangle}_{=|0\rangle} = \hat{U}\hat{a}^\dagger_{-1,1} \ldots \hat{a}^\dagger_{-1,N}|0\rangle, \end{aligned}$$

which finally reads

$$|\Phi(\theta,\phi)\rangle = \hat{U}(\theta,\phi)|\Phi_0\rangle = \exp(\mathrm{i}\phi\hat{J}_z)\exp(\mathrm{i}\theta\hat{J}_y)|\Phi_0\rangle. \tag{7.13c}$$

A word is in order about the nature of this transformation. The basic operators $\hat{J}_i$, as given in (7.1) and (7.4), are composed as

$$\hat{J}_z = \frac{1}{2}\sum_{\alpha=1}^{N} \left[\hat{a}^\dagger_{+1,\alpha}\hat{a}_{-1,\alpha} + \hat{a}^\dagger_{-1,\alpha}\hat{a}_{+1,\alpha}\right].$$

Each submode α contributes identically to the whole operator. Correspondingly, the unitary transformation, e.g., with $\exp(\mathrm{i}\theta\hat{J}_y)$ produces a collective deformation. This becomes apparent in the limit of small θ, i.e.,

$$\begin{aligned}\exp(\mathrm{i}\theta\hat{J}_y)|\Phi_0\rangle &\approx |\Phi_0\rangle - \frac{\mathrm{i}\theta}{2}\sum_\alpha\left[\hat{a}^\dagger_{+1,\alpha}\hat{a}_{-1,\alpha} - \hat{a}^\dagger_{-1,\alpha}\hat{a}_{+1,\alpha}\right]|\Phi_0\rangle \\ &= |\Phi_0\rangle - \frac{\mathrm{i}\theta}{2}\sum_\alpha \hat{a}^\dagger_{+1,\alpha}\hat{a}_{-1,\alpha}|\Phi_0\rangle.\end{aligned}$$

This describes a coherent superposition of $1ph$ states, where each $1ph$ configuration contributes the same small amount. The N available $1ph$ excitations together cooperate coherently to produce a large change in the wave function. This can be read off from the factor $N/2$ in the expectation value $\overline{J}_z = \sin(\theta)\frac{N}{2}$, see (7.22). It often happens that the residual two-body interaction acts particularly strongly between collective excitations which, in turn, leads to the appearance of collective modes with low excitation energies and/or large transition strengths. This aspect of collective dynamics will be discussed for the example of the LMG model in Chap. 8.

7.6.2 Expectation Values with the Transformed State

In this section, we are going to evaluate the expectation values of the basic operators $\hat{J}_i$ and their products for the transformed state (7.13). The general problem is to compute the expectation value for any observable $\hat{A}$, i.e.,

$$\overline{A} = \langle\Phi(\theta,\phi)|\hat{A}|\Phi(\theta,\phi)\rangle = \langle\Phi_0|\hat{U}^\dagger(\theta,\phi)\hat{A}\hat{U}(\theta,\phi)|\Phi_0\rangle, \tag{7.14}$$

with $\hat{U} = \hat{U}(\theta,\phi)$ as given in (7.13). There are several ways to perform this task. The most efficient is to evaluate the unitary transformation on the observable. One starts from the expression of $\hat{A}$ in terms of $\hat{J}_\nu$, transforms it with $\hat{U}^\dagger\hat{J}_\nu\hat{U}$, and then evaluates it with the known expectation values over $|\Phi_0\rangle$. To that end, we first compute the back-transformed basic operators of the LMG algebra. These become

$$\hat{U}^\dagger(\theta,\phi)\hat{J}_x\hat{U}(\theta,\phi) = \cos(\theta)\hat{J}_x - \mathrm{i}\sin(\theta)\sin(\phi)\hat{J}_y - \sin(\theta)\cos(\phi)\hat{J}_z, \tag{7.15a}$$
$$\hat{U}^\dagger(\theta,\phi)\hat{J}_y\hat{U}(\theta,\phi) = \cos(\phi)\hat{J}_y + \mathrm{i}\sin(\phi)\hat{J}_z, \tag{7.15b}$$
$$\hat{U}^\dagger(\theta,\phi)\hat{J}_z\hat{U}(\theta,\phi) = \sin(\theta)\hat{J}_x + \mathrm{i}\cos(\theta)\sin(\phi)\hat{J}_y + \cos(\theta)\cos(\phi)\hat{J}_z, \tag{7.15c}$$

with $\hat{U}(\theta,\phi)$ as given in (7.13).

This result will now be proven in detail. In a first step, we provide the transformation for one single spin using the properties $[\hat{s}_x,\hat{s}_y] = \mathrm{i}\hat{s}_z$ and $\hat{s}_y\hat{s}_x\hat{s}_y = -\hat{s}_x$ and similarly for the other combinations. This yields

$$
\begin{aligned}
e^{-\mathrm{i}\theta\hat{s}_y}\hat{s}_x e^{\mathrm{i}\theta\hat{s}_y} &= (\cos(\tfrac{\theta}{2}) - \mathrm{i}\sin(\tfrac{\theta}{2})\hat{s}_y)\hat{s}_x(\cos(\tfrac{\theta}{2}) + \mathrm{i}\sin(\tfrac{\theta}{2})\hat{s}_y) \\
&= \hat{s}_x\left(\cos^2(\tfrac{\theta}{2}) - \sin^2(\tfrac{\theta}{2})\right) - \hat{s}_z\sin(\tfrac{\theta}{2})\cos(\tfrac{\theta}{2}) = \hat{s}_x\cos(\theta) - \hat{s}_z\sin(\theta), \\
e^{-\mathrm{i}\theta\hat{s}_y}\hat{s}_y e^{\mathrm{i}\theta\hat{s}_y} &= \hat{s}_y, \\
e^{-\mathrm{i}\theta\hat{s}_y}\hat{s}_z e^{\mathrm{i}\theta\hat{s}_y} &= (\cos(\tfrac{\theta}{2}) - \mathrm{i}\sin(\tfrac{\theta}{2})\hat{s}_y)\hat{s}_z(\cos(\tfrac{\theta}{2}) + \mathrm{i}\sin(\tfrac{\theta}{2})\hat{s}_y) = \hat{s}_z\cos(\theta) + \hat{s}_x\sin(\theta).
\end{aligned}
$$

This allows to evaluate the full back transformation for θ-rotation as

$$
\begin{aligned}
\exp(-\mathrm{i}\theta\hat{J}_y)\hat{J}_x\exp(\mathrm{i}\theta\hat{J}_y) &= \cos(\theta)\hat{J}_x - \sin(\theta)\hat{J}_0, \\
\exp(-\mathrm{i}\theta\hat{J}_y)\hat{J}_y\exp(\mathrm{i}\theta\hat{J}_y) &= 0, \\
\exp(-\mathrm{i}\theta\hat{J}_y)\hat{J}_z,\exp(\mathrm{i}\theta\hat{J}_y) &= \cos(\theta)\hat{J}_0 + \sin(\theta)\hat{J}_x,
\end{aligned}
$$

$$
\Longrightarrow
$$

$$
\mathrm{e}^{-\mathrm{i}\theta\hat{J}_y}\begin{pmatrix}\hat{J}_x\\ \hat{J}_y\\ \hat{J}_z\end{pmatrix}\mathrm{e}^{\mathrm{i}\theta\hat{J}_y} = \begin{pmatrix}\cos(\theta) & 0 & -\sin(\theta)\\ 0 & 1 & 0\\ \sin(\theta) & 0 & \cos(\theta)\end{pmatrix}\begin{pmatrix}\hat{J}_x\\ \hat{J}_y\\ \hat{J}_z\end{pmatrix}.
$$

With very similar steps (not detailed here) the $\hat{J}_x$ transformation about rotation angle ϕ is found to be

$$
\mathrm{e}^{-\mathrm{i}\phi\hat{J}_z}\begin{pmatrix}\hat{J}_x\\ \hat{J}_y\\ \hat{J}_z\end{pmatrix}\mathrm{e}^{\mathrm{i}\phi\hat{J}_z} = \begin{pmatrix}\cos(\phi) & \sin(\phi) & 0\\ -\sin(\phi) & \cos(\phi) & 0\\ 0 & 0 & 1\end{pmatrix}\begin{pmatrix}\hat{J}_x\\ \hat{J}_y\\ \hat{J}_z\end{pmatrix}.
$$

The two results are concatenated to the full transformation with the full $\hat{U}$ as given in (7.13)

$$
\begin{aligned}
\hat{U}^\dagger\begin{pmatrix}\hat{J}_x\\ \hat{J}_y\\ \hat{J}_z\end{pmatrix}\hat{U} &= \begin{pmatrix}\cos(\phi) & \sin(\phi) & 0\\ -\sin(\phi) & \cos(\phi) & 0\\ 0 & 0 & 1\end{pmatrix}\begin{pmatrix}\cos(\theta) & 0 & -\sin(\theta)\\ 0 & 1 & 0\\ \sin(\theta) & 0 & \cos(\theta)\end{pmatrix}\begin{pmatrix}\hat{J}_x\\ \hat{J}_y\\ \hat{J}_z\end{pmatrix} \\
&= \begin{pmatrix}\cos(\vartheta)\cos(\phi) & \sin(\phi) & -\sin(\vartheta)\cos(\phi)\\ -\cos(\vartheta)\sin(\phi) & \cos(\phi) & \sin(\vartheta)\sin(\phi)\\ \sin(\vartheta) & 0 & \cos(\vartheta)\end{pmatrix}\begin{pmatrix}\hat{J}_x\\ \hat{J}_y\\ \hat{J}_z\end{pmatrix},
\end{aligned}
$$

This result immediately leads to the form given in (7.15).

The expectation values of the $\hat{J}_i$ can now be evaluated very simply using (7.15) and the basic relations $\langle\Phi_0|\hat{J}_x|\Phi_0\rangle = 0$, $\langle\Phi_0|\hat{J}_y|\Phi_0\rangle = 0$, $\langle\Phi_0|\hat{J}_0|\Phi_0\rangle = -N/2$, as can be seen from (7.7) and (7.8b). For the transformed state this results in $|\Phi\rangle = |\Phi(\theta,\phi)\rangle$ and

$$\overline{J}_x = \langle\Phi|\hat{J}_x|\Phi\rangle = \sin(\theta)\cos(\phi)\frac{N}{2}, \tag{7.16a}$$

$$\overline{J}_y = \langle\Phi|\hat{J}_y|\Phi\rangle = -\sin(\vartheta)\sin(\phi)\frac{N}{2}, \tag{7.16b}$$

$$\overline{J}_0 = \langle\Phi|\hat{J}_0|\Phi\rangle = -\cos(\theta)\frac{N}{2}. \tag{7.16c}$$

For the total energy, we will also need the expectation value of $\hat{J}_x^2$. We decompose $\langle\Phi_0|\hat{U}^\dagger\hat{J}_x^2\hat{U}|\Phi_0\rangle = \langle\Phi_0|(\hat{U}^\dagger\hat{J}_x\hat{U})^2|\Phi_0\rangle$, use (7.15a), the expectation values (7.8c), $\langle\Phi_0|\{\hat{J}_x,\hat{J}_y\}|\Phi_0\rangle = 0$, and $\langle\Phi_0|\hat{J}_0^2|\Phi_0\rangle = N^2/4$. This yields

$$\overline{J_x^2} = \langle\Phi|\hat{J}_x^2|\Phi\rangle = \frac{N}{4}\left(\cos^2(\theta) - \sin^2(\theta)\sin^2(\phi)\right) + \frac{N^2}{4}\sin^2(\theta)\cos^2(\phi). \tag{7.17}$$

Note that the leading term $\propto N^2$ is eliminated when computing the variance $\overline{\Delta^2 J_x} = \overline{J_x^2} - \overline{J_x}^2$.

7.6.3 The Energy Landscape and the HF Minimum

The aim now is to compute the total energy for the transformed state (7.13), i.e.,

$$E(\theta,\phi) = \langle\Phi|\hat{H}|\Phi\rangle = \langle\Phi_0|\hat{U}^\dagger\hat{H}\hat{U}|\Phi_0\rangle,$$

for the LMG Hamiltonian

$$\hat{H} = \varepsilon\hat{J}_0 + 2G\hat{J}_x^2. \tag{7.18}$$

All necessary expectation values have been evaluated in the previous section, see (7.16) and (7.17). The energy thus becomes immediately

$$E = \frac{\varepsilon N}{2}\left[-\cos(\theta) + \chi\frac{\cos^2(\theta) - \sin^2(\theta)}{2(N-1)} + \frac{\chi}{2}\sin^2(\theta)\cos^2(\phi)\right], \tag{7.19a}$$

$$\chi = \frac{2G(N-1)}{\varepsilon}, \tag{7.19b}$$

where χ is an effective interaction strength, the two-body interaction strength G properly scaled in the units of the system. It is interesting to note that the effective strength is a product of G and particle number $N-1$. The relation between the zeroth-order contribution ($\propto \hat{H}_0$) and that of the interaction can be adjusted by changing G and/or changing the system size N.

The solution of the HF problem becomes particularly simple in this model. It amounts to finding the minimum of the energy (7.19) within the limits $-\pi < \theta \le \pi$,

$-\frac{\pi}{2} < \phi \leq \frac{\pi}{2}$. The minimum always occurs at $\phi = 0$. This is obvious for $G \leq 0$, i.e., $\chi \leq 0$, because both $\cos(\phi)$ and $\cos^2(\phi)$ have their maxima there with value 1. The arguments are more involved for large positive couplings $G > 0$. Without going into the details we simply assert that here also $\phi = 0$ is related to minimal energy. What remains is basically a one-dimensional energy landscape,

$$E = \frac{\varepsilon N}{2}\left[-\cos(\theta) + \frac{\chi}{2}\sin^2(\theta) + \chi\frac{\cos^2(\theta) - \sin^2(\theta)}{2(N-1)}\right], \tag{7.20}$$

which is well suited for a detailed discussion of the HF equations. We usually assume large N and drop the third term. Figure 7.3 illustrates it for a variety of effective coupling strengths. The energy function is 2π periodic and so are the Slater states, i.e., $|\Phi(\theta + 2\pi, \phi)\rangle = |\Phi(\theta, \phi)\rangle$. The figure shows one relevant interval for θ in which all states are truly different. The unperturbed case ($\chi = 0$) is a simple cosine function with one minimum at $\theta = 0$ which represents, of course, the unperturbed ground state. Positive couplings $\chi > 0$ maintain $\theta = 0$ as the minimum, even enhance the curvature there. The HF solution still remains $|\Phi_0\rangle$. Negative χ start to soften the minimum until a curvature of zero is reached at a critical coupling strength $\chi = -1$. Below this, a new minimum forms at finite angle θ. From a formal point of view, the HF equation for the LMG model is the variational equation

$$\partial_\theta E = 0 \quad \Longrightarrow \quad \sin(\theta)\big(1 + \chi\cos(\theta)\big) = 0\,, \tag{7.21}$$

which determines the stationary points in the energy landscape. There is always the solution $\sin(\theta) = 0 \Longrightarrow \theta = 0$ or π. We see from Fig. 7.3 that $\theta = 0$ indeed represents the HF minimum for $\chi > -1$ and becomes a maximum (i.e., an unstable solution) for $\chi < -1$. The other solution $\theta = \pi$ is a maximum for $\chi < 1$ and turns into a minimum for $\chi > 1$. This implies that for $\chi > 1$ there is a second,

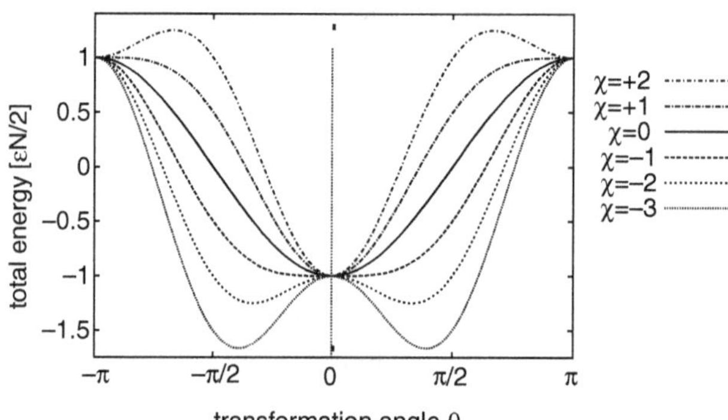

Fig. 7.3 The total energy (7.19) of the LMG model versus transformation angle θ for various choices of the effective coupling strength χ as indicated

stable minimum which, however, never becomes the deepest one. Such a minimum is called an isomer. A second branch of solutions emerges from the second factor in condition (7.21), i.e., $1 + \chi\cos(\theta) = 0 \Longrightarrow \cos(\theta) = -1/\chi$. This condition cannot be met for $|1/\chi| > 1$, i.e., $|\chi| < 1$. In this regime, there remain only the two solutions $\theta = 0$ and π as discussed above. Two new solutions come up for $|\chi| > 1$ at the finite angles $\theta = \pm\arccos(-\chi^{-1})$. These are minima for $\chi < -1$. As soon as these minima appear, the formerly deepest minimum $\theta = 0$ turns into a maximum, see Fig. 7.3. For $\chi > 1$, we obtain two new maxima while the former maximum at $\theta = \pm\pi$ is turned into an (isomeric) minimum.

The very simple LMG model thus already shows a rich variety of solutions of the HF equation (7.21). We see that this equation is only a necessary condition for finding a HF minimum. It embraces all stationary states. Maxima (or saddle points) are unstable solutions, and isomeric minima have to be kept in mind as well. The unstable solutions are rarely a problem in practical numerical solution schemes which can only end up in a stable solution. The isomeric minima remain a source of uncertainty. Realistic systems with their huge number of degrees of freedom have very complex energy landscapes in which it is never certain that the deepest minimum has in fact been found. The safeguards against this irreducible uncertainty are physical intuition, experience, diligence, and a healthy amount of residual skepticism.

Figure 7.4 compares the HF solution with the exact one as function of the dimensionless coupling strength χ for a small system $N = 8$ and a medium large one with $N = 32$. The comparison concentrates on the region of the critical point near $\chi = -1$ where the HF minimum starts to develop a finite deformation θ. The lower panels of Fig. 7.4 compare the total energies. HF turns out to generally be a good approximation. The largest deviations can be seen near the critical point and these shrink with increasing system size N. The exact energy has a much smoother trend than the HF energy, which has, in fact, a non-analytical point at the critical χ, namely, a discontinuous derivative. This becomes more apparent in the two detailed observables shown in the other panels. The expectation value of $\hat{J}_x$ in the middle panels exhibits a dramatic difference: the exact solution maintains $\langle\hat{J}_x\rangle = 0$ throughout, while HF develops a large finite value. This is due to a spontaneous symmetry breaking induced by the mean-field approximation and will be discussed further in Sect. 7.7.1. The upper panel shows the expectation value of the squared operator $\hat{J}_x^2$. This quantity develops large values in the critical regime; trends and values are remarkably similar for the exact solution and HF. Differences are largest near the critical point where HF has a discontinuous first derivative while the exact result is smooth. It is instructive to decompose $\langle\hat{J}_x^2\rangle = \langle\hat{J}_x\rangle^2 + \langle\Delta\hat{J}_x^2\rangle$. The exact solution has $\langle\hat{J}_x\rangle = 0$ and all $\langle\hat{J}_x^2\rangle$ are produced from the (large) variance $\langle\Delta\hat{J}_x^2\rangle$, while the mean-field solution has the dominant contribution from $\langle\hat{J}_x\rangle^2$. This suggests that the finite deformation with accompanying finite $\langle\hat{J}_x\rangle$ carries physical information hinting at the exact result for the squared expectation value $\langle\hat{J}_x^2\rangle$.

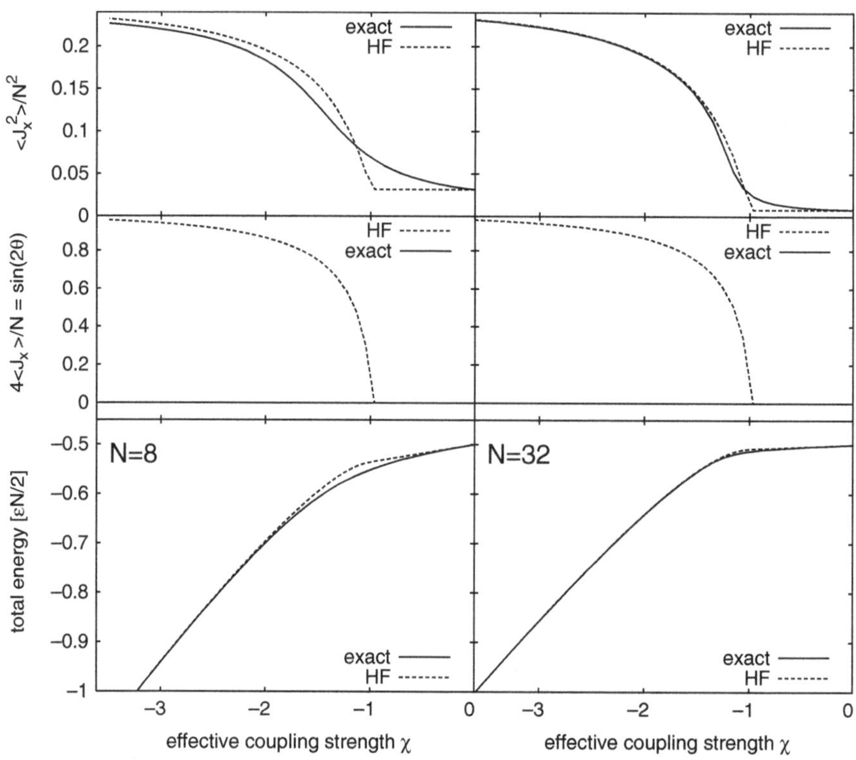

Fig. 7.4 Comparison of the exact solution with HF for the LMG model and for two different particle numbers, $N = 8$ on the *left panels* and $N = 32$ to the *right*. The *upper most panels* show the expectation value $\overline{J_x^2}$, the *middle panels* $\overline{J_x}$, and the *lowest* ones the total energy

7.7 Spontaneous Symmetry Breaking

7.7.1 Violation of "Reflection Symmetry" in the LMG Model

The LMG model is extremely rich and allows much more than just a discussion of the structure of a HF solution. It also comprises concepts such as collective deformation (see Sect. 7.6.1), spontaneous symmetry breaking, symmetry restoration, or various aspects of dynamics. The dynamical aspects will be taken up in Chap. 8. Here, we will continue with the structural features of symmetry breaking. For this purpose, it suffices to consider only Slater states (7.13) with $\phi = 0$, as done already when discussing the HF solutions.

We start again by having a quick look at Fig. 7.3 showing the energy as function of deformation angle θ for various coupling strengths. The patterns are symmetric under "reflection," i.e., under the transformation $\theta \longrightarrow -\theta$. It is important to point out that the Slater states change under reflection, i.e., $|\Phi(-\theta)\rangle \neq |\Phi(+\theta)\rangle$, while both states have the same energy $E(-\theta) = E(+\theta)$. We already know the $\alpha \longleftrightarrow \alpha'$-

symmetry of the LMG model and have exploited it extensively. Now there appears a further symmetry in the LMG Hamiltonian related to reflection $\theta \longrightarrow -\theta$. First recall the expectation values of the basic observables $\hat{J}_i$ for the case $\phi = 0$. Specializing the expressions (7.16) yields

$$\overline{J}_x = \sin(\theta)\frac{N}{2}, \quad \overline{J}_y = 0, \quad \overline{J}_0 = -\cos(\theta)\frac{N}{2}. \tag{7.22}$$

The operation $\theta \longrightarrow -\theta$ is obviously related to reflection of the $\hat{J}_x$ observable, i.e., $\hat{J}_x \longrightarrow -\hat{J}_x$, while $\hat{J}_0$ remains unchanged. A quick glance at the LMG Hamiltonian (7.18) shows that it has precisely this reflection symmetry because $\hat{J}_x$ appears squared. The symmetry property can thus be formalized by introducing the operation $\hat{\Pi}$ of reflection which is defined via

$$\hat{\Pi}\hat{J}_x\hat{\Pi} = -\hat{J}_x , \ \hat{\Pi}\hat{J}_y\hat{\Pi} = -\hat{J}_y , \ \hat{\Pi}\hat{J}_0\hat{\Pi} = \hat{J}_0 , \ \hat{\Pi}^\dagger = \hat{\Pi} . \tag{7.23a}$$

The action on a Slater state is

$$\hat{\Pi}|\Phi(\theta)\rangle = |\Phi(-\theta)\rangle. \tag{7.23b}$$

Note that the Slater states are generally not symmetric under reflection, except for the unperturbed ground state $|\Phi_0\rangle$. The reflection symmetry of the Hamiltonian is formulated compactly as

$$\hat{\Pi}\hat{H}\hat{\Pi} = \hat{H}. \tag{7.23c}$$

Whatever the detailed form of $\hat{H}$ might be, the symmetry property (7.23c) suffices to obtain the feature

$$E(-\theta) = \langle\Phi(-\theta)|\hat{H}|\Phi(-\theta)\rangle = \langle\Phi(\theta)|\underbrace{\hat{\Pi}\hat{H}\hat{\Pi}}_{\hat{H}}|\Phi(\theta)\rangle = E(\theta). \tag{7.24}$$

There are now two basically different scenarios conceivable: the first one has the energy minimum occurring at $\theta = 0$ (as it does in the LMG model for undercritical coupling). In this case the ground state is unique and symmetric (because of $\hat{\Pi}|\Phi_0\rangle = |\Phi_0\rangle$). More surprising is the second scenario. The energy minimum appears at a finite angle $\theta_{\min} \neq 0$. Now there is a pair of HF states each of which violates reflection symmetry but which are related to each other by the symmetry operation

$$|\Phi(-\theta_{\min})\rangle \neq |\Phi(+\theta_{\min})\rangle \quad \longleftrightarrow \quad E_{\mathrm{HF}} = E(+\theta_{\min}) = E(-\theta_{\min}). \tag{7.25}$$

They are thus necessarily degenerate in energy. Each single HF state $|\Phi(\pm\theta_{\min})\rangle$ breaks the reflection symmetry. The transition from the first (and symmetric) scenario to the second, asymmetric case is triggered by a change in the coupling strength which turns the stable minimum at $\theta = 0$ to a maximum, i.e., to an unstable point in the energy landscape (see Fig. 7.3 in the step from $\chi > -1$ to $\chi < -1$). This is the typical mechanism for spontaneous symmetry breaking which occurs in many different places in physics. In the following we will briefly discuss the Jahn–Teller effect for clusters and molecules (Sect. 7.7.3), the competition between spherical and deformed nuclei (Sect. 7.7.2) as examples. A further important example is the pairing transition to a condensate of pairs of fermions which will be considered extensively in Sect. 9.4. Further examples are the laser transition from stochastic to coherent light [45], the Higgs mechanism, and chiral symmetry breaking in field theory [116].

7.7.2 Transition from Spherical to Deformed Nuclei

The trend of nuclear observables with changing proton or neutron number shows that there is often a transition from spherical shape to stable ground-state deformation. This is a shell effect as discussed extensively in Chap. 3. Magic nucleon numbers, which form a closed shell for a spherical mean field, lead to a spherical ground state, while any non-magic nucleon number drives to deformation. Reality is more involved than this simple picture. Figure 7.5 shows quadrupole excitation energies (lower panel) and quadrupole ground-state deformations (upper panel) along a chain of Sr isotopes (proton number $Z = 38$). The deformations are expressed in terms of the axis ratio $\langle z^2 \rangle$ relative to $\langle x^2 \rangle$. A value of $\eta = 0$ signifies spherical shape, $\eta > 1$ prolate deformations, and $\eta < 0$ oblate ones. The deformation should be viewed similar to $\langle \hat{J}_x^2 \rangle$ in the LMG model (upper panel of Fig. 7.4).

The pattern in Fig. 7.5 clearly indicates a region of spherical nuclei and a transition to deformed ones, which is corroborated by the trend of the quadrupole

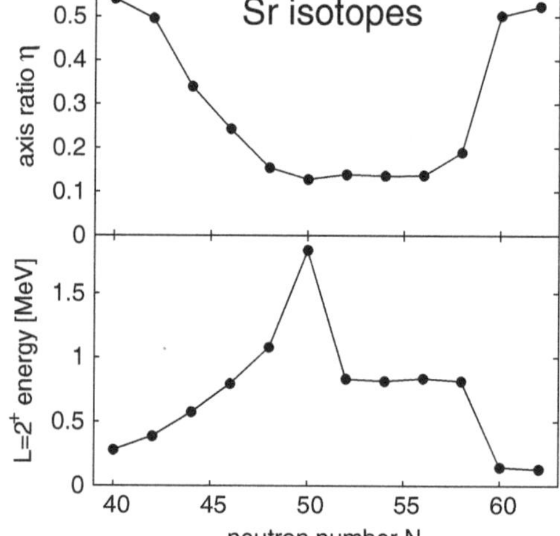

Fig. 7.5 Experimental data in Sr isotopes [86]. *Lower*: excitation energies of the lowest 2^+ modes. *Upper*: nuclear deformations expressed as axis ratio $\eta = \langle z^2 \rangle / \langle x^2 \rangle - 1$

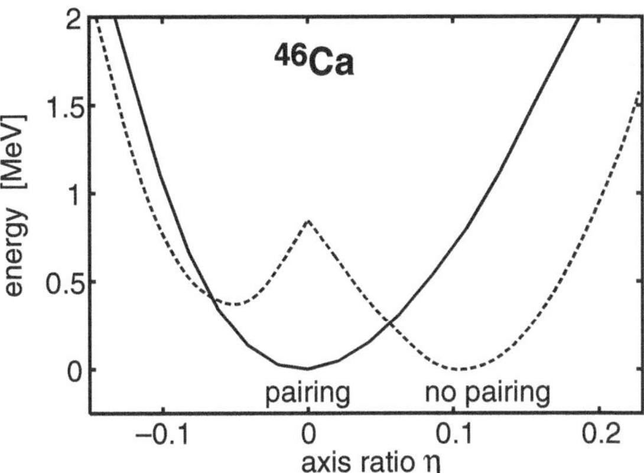

Fig. 7.6 Quadrupole deformation energy curves for the nucleus ^{46}Ca computed with quadrupole constrained Skyrme–Hartree–Fock using the parametrization SLy6 [22]. Compared are a calculation using standard BCS pairing (*solid line*) with one without pairing (*dashed*), see Chap. 9. The quadrupole deformation is characterized by the axis ratio $\eta = \langle z^2 \rangle / \langle x^2 \rangle - 1$

excitation energies in the lower panel. A transition to deformed nuclei is expected in view of the shell effects discussed in Sect. 3. Pure shell effects, however, would leave only the magic $N = 50$ nuclei spherical and induce deformation immediately in the vicinity. What is seen instead is a whole region of spherical nuclei next to the magic point. The situation is qualitatively similar to the trend in the LMG model where the minimum at $\theta = 0$ stays stable for a long time until the interaction becomes strong enough to overrule the spectral gap. From analogy we conclude that there must be a basic gap in the nuclear case which stabilizes sphericity for a while before the shell effects take over. This is the pairing gap (see Sect. 9.4). This situation is called a *dynamical Jahn–Teller effect* [32, 88] where a residual interaction (here pairing) can counterweight the spontaneous symmetry breaking (by shell effects).

The stabilization through the pairing gap was deduced in Fig. 7.5 from an isotopic trend. A direct observation is provided in Fig. 7.6 where we show the deformation energy curves for one nucleus, ^{46}Ca, with and without pairing interaction. The deformation energies correspond to Born–Oppenheimer energy curves (see Sect. 4.2) and are produced here by adding a quadrupole constraint $-\lambda \hat{Q}_{20}$ to the mean-field Hamiltonian (Chap. 5). ^{46}Ca has magic proton number $Z = 20$ but non-magic neutron number $N = 26$. The effect of pairing seen in Fig. 7.6 is dramatic. Without pairing, the shell model drives to a well-developed deformation (as the examples in Chap. 3). The pairing interaction restores the spherical shape for this nucleus, which is still semi-magic and whose neutron number is still close to the next magic shell at $N = 28$.

Summarizing, we see that the LMG model provides a realistic mapping of the counteracting effects taking place in the nuclear phase transition.

7.7.3 The Jahn–Teller Effect

The Jahn–Teller effect in molecular and solid-state physics also deals with this sort of spontaneous symmetry breaking [32]. The typical situation is that an electronic configuration is degenerate for a symmetric shape, so that the system is driven to a lower symmetry. Electronic correlations can counteract this deformation, like pairing does in the nuclear example above, but they are usually too weak for substantial symmetry restoration. A stronger counteracting effect here is thermal excitation which also serves to produce a unique electronic configuration. We exemplify this for the simple case of a Na_{14} cluster, a type of system which has already been discussed in Chap. 3. We have learned from Fig. 3.4 that Na_{14} has an open shell (i.e., degenerate) electronic configuration at spherical shape and thus develops a substantial ground-state deformation.

Figure 7.7 shows the deformation energy curves for Na_{14}, the total energy as function of ionic deformation, computed with density functional theory at the level of LDA (see Sect. 6.1). In order to have a simple description of symmetry and symmetry breaking, we use the deformed jellium model for the ionic background. All shapes are axially symmetric and are characterized by the driving quadrupole moment α_{20} in the jellium model (1.1). The point $\alpha_{20} = 0$ corresponds to perfect spherical symmetry and $\alpha_{20} \neq 0$ to broken symmetry. The curve at temperature $T = 0$ clearly shows the shell effect, producing a deep minimum at well-deformed shapes, as discussed in connection with Fig. 3.4. The temperature $T = 0.14$ eV does already soften the trend to deformation and $T = 0.27$ eV finally produces a (shallow) minimum at spherical shape. This symmetry restoration through thermal excitation temperature is also known from the pairing transition where the pairing gap disappears at a critical temperature (see Sect. 9.4). The temperatures in the present example Na_{14} are very high, indicating that the restoration of a high symmetry, here sphericity, can be costly and corresponding to the high reaction barrier of more than 0.6 eV. Realistic examples deal with lower symmetries [32] and relate to more moderate temperatures but require a deep knowledge of the various point symmetries as they occur in molecules.

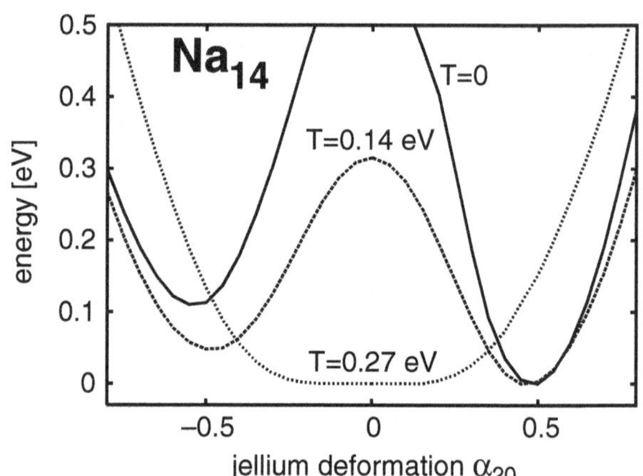

Fig. 7.7 Quadrupole deformation energy curves for the metal cluster Na_{14} computed with LDA (see Sect. 6.1) and the deformed jellium model (1.1) for the ionic background. The total energy is drawn versus the dimensionless quadrupole momentum of the jellium distribution. Results for three different temperatures are shown

7.7.4 Symmetry Restoration by Projection

In the critical regime, the HF ground state of the LMG model breaks reflection symmetry and appears as two degenerate states. The exact ground state as constructed in Sect. 7.5 is always unique and maintains symmetry. The symmetry breaking and degeneracy arises from the HF approximation. The energetic degeneracy suggests a strategy to restore at least the symmetry and so come closer to the exact ground state. It is known from quantum-mechanical perturbation theory that perturbations become arbitrarily important in case of degeneracies and that the full interaction must be diagonalized in the degenerate subspace [24]. Similarly, we here build a unique ground state from a coherent superposition of the two equivalent HF states, i.e.,

$$|\Psi_{\text{restor}}\rangle = c_+|\Phi(\theta_{\min})\rangle + c_-|\Phi(-\theta_{\min})\rangle. \tag{7.26a}$$

Symmetry arguments fix the expansion coefficients $c_\pm$ without painful diagonalization. The symmetry-restored state $|\Psi_{\text{restor}}\rangle$ should be symmetric under reflections, i.e.,

$$\hat{\Pi}|\Psi_{\text{restor}}\rangle = +|\Psi_{\text{restor}}\rangle \quad \Longrightarrow \quad c_+ = c_- . \tag{7.26b}$$

What remains is to compute the overall normalization. The energy of the symmetry-restored state becomes

$$\begin{aligned} E_{\text{restor}}(\theta) &= \frac{\langle\Psi_{\text{restor}}|\hat{H}|\Psi_{\text{restor}}\rangle}{\langle\Psi_{\text{restor}}|\Psi_{\text{restor}}\rangle} = \frac{E(\theta_{\min}) + \langle\Phi(-\theta_{\min})|\hat{H}|\Phi(\theta_{\min})\rangle}{1+\langle\Phi(-\theta_{\min})|\Phi(\theta_{\min})\rangle} \\ &= \frac{E_{\text{HF}}(\theta) - \dfrac{N\varepsilon}{2}\left(\cos^{N-1}(\theta) + \dfrac{\chi}{2(N-1)}\cos^N(\theta)\right)}{1+\cos^N(\theta)} . \end{aligned} \tag{7.26c}$$

The evaluation is done most simply by mapping into spin space. The overlap separates into single system ($\equiv$ particle) overlaps because operators for different α commute. We start with the norm overlap

$$\begin{aligned} \langle\Phi(-\theta)|\Phi(\theta)\rangle &= \langle 0|\hat{b}_{-1,1}(-\theta)\ldots\hat{b}_{-1,N}(-\theta)\hat{b}^\dagger_{-1,1}(\theta)\ldots\hat{b}^\dagger_{-1,N}(\theta)|0\rangle \\ &= \prod_{\alpha=1}^{N}\langle 0|\hat{b}_{-1,\alpha}(-\theta)\hat{b}^\dagger_{-1,\alpha}(\theta)|0\rangle \\ &\equiv \langle\chi_\downarrow|\exp(2\mathrm{i}\theta\hat{s}_y)|\chi_\downarrow\rangle^N = \langle\chi_\downarrow|(\cos(\theta)+\mathrm{i}\sin(\theta)\hat{s}_y)|\chi_\downarrow\rangle^N \\ &= \cos(\theta)^N . \end{aligned}$$

Here the well-known relations for Pauli spin matrices $\exp(2\mathrm{i}\theta\hat{s}_y) = \cos(\theta) + \mathrm{i}\sin(\theta)\hat{s}_y$ and $\langle\chi_\downarrow|\hat{s}_y|\chi_\downarrow\rangle = 0$ were used. The expectation value of $\hat{J}_0$ with a similar decomposition becomes

$$\begin{aligned}\langle\Phi(-\theta)|\hat{J}_0|\Phi(\theta)\rangle &= \sum_{\alpha=1}^{N}\langle\Phi(-\theta)|\hat{s}_0^{(\alpha)}|\Phi(\theta)\rangle \\ &\equiv N\langle\chi_\downarrow|\exp(\mathrm{i}\theta\hat{s}_y)\hat{s}_0\exp(\mathrm{i}\theta\hat{s}_y)|\chi_\downarrow\rangle\ \langle\chi_\downarrow|\exp(2\mathrm{i}\theta\hat{s}_y)|\chi_\downarrow\rangle^{N-1} \\ &= N\cos(\theta)^{N-1}\langle\chi_\downarrow|(\cos(\tfrac{\theta}{2}) + \mathrm{i}\sin(\tfrac{\theta}{2})\hat{s}_y)\hat{s}_0(\cos(\tfrac{\theta}{2}) \\ &\quad +\mathrm{i}\sin(\tfrac{\theta}{2})\hat{s}_y)|\chi_\downarrow\rangle \\ &= -\frac{N}{2}\cos(\theta)^{N-1}\left(\cos^2(\tfrac{\theta}{2}) + \sin^2(\tfrac{\theta}{2})\right) = -\frac{N}{2}\cos(\theta)^{N-1},\end{aligned}$$

where $\{\hat{s}_y, \hat{s}_0\} = 0$ and $\hat{s}_y\hat{s}_0\hat{s}_y = -\hat{s}_0$ were used. The operator $\hat{J}_x^2$ requires a few more steps. Separating

$$\begin{aligned}\langle\Phi(-\theta)|\hat{J}_x^2|\Phi(\theta)\rangle &\equiv \sum_{\alpha\neq\alpha'}\underbrace{\langle\Phi(-\theta)|\hat{s}_x^{(\alpha)}\hat{s}_x^{(\alpha')}|\Phi(\theta)\rangle}_{=0} + \sum_{\alpha}\langle\Phi(-\theta)|\underbrace{(\hat{s}_x^{(\alpha)})^2}_{=1}|\Phi(\theta)\rangle \\ N\langle\Phi(-\theta)|\Phi(\theta)\rangle &= N\cos(\theta)^N,\end{aligned}$$

the first term vanishes because of

$$\begin{aligned}\langle\chi_\downarrow|e^{\mathrm{i}\theta\hat{s}_y}\hat{s}_x e^{\mathrm{i}\theta\hat{s}_y}|\chi_\downarrow\rangle &= \langle\chi_\downarrow|(\cos(\tfrac{\theta}{2}) + \mathrm{i}\sin(\tfrac{\theta}{2})\hat{s}_y)\hat{s}_x(\cos(\tfrac{\theta}{2}) + \mathrm{i}\sin(\tfrac{\theta}{2})\hat{s}_y)|\chi_\downarrow\rangle \\ &= \underbrace{\langle\chi_\downarrow|\hat{s}_x|\chi_\downarrow\rangle}_{=0}\left(\cos^2(\tfrac{\theta}{2}) + \sin^2(\tfrac{\theta}{2})\right).\end{aligned}$$

Putting all pieces together leads to the closed expression for the projected energy as given in (7.26c).

In a first step, it seems natural to project the HF state at a given deformation $\theta_{\min}$. This is called *Projection After Variation (PAV)* [92], i.e., $E_{\mathrm{PAV}} = E_{\mathrm{restor}}(\theta_{\min})$ with the energy expression as given in (7.26c). Figure 7.8 compares the energies. As the HF result already provides reasonable quality (see Fig. 7.4), we enhance the graphical resolution by comparing the error relative to the exact result. The HF result in the figure shows, of course, the largest error and this error has a maximum near the critical point $\chi = -1$. where the optimal angle $\theta_{\min}$ starts to deviate from zero. The PAV energy follows the HF result from $\chi = 0$ up to the critical point because there is nothing to gain for $\theta_{\min} = 0$, but the PAV-projected energy yields substantial improvements near $\theta_{\min}$. There is, in fact, no need to stick to the HF angle $\theta_{\min}$. The energy (7.26c) can be considered as a freely varying function of θ and it is obvious that the optimum choice is the deformation θ which minimizes the energy. This is called *Variation After Projection (VAP)* with the energy $E_{\mathrm{VAP}} = \min_\theta\{E_{\mathrm{restor}}(\theta)\}$. VAP allows to deal with finite θ already in the under-critical regime and so can provide improvements in all regimes. That is obviously the case, as can be seen

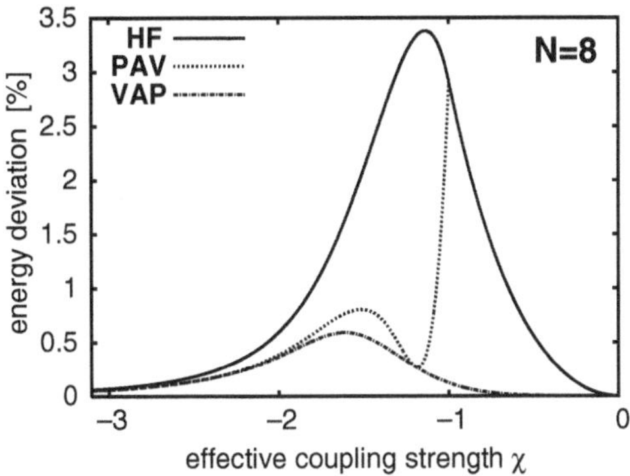

Fig. 7.8 Energy error relative to the exact solution of the LMG model for particle number $N = 8$ for three approximations: HF, parity projected HF energy state (PAV), and minimized projected energy (VAP)

from Fig. 7.8. In all regimes VAP is a big step forward toward the exact result. It incorporates a great deal of correlations, namely, those related to soft modes of the system. Practical examples are rotational projection for deformed nuclei or parity projection in symmetry-broken molecules.

7.8 Concluding Remarks

Quasispin models provide ideal simple models for the description of many physical situations in a rather realistic way, at least what concerns the understanding of shortcomings of mean-field models. Because they can be worked out analytically at a high level of detail thanks to the simplifications they are constructed on, these difficulties can be identified in an explicit and transparent manner. We have illustrated these aspects in quite some detail in this chapter, basing our discussion on the prototypical LMG model. We have discussed in detail its analytical mean-field solutions and its shortcomings and have in particular detailed the mechanism of symmetry breaking and seen how such mechanisms are observed in the case of molecules and nuclei. The LMG model can also be used in dynamical situations and again serve as a prototypical model to test linearized dynamics. These aspects will be discussed in Chap. 8.

Chapter 8
Excitation Spectra

One of the most common ways to analyze the properties of a many-fermion system is to study its excitation spectrum, which can largely be described in terms of small-amplitude oscillations. Such oscillations constitute the response of the system to a small perturbation and thus help to understand linear response dynamics. The small amplitude allows a handling by linearized analysis of motion, a generalization of the harmonic motion around the ground state in any simple system [40]. In a many-fermion system, the harmonic analysis is similar but a bit more complex because of the quantum nature and the large number of degrees of freedom involved [10]. It amounts to exploring the space of one-particle–one-hole ($1ph$) excitations about the ground state. A particularly interesting feature is that individual perturbations couple to collective (harmonic oscillations) similarly to a set of coupled oscillators. This collective (small-amplitude) motion has been widely studied in self-bound fermion systems such as nuclei, metal clusters, quantum dots, or atomic traps.

In nuclei, small-amplitude collective oscillations are known as Giant Resonances (GR) and were identified in the late 1940s [92]. They can be characterized by a multipole order depending on the global characteristics of the oscillations. The most widely studied GR is the Giant Dipole Resonance (GDR), in which the protons oscillate as a whole against the neutrons. It directly reflects the strong interaction between neutrons and protons, a key constituent of the nuclear interaction. One should also mention the Giant monopole resonance which corresponds to a breathing mode of the nucleus and gives access to the compressibility of nuclei and ultimately nuclear matter. The next higher modes are the quadrupole and octupole modes, in which the nucleus explores prolate and oblate deformations for the quadrupole and pear-shape deformations for the octupole mode, respectively. Note that in GR protons and neutrons can vibrate in phase (isoscalar) or in opposite phase (isovector), the two modes leading to different frequencies.

In metal clusters, collective motion is dominantly associated with the Mie plasmon, following the early work of Mie in 1908 in which the response of a metal sphere to light was explored [71]. The Mie plasmon corresponds to a collective oscillation of the center of mass of the electron cloud with respect to the ionic background. It is the dominant response mechanism of a metal cluster subject to an electromagnetic perturbation such as delivered by a laser or a charged projectile [89].

J.A. Maruhn et al., *Simple Models of Many-Fermion Systems*,
DOI 10.1007/978-3-642-03839-6_8,

Because lasers usually have a well-defined frequency, the response mechanisms can furthermore become resonant (resonance between collective oscillation frequency and laser frequency) and thus lead to complex dynamical scenarios. The analysis of the eigenfrequencies of the system through the Mie plasmon is thus crucial in this field. Note also that the Mie plasmon is the analogue of the nuclear GDR with which it shares many properties, such as the access it gives to the system's shape (the oscillation frequency depends on the spatial extension of the system: a long axis implies a low frequency, a small axis a high one) [117, 19].

Even if collective dynamics may provide a doorway to the understanding of the dynamical response of a system, it primarily characterizes low-energy phenomena, since by construction it is confined to small-amplitude motion. It thus also offers a unique access to ground-state properties of the system to the extent that it characterizes small vibrations around it. Not surprisingly the theory of such small-amplitude motion is thus formulated in the approximation of harmonic motion in which real time is merely contained in the actual frequencies involved. The aim of this chapter is to formulate a theory for the analysis of such small-amplitude motion. We shall derive it starting from a dynamic equation, the direct time-dependent generalization of the static Hartree–Fock approach, which then is linearized.

8.1 Collective Effects in Excitation Spectra

In previous chapters, we have presented several stages of approach to describe the ground-state structure of fermion systems in a mean-field picture, simple shell models in Chap. 3, and self-consistent models in Chap. 5. A key feature was the single-particle levels, which determine crucial quantum effects like magic numbers or Jahn–Teller deformation, see e.g., Sect. 3.2.4. One expects that the single-particle levels also play a role in excitations. This was discussed in Sect. 5.3.3 and these formal considerations point out the importance of the difference of single-particle energies but also indicate a remaining residual two-particle interaction whose effects on excitation spectra have yet to be explored. This will be done in the present chapter.

The left part of Fig. 8.1 shows as example the level scheme for the cluster Na_{93}^+. A few possible $1ph$ excitations staying close to the Fermi surface are indicated. From a pure $1ph$ picture we thus expect a fuzzy multitude of excitations with comparable (small) strength and energy of about 1 eV. The right panels of the figure show various stages for computing the optical absorption strength (for its definition see Sect. 8.3.5). The experimental strength in panel (a) exhibits a totally different pattern, a single strong peak at about 3 eV. On the other hand, the pure spectrum of $L = 1$ excitations (without dipole weight) in panel (d) confirms the naive expectation. A block of $1ph$ states around 1 eV and several others at higher energies. Panel (c) shows the dipole transition strength $S_1 = \sum_{ph} |d_{ph}|^2(\varepsilon_p - \varepsilon_h)$ for the pure $1ph$ picture. It is concentrated at low energies because higher transitions have only small dipole moments. We now take into account the residual interaction with the RPA

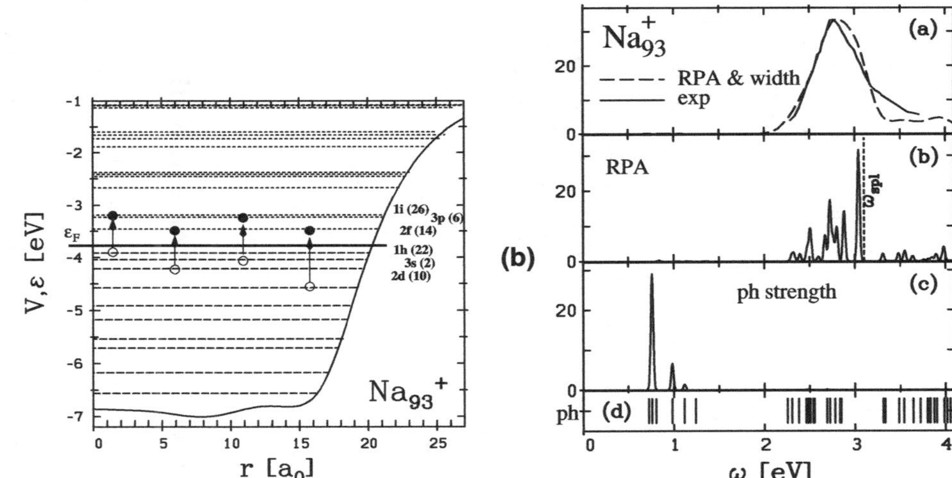

Fig. 8.1 Demonstration of the coherent collective shift in the excitation spectra for the cluster Na_{93}^{+}, computed with DFT and the spherical jellium model (1.1) for the ionic background charge. The *left* part shows the ground-state Kohn–Sham potential and corresponding single-electron levels. Occupied levels are drawn with *long dashes* and unoccupied ones with *short dashes*. The *heavy line* indicates the Fermi energy. A few $1ph$ excitations near the Fermi surface are illustrated by *arrows* and *circles*. The *right* part shows the excitation spectrum for angular momentum $L = 1$ at different levels of approach. (**d**) Pure $1ph$ energies $\varepsilon_p - \varepsilon_h$. (**c**) The dipole transition strength (optical absorption) in a pure $1ph$ picture. (**b**) The dipole strength after taking into account the residual interaction (by means of RPA as outlined below). (**a**) A thermally averaged RPA optical absorption strength (*dashed*) compared with the experimental result. Adapted from [89]

method as outlined below. This leads to the result in panel (b) which on average is already very close to experiment. In a final step, thermal shape fluctuations of the cluster are taken into account (measurements are done at about room temperature). This smoothes the calculated results to the pattern shown in panel (a), very nicely agreeing with the experimental strength.

Although the background and details of the example can only be understood at the end of this chapter, we see here that there are situations where the pure $1ph$ picture fails badly. The necessary complete treatment will be described in this section. A few more words are in order for the present example. Closer analysis will show that the strong resonance peak, called the Mie plasmon resonance (Sect. 8.4.2.2), corresponds to collective motion where all electrons close to the Fermi surface perform dipole oscillations in phase with each other. Such collective motion coherently collects all contributions from the residual interaction to eventually produce a large collective shift away from the $1ph$ energies. This appearance of collective resonances is ubiquitous in almost all fermion systems. We have seen here the surface plasmon resonance in metal clusters. In nuclei, one has the multitude of giant resonances of various multipolarities and the low-energy collectivity in rotations and vibrations. Quantum dots can show collectivity of the plasmon type as well as in spin modes. Atom clouds in traps can also rotate and vibrate as a whole. The collective energy shift can go both directions, as compared to pure ph transitions, depending on the sign of the residual interaction. Surface plasmons and nuclear giant resonances show upward shifts. Nuclear vibrations and spin modes in electronic systems are related to downward shifts. A further distinction concerns the amplitude of the motion. Modes which are up-shifted usually oscillate with small amplitudes (for simple energetic reasons) and can thus be described by linearized TDHF (=RPA) as outlined in this chapter. Note, though, that there are also some excitation modes with large amplitudes which require methods beyond RPA. These are, e.g., nuclear rotations or vibrations and in the extreme nuclear fission, see Sect. 3.3.3.3.

8.2 The Time-Dependent Hartree–Fock Approximation

The time-dependent Hartree–Fock approximation, widely known by its abbreviation TDHF appears like a simple generalization of static Hartree–Fock. We will introduce it in an intuitive way and subsequently derive it through a time-dependent variational principle. TDHF will be used as a formal basis for deriving RPA, the linearized version of TDHF. RPA can be derived in many ways but the path through TDHF provides a rather transparent access to the theory.

8.2.1 Intuitive Formulation

The TDHF approximation can be superficially obtained by replacing the stationary Schrödinger-like equation (5.15) by a time-dependent one

$$i\hbar\frac{\partial}{\partial t}\varphi_\alpha(x,t) = \hat{h}\varphi_\alpha(x,t), \tag{8.1}$$

where the mean-field operators remain the same as in Eqs. (5.15c) and (5.15d). While this may appear quite natural, it should of course be legitimated from underlying principles.

8.2.2 Time-Dependent Variational Principle

A better founded formulation of TDHF is obtained from the stationary variational principle (5.6) by replacing the energy by the time derivative

$$\delta \int dt \big(\langle \Phi | i\hbar\frac{\partial}{\partial t} - \hat{H} | \Phi \rangle\big) = 0, \tag{8.2}$$

where $\hat{H}$ is the total many-body Hamiltonian. This does reproduce the time-dependent Schrödinger equation for an unrestricted wave function. For a restricted wave function it seems to imply only that the time development should in some sense be close to that of the unrestricted one, so that there is no assurance that the approximate solution will not go totally bad as time progresses.

We can now carry through the same variation as in Chap. 5. The time derivative at first seems not to be a one-body operator, but applied to a product wave function it effectively reduces to a sum over time derivatives of the individual wave functions, so that the matrix elements can be calculated just as for the kinetic energy. As a result the Hartree–Fock conditions are generalized to (we write them immediately using the single-particle Hamiltonian)

$$\left(i\hbar\frac{\partial}{\partial t} - \hat{h}\right)_{ph} = 0 \quad \text{or} \quad \langle \Phi(t)|\hat{a}_h^\dagger \hat{a}_p \left(\mathrm{i}\partial_t - \hat{H}\right)|\Phi(t)\rangle = 0, \tag{8.3}$$

where p and h again refer to unoccupied (particle) and occupied (hole) states, respectively. The meaning of these conditions is that time development using the single-particle operator $\hat{h}$ should not mix occupied and unoccupied states. As in the static case, the simplest way to assure this is to use the single-particle states propagated by this operator, i.e., to calculate their time dependence using (8.1). Since $\hat{h}$ is a Hermitian operator, the subspaces will nicely remain orthogonal.

In practice, only the occupied states have to be followed in time, as the unoccupied ones simply span the space orthogonal to these. Note, however, that as in the static case, there is an arbitrariness: a time-dependent unitary transformation mixing the occupied states among themselves would not change the physical solution.

8.2.3 TDHF Beyond the Linear Domain

In the following the TDHF approximation will serve as a starting point for deriving the RPA which corresponds to the linearized version of TDHF. As always in linearized theories the explicit time dependence of the processes under study disappears to the benefit of an analysis of motion in terms of harmonic modes. The full TDHF approximation, however, covers a much wider range of physical situations than the ones explored in the sequel of this chapter. In particular, it allows to describe large amplitude motion and higher excitations. Typical applications for large amplitudes are nuclear fission, fusion, and fragmentation [74]. Higher excitations play a large role in laser-induced dynamics of electronic systems, e.g., in cluster dynamics [89].

It should also be noted that TDHF may serve as a basis for the development of other approximations, especially in the semi-classical domain. One major case is the Vlasov equation, which corresponds to the classical analogue of TDHF and which can be derived from TDHF for applications to fermion systems. Going to the classical (or semi-classical when properly including fermion statistics) limit implies a loss of some quantal effects (especially the ones associated with shell structure) which means that such a theory can mostly be applied to energetic situations. The Vlasov strategy is especially suited to that domain, not per se, but because it provides the most efficient basis for the inclusion of dynamical correlations through the addition of collisions terms "à la Boltzmann" [11, 89], a step which has not yet been convincingly (and practically) achieved in the fully quantal domain. Another typical approximation possibly derived from TDHF is the Time-Dependent Thomas–Fermi Approximation (TDTF) in which one imposes a specific collective form for the many-body wave function which finally reduces the dynamical equation to a fluid dynamics description of the system. This TDTF approximation has been used in many domains of physics, in nuclear physics [92], and more recently in the case of fermionic traps [38]. We will not use these approaches (Vlasov and beyond or TDTF) here and focus the forthcoming analysis on the quantal linear domain.

8.3 RPA – the Algebra of Small-Amplitude Oscillations

Small amplitude means small deviations from the mean-field ground state. We have learned in Chap. 5 in the context of the Hartree–Fock variation that the space of small-amplitude deviations is spanned by the complete set of $1ph$ excitations. The aim is thus to develop an approach which takes care of the residual interaction in $1ph$ space. Motion has a natural connection with TDHF and we will develop that method as the small-amplitude limit of TDHF. The same result can be obtained in many different ways (e.g., with diagrammatic perturbation theory [13] or equations of motion techniques [93]). From its early diagrammatic derivation [16] it inherited the name Random-Phase Approximation (RPA) because a cancellation of complicated diagrams due to random-phase fluctuations was assumed. We ought also to mention that actual content of RPA may differ in different fields and applications. For example, RPA in solid-state physics often embodies only the direct part of the Coulomb residual interaction. In finite systems (nuclei, clusters), usually the fully consistent approach as outlined below is intended.

8.3.1 The Starting Point: TDHF

The small-amplitude limit of TDHF and the subsequent derivation of the random-phase approximation (RPA) becomes particularly transparent in a representation of Slater states and generating operators. This establishes a great similarity to the quantum-mechanical oscillator problem in Heisenberg representation. We therefore chose that line of development for the presentation here.

The one key ingredient for the description now becomes the time-dependent Slater state $|\Phi(t)\rangle$, given in detail as

$$|\Phi(t)\rangle = \hat{a}_1^\dagger(t)\hat{a}_2^\dagger(t)\cdots\hat{a}_N^\dagger(t)|0\rangle \equiv \mathcal{A}\{\varphi_1(x_1,t)\ldots\varphi_N(x_N,t)\}\,, \tag{8.4}$$

where the $\hat{a}_i^\dagger(t)$, or $\varphi_i(x,t)$ correspondingly, constitute the set of occupied time-dependent single-particle states. The time evolution of these single-particle states is optimized by the TDHF equations (8.3), which here are used in the second form given there. We are now interested in types of motion which oscillate only with small amplitude about a stationary Hartree–Fock state $|\Phi_0\rangle$. The HF state fulfills the stationary HF equations (5.10), i.e.,

$$\langle\Phi_0|\hat{a}_h^\dagger\hat{a}_p\hat{H}|\Phi_0\rangle = 0\,, \quad \langle\Phi_0|\hat{H}\hat{a}_p^\dagger\hat{a}_h|\Phi_0\rangle = 0\,.$$

This motivates making the ansatz

$$|\Phi(t)\rangle \approx \big[|\Phi_0\rangle + |\delta\Phi(t)\rangle\big]\mathrm{e}^{-\mathrm{i}E_0 t}\,, \tag{8.5a}$$

where E_0 is the HF ground-state energy, and then linearizing the TDHF equations with respect to $|\delta\Phi(t)\rangle$. As already worked out in Chap. 5, the manifold of small changes of a Slater state spans the space of $1ph$ states (5.8). It is important to emphasize what $1ph$ means in this context of linearization. As we consider variations about the basis state $|\Phi_0\rangle$, these are the $1ph$ states with respect to $|\Phi_0\rangle$ and the single-particle states associated with $\hat{a}_\alpha^\dagger$ or $\hat{a}_\alpha$ are the single-particle states of the HF ground state $|\Phi_0\rangle$. This is parametrized here as

$$|\delta\Phi(t)\rangle = \mathrm{i}\eta\hat{G}|\Phi_0\rangle\,, \quad \hat{G} = \sum_{p>N,h\leq N}\left\{G_{ph}\hat{a}_p^\dagger\hat{a}_h + G_{hp}\hat{a}_h^\dagger\hat{a}_p\right\}\,, \tag{8.5b}$$

where η is a small number and the matrix elements G_{ph} and G_{hp} are still arbitrary. We take the liberty to chose $\hat{G}$ as being Hermitian, so that $G_{hp} = G^*_{ph}$ and consequently

$$\langle\delta\Phi(t)| = -\mathrm{i}\eta\langle\Phi_0|\hat{G}\,, \quad \hat{G}^\dagger = \hat{G}\,. \tag{8.5c}$$

8.3.2 Linearization

The aim is to expand the TDHF equation (8.3) in first order using the expanded state (8.5). We start with the time derivative

$$\begin{aligned}\langle\Phi|\hat{a}^\dagger_h\hat{a}_p\mathrm{i}\partial_t|\Phi\rangle \approx\; & E_0\underbrace{\langle\Phi_0|\hat{a}^\dagger_h\hat{a}_p|\Phi_0\rangle}_{=0} + \mathrm{i}\eta\langle\Phi_0|\hat{a}^\dagger_h\hat{a}_p\mathrm{i}\partial_t\hat{G}|\Phi_0\rangle \\ & + \mathrm{i}\eta E_0\left[\langle\Phi_0|\hat{a}^\dagger_h\hat{a}_p\hat{G}|\Phi_0\rangle - \langle\Phi_0|\hat{G}\hat{a}^\dagger_h\hat{a}_p|\Phi_0\rangle\right].\end{aligned}$$

The term with the action of $\hat{H}$ becomes

$$\begin{aligned}\langle\Phi|\hat{a}^\dagger_h\hat{a}_p\hat{H}|\Phi\rangle &\approx \underbrace{\langle\Phi_0|\hat{a}^\dagger_h\hat{a}_p\hat{H}|\Phi_0\rangle}_{=0} + \mathrm{i}\eta\langle\Phi_0|\hat{a}^\dagger_h\hat{a}_p\hat{H}\hat{G}|\Phi_0\rangle - \mathrm{i}\eta\langle\Phi_0|\hat{G}\hat{a}^\dagger_h\hat{a}_p\hat{H}|\Phi_0\rangle \\ &= \mathrm{i}\eta\langle\Phi_0|\big[\hat{a}^\dagger_h\hat{a}_p\hat{H},\hat{G}\big]|\Phi_0\rangle \\ &= \mathrm{i}\eta\langle\Phi_0|\hat{a}^\dagger_h\hat{a}_p\big[\hat{H},\hat{G}\big]|\Phi_0\rangle + \mathrm{i}\eta\langle\Phi_0|\big[\hat{a}^\dagger_h\hat{a}_p,\hat{G}\big]\hat{H}|\Phi_0\rangle\,.\end{aligned}$$

The second term is disentangled as

$$\begin{aligned}\langle\Phi_0|\big[\hat{a}^\dagger_h\hat{a}_p,\hat{G}\big]\hat{H}|\Phi_0\rangle &= \sum_{h'} G_{ph'}\underbrace{\langle\Phi_0|\hat{a}^\dagger_h\hat{a}_{h'}}_{=\delta_{hh'}\langle\Phi_0|}\hat{H}|\Phi_0\rangle - \sum_{p'} G_{p'h}\underbrace{\langle\Phi_0|\hat{a}^\dagger_{p'}\hat{a}_p}_{=0}\hat{H}|\Phi_0\rangle \\ &= G_{ph}\langle\Phi_0|\hat{H}|\Phi_0\rangle = E_0\langle\Phi_0|\hat{a}^\dagger_h\hat{a}_p\hat{G}|\Phi_0\rangle\end{aligned}$$

and thus

$$\langle\Phi|\hat{a}^\dagger_h\hat{a}_p\hat{H}|\Phi\rangle \approx \mathrm{i}\eta\langle\Phi_0|\hat{a}^\dagger_h\hat{a}_p\big[\hat{H},\hat{G}\big]|\Phi_0\rangle + \mathrm{i}\eta E_0\langle\Phi_0|\hat{a}^\dagger_h\hat{a}_p\hat{G}|\Phi_0\rangle\,.$$

Combining these results we find the linearized TDHF equation as

$$0 = \langle\Phi_0|\hat{a}^\dagger_h\hat{a}_p\big(\mathrm{i}\partial_t\hat{G} - [\hat{H},\hat{G}]\big)|\Phi_0\rangle = \langle\Phi_0|\big[\hat{a}^\dagger_h\hat{a}_p,\mathrm{i}\partial_t\hat{G} - [\hat{H},\hat{G}]\big]|\Phi_0\rangle,$$

where a zero was inserted, namely $\hat{a}_h^\dagger \hat{a}_p|\Phi_0\rangle = 0$, to reshape the expression in terms of commutators throughout. That will simplify the treatment later on. The conjugate equation involving $\hat{a}_p^\dagger \hat{a}_h$ can be derived similarly.

Altogether the linearized TDHF equations are thus

$$\langle\Phi_0|[\hat{a}_p^\dagger \hat{a}_h, \mathrm{i}\partial_t \hat{G} - [\hat{H}, \hat{G}]]|\Phi_0\rangle = 0 = \langle\Phi_0|[\hat{a}_h^\dagger \hat{a}_p, \mathrm{i}\partial_t \hat{G} - [\hat{H}, \hat{G}]]|\Phi_0\rangle, \tag{8.6}$$

to be considered for all conceivable combinations of $h \leq N$ and $p > N$. The complete manifold of $1ph$ operators projects out the $1ph$ part of the other operators in the commutator. $\partial_t \hat{G}$ is already purely $1ph$ by construction, while $[\hat{H}, \hat{G}]$ as such contains much more. The equation of motion for $\hat{G}$ could thus be written more compactly as

$$\mathrm{i}\partial_t \hat{G} = [\hat{H}, \hat{G}]_{ph} ,$$

where the suffix ph denotes the $1ph$ part of the commutator.

8.3.3 Eigenmodes and Eigenfrequencies

The equation of motion (8.6) poses an initial value problem like TDHF: giving an initial state $|\Phi(t = 0)\rangle \approx \left[1 + \mathrm{i}\hat{G}(t = 0)\right]|\Phi_0\rangle$ also determines the corresponding initial generator $\hat{G}(t = 0)$. Equation (8.6) then allows determining $\hat{G}(t)$ at later times. Equation (8.6) is a linear equation for $\hat{G}(t)$. This means it is an equation for coupled harmonic modes. From the theory of coupled harmonic oscillators, it is known that every linear propagation can be decomposed in terms of eigenmodes m

$$\hat{G} = \sum_m \left[\hat{C}_m^\dagger e^{-\mathrm{i}\omega_m t} - \hat{C}_m e^{\mathrm{i}\omega_m t}\right] , \quad \omega_m > 0 . \tag{8.7}$$

The positive definiteness of the ω_m is needed to render the decomposition unambiguous. The time derivative is then reduced to a simple algebraic operation

$$\mathrm{i}\partial_t \hat{G} = \sum_m \omega_m \left[\hat{C}_m^\dagger e^{-\mathrm{i}\omega_m t} + \hat{C}_m e^{\mathrm{i}\omega_m t}\right] . \tag{8.8}$$

We insert this expansion into the linearized TDHF equation (8.6) and sort the terms according to the frequency factor $e^{-\mathrm{i}\omega_m t}$. Each factor is linearly independent of the others, which implies that the equation for each m has to be fulfilled separately, leading to the eigenvalue equations

$$\langle\Phi_0|[\left\{\begin{matrix}\hat{a}_h^\dagger\hat{a}_p\\ \hat{a}_p^\dagger\hat{a}_h\end{matrix}\right\},[\hat{H},\hat{C}_m^\dagger]]|\Phi_0\rangle=\omega_m\langle\Phi_0|[\left\{\begin{matrix}\hat{a}_h^\dagger\hat{a}_p\\ \hat{a}_p^\dagger\hat{a}_h\end{matrix}\right\},\hat{C}_m^\dagger]|\Phi_0\rangle\,, \tag{8.9}$$

and the conjugate equations for $\hat{C}_m$. A particularly instructive form of the equations is obtained by realizing that the space of variations $\hat{a}_h^\dagger\hat{a}_p$, $\hat{a}_p^\dagger\hat{a}_h$ is equally well spanned by the set of eigenmode operators $\hat{C}_m^\dagger$ and $\hat{C}_m$, i.e.,

$$\left\{\hat{a}_h^\dagger\hat{a}_p,\hat{a}_p^\dagger\hat{a}_h\right\}\equiv\left\{\hat{C}_m^\dagger,\hat{C}_m\right\}, \tag{8.10}$$

where p, h runs over all $1ph$ combinations and m over the full spectrum of eigenmodes of the linearized equations. Running through all these m is equivalent to checking all conceivable variations in ph space. This yields the eigenmode equations as

$$\langle\Phi_0|[\hat{C}_n,[\hat{H},\hat{C}_m^\dagger]]|\Phi_0\rangle=\omega_m\langle\Phi_0|[\hat{C}_n,\hat{C}_m^\dagger]|\Phi_0\rangle\,,$$
$$\langle\Phi_0|[\hat{C}_n^\dagger,[\hat{H},\hat{C}_m^\dagger]]|\Phi_0\rangle=\omega_m\langle\Phi_0|[\hat{C}_n^\dagger,\hat{C}_m^\dagger]|\Phi_0\rangle\,.$$

The commutators on the right-hand sides can be simplified. To that end, we consider the complex conjugate of the above equations:

$$\langle\Phi_0|[\hat{C}_m,[\hat{H},\hat{C}_n^\dagger]]|\Phi_0\rangle=\omega_m\langle\Phi_0|[\hat{C}_m,\hat{C}_n^\dagger]|\Phi_0\rangle,$$
$$\langle\Phi_0|[\hat{C}_m^\dagger,[\hat{H},\hat{C}_n^\dagger]]|\Phi_0\rangle=\omega_m\langle\Phi_0|[\hat{C}_m^\dagger,\hat{C}_n^\dagger]|\Phi_0\rangle\,,$$

and exchange labels $n\longleftrightarrow m$, yielding

$$\langle\Phi_0|[\hat{C}_n,[\hat{H},\hat{C}_m^\dagger]]|\Phi_0\rangle=\omega_n\langle\Phi_0|[\hat{C}_n,\hat{C}_m^\dagger]|\Phi_0\rangle\,,$$
$$\langle\Phi_0|[\hat{C}_n^\dagger,[\hat{H},\hat{C}_m^\dagger]]|\Phi_0\rangle=\omega_n\langle\Phi_0|[\hat{C}_n^\dagger,\hat{C}_m^\dagger]|\Phi_0\rangle\,.$$

We assume that all eigenfrequencies are different and subtract the two equations. The left-hand sides cancel and we are left with

$$0=(\omega_n-\omega_m)\langle\Phi_0|[\hat{C}_n,\hat{C}_m^\dagger]|\Phi_0\rangle\,,$$
$$0=(\omega_n+\omega_m)\langle\Phi_0|[\hat{C}_n^\dagger,\hat{C}_m^\dagger]|\Phi_0\rangle\,.$$

The second equation requires the commutator part to vanish because $\omega_n+\omega_m>0$. The first equation acts similarly for $n\neq m$, but leaves a choice for $n=m$. Linear equations do not determine the normalization of an eigenmode. This freedom can be exploited by normalizing the commutator to one.

This yields the eigenmode equations as

$$\langle\Phi_0|[\hat{C}_n, [\hat{H}, \hat{C}_m^\dagger]]|\Phi_0\rangle = \omega_m \delta_{nm}, \tag{8.11a}$$

$$\langle\Phi_0|[\hat{C}_n^\dagger, [\hat{H}, \hat{C}_m^\dagger]]|\Phi_0\rangle = 0, \tag{8.11b}$$

$$\langle\Phi_0|[\hat{C}_n, \hat{C}_m^\dagger]|\Phi_0\rangle = \delta_{nm}, \tag{8.11c}$$

$$\langle\Phi_0|[\hat{C}_n^\dagger, \hat{C}_m^\dagger]|\Phi_0\rangle = 0\,. \tag{8.11d}$$

They are homogeneous linear equations for the generators $\hat{C}_m^\dagger$ of the eigenmodes with eigenfrequency ω_m. They are often called the equations of the *random-phase approximation* (RPA). The notion "random phase" becomes apparent in a derivation from diagrammatic analysis where all diagrams not included in RPA are assumed to disappear by phase averaging [16].

The algebraic formulation is intuitive because one merely has to skip the "bra" and the "ket" and to consider the commutators as such, obtaining the quantum-mechanical algebra of coupled harmonic oscillators. That algebra cannot, however, be exactly matched for a many-fermion problem. With the expectation values as given here, we have what is called a weak operator algebra. It keeps closeness to the basis state $|\Phi_0\rangle$ in accordance with the underlying linearization, and it is an oscillator algebra because every system in the limit of small amplitudes becomes a system of coupled oscillators. The above algebraic formulation is also very useful in developing approximation schemes for which one has simply to invent a convenient subset of $1ph$ operators for the expansion of the $\hat{C}_m^\dagger$. Moreover, it is a very elegant formulation in connection with model Hamiltonians which comply very well with algebraic treatment like the quasispin, or LMG, model, see Sect. 7.2.

8.3.4 The RPA Equations in Matrix Form

The RPA equations in the form (8.11) require some more steps to develop a practical solution scheme. There are different ways to do this. An example for a purely algebraic solution is given in Sect. 8.3.7. Here we present the conceptually simplest (not necessarily most efficient) representation as a matrix equation. A complete basis of the space of $1ph$ operators is provided trivially by the set of all conceivable $1ph$ operators, i.e., $\left\{\hat{a}_p^\dagger\hat{a}_h, \hat{a}_h^\dagger\hat{a}_p, h \le N,\ p > N\right\}$. Expanding the sought-for eigenmode operators in terms of this $1ph$ basis yields

$$\hat{C}_m^\dagger = \sum_{ph}\left\{x_{ph}^{(m)}\hat{a}_p^\dagger\hat{a}_h - y_{hp}^{(m)}\hat{a}_h^\dagger\hat{a}_p\right\}, \tag{8.12a}$$

$$\hat{C}_m = \sum_{ph}\left\{x_{ph}^{(m)*}\hat{a}_h^\dagger\hat{a}_p - y_{hp}^{(m)*}\hat{a}_p^\dagger\hat{a}_h\right\}. \tag{8.12b}$$

We insert this expansion into the RPA equations (8.9)

$$\sum_{p'h'}\left(\langle\Phi_0|[\begin{Bmatrix}\hat{a}_h^\dagger\hat{a}_p\\ \hat{a}_p^\dagger\hat{a}_h\end{Bmatrix},[\hat{H},\hat{a}_{p'}^\dagger\hat{a}_{h'}]]|\Phi_0\rangle x_{p'h'}^{(m)}-\langle\Phi_0|[\begin{Bmatrix}\hat{a}_h^\dagger\hat{a}_p\\ \hat{a}_p^\dagger\hat{a}_h\end{Bmatrix},[\hat{H},\hat{a}_{h'}^\dagger\hat{a}_{p'}]]|\Phi_0\rangle y_{p'h'}^{(m)}\right)$$
$$=\omega_m\sum_{p'h'}\left(\langle\Phi_0|[\begin{Bmatrix}\hat{a}_h^\dagger\hat{a}_p\\ \hat{a}_p^\dagger\hat{a}_h\end{Bmatrix},\hat{a}_{p'}^\dagger\hat{a}_{h'}]|\Phi_0\rangle x_{p'h'}^{(m)}-\langle\Phi_0|[\begin{Bmatrix}\hat{a}_h^\dagger\hat{a}_p\\ \hat{a}_p^\dagger\hat{a}_h\end{Bmatrix},\hat{a}_{h'}^\dagger\hat{a}_{p'}]|\Phi_0\rangle y_{p'h'}^{(m)}\right),$$

and employ the key commutators

$$\langle\Phi_0|[\hat{a}_h^\dagger\hat{a}_p,\hat{a}_{p'}^\dagger\hat{a}_{h'}]|\Phi_0\rangle=\delta_{ph,p'h'},$$
$$\langle\Phi_0|[\hat{a}_h^\dagger\hat{a}_p,\hat{a}_{h'}^\dagger\hat{a}_{p'}]|\Phi_0\rangle=0\,.$$

This yields the RPA equation in matrix representation:

$$\omega_m x_{ph}^{(m)}=\sum_{p'h'}A_{ph;p'h'}x_{p'h'}^{(m)}+\sum_{p'h'}B_{ph;p'h'}y_{p'h'}^{(m)}, \tag{8.13a}$$

$$-\omega_m y_{ph}^{(m)}=\sum_{p'h'}B_{ph;p'h'}x_{p'h'}^{(m)}+\sum_{p'h'}A_{ph;p'h'}y_{p'h'}^{(m)}, \tag{8.13b}$$

$$A_{ph;p'h'}=\langle\Phi_0|[\hat{a}_h^\dagger\hat{a}_p,[\hat{H},\hat{a}_{p'}^\dagger\hat{a}_{h'}]]|\Phi_0\rangle, \tag{8.13c}$$

$$B_{ph;p'h'}=\langle\Phi_0|[\hat{a}_p^\dagger\hat{a}_h,[\hat{H},\hat{a}_{p'}^\dagger\hat{a}_{h'}]]|\Phi_0\rangle\,. \tag{8.13d}$$

This is obviously a set of linear homogeneous equations for the coefficients $x_{ph}^{(m)}$ and $y_{ph}^{(m)}$. It determines the eigenfrequencies ω_m and the corresponding eigenvectors $x_{ph}^{(m)}$, $y_{ph}^{(m)}$. The matrices A and B are determined by the underlying Hamiltonian. Further evaluation depends on the details of the model. The orthonormality conditions (8.11c) and (8.11d) for the operators $\hat{C}_m^\dagger$, $\hat{C}_m$ become conditions for the expansion coefficients as

$$\sum_{ph}x_{ph}^{(m)*}x_{ph}^{(n)}-y_{ph}^{(m)*}y_{ph}^{(n)}=\delta_{mn}, \tag{8.14a}$$

$$\sum_{ph}x_{ph}^{(m)*}y_{ph}^{(n)}-y_{ph}^{(m)*}x_{ph}^{(n)}=0\,. \tag{8.14b}$$

In order to exemplify the evaluation of the RPA equations in more detail, consider the example of a many-particle system with a local two body interaction $V(|\mathbf{r}-\mathbf{r}'|)$

$$\omega_m x_{ph}^{(m)} = (\varepsilon_p - \varepsilon_h) x_{ph}^{(m)} + \sum_{p'h'} V_{ph;p'h'} \left[x_{p'h'}^{(m)} + (y_{p'h'}^{(m)})^* \right] , \tag{8.15a}$$

$$-\omega_m y_{ph}^{(m)} = (\varepsilon_p - \varepsilon_h) y_{ph}^{(m)} + \sum_{p'h'} V_{ph;p'h'} \left[(x_{p'h'}^{(m)})^* + y_{p'h'}^{(m)} \right] , \tag{8.15b}$$

$$V_{ph;p'h'} = \int d^3r\, d^3r'\, \varphi_p^{(0)\dagger}(\mathbf{r}) \varphi_{h'}^{(0)\dagger}(\mathbf{r}') V(|\mathbf{r}-\mathbf{r}'|) \varphi_h^{(0)}(\mathbf{r}) \varphi_{p'}^{(0)}(\mathbf{r}') , \tag{8.15c}$$

$$h, h' \leq N , \quad p, p' > N . \tag{8.15d}$$

8.3.5 Transition Moments and Strength Distributions

The RPA equations (8.11) or (8.13) provide the spectrum of eigenfrequencies ω_m and corresponding eigenmodes $\hat{C}_m^\dagger$. The latter allow to compute the transition strength for a given one-body observable $\hat{Q}$ where $\hat{Q}$ is typically some multipole moment. Most important is the case of the dipole moment, which is related to the photo-absorption strength

$$\mathcal{S}_Q(\omega) = 2\pi \sum_m \omega \left| \langle \Phi_0 | [\hat{Q}, \hat{C}_m^\dagger] | \Phi_0 \rangle \right|^2 [\delta(\omega - \omega_m) + \delta(\omega + \omega_m)] . \tag{8.16}$$

RPA provides discrete spectra. Realistic strength distributions are smoothed by several effects (particle emission, two-body correlations, thermal averaging). This is often accounted for in practice by replacing the δ-distribution by a smooth Lorentzian or Gaussian of small, but finite, width.

The derivation of the strength (8.16) proceeds quite similarly to the derivation of Fermi's golden rule [24]. We add an external time dependent dipole field as a perturbation $\eta f(t)\hat{D}$ where $f(t)$ is the temporal profile and η contains the field strength together with other factors. The solution to the now time-dependent Hamiltonian $\hat{H} - \eta f(t)\hat{Q}$ is sought again in the small amplitude limit by the superposition (8.5). The steps of linearization are run through as in Sect. 8.3.2. A new feature is now that the perturbation $\hat{W}$ does not belong to the HF optimization and thus, there is a non-vanishing $\langle \Phi_0 | \hat{a}_h^\dagger \hat{a}_p \hat{W} | \Phi_0 \rangle$. We keep this term as the leading order of perturbation and obtain the modified linearized equations as

$$\mathrm{i}\partial_t \hat{G} = [\hat{H}, \hat{G}]_{ph} - \eta f(t) Q_{ph} .$$

The dynamic generator $\hat{G}$ is expanded as

$$\hat{G} = \sum_m \left[\hat{C}_m^\dagger e^{-\mathrm{i}\omega_m t} c_m(t) - \hat{C}_m e^{\mathrm{i}\omega_m t} c_m^*(t)\right] , \quad \omega_m > 0,$$

with expansion coefficients c_m yet to be determined. Inserting this expansion into the above dynamic equation for $\hat{G}$ yields the equations of motion for the coefficients

$$\mathrm{i}\hbar\dot{c}_m = \eta f(t)\langle\Phi_0|[\hat{Q},\hat{C}_m^\dagger]|\Phi_0\rangle e^{\mathrm{i}\omega_m t} \quad \Longrightarrow$$

$$c_m = -\frac{\mathrm{i}\eta}{\hbar}\langle\Phi_0|[\hat{Q},\hat{C}_m^\dagger]|\Phi_0\rangle \int_{-\infty}^{t} dt' f(t') e^{\mathrm{i}\omega_m t'} ,$$

where final transitions are evaluated at $t \longrightarrow \infty$. We consider a perturbation of finite duration $f(t) = \theta(T/2 - |t|)e^{-\mathrm{i}\omega t}$ in the limit of large T, which yields the asymptotic transition probability

$$|c_m(\infty)| \propto T\frac{\sin^2(\omega T/2)}{(\omega T/2)^2} \longrightarrow T\, 2\pi\delta(\omega - \omega_m) .$$

The photo-absorption cross section is related to the transition rate, which is the transition probability per time unit. Thus dividing by the factor T in the probability and evaluating the appropriate photon normalization factors [24, 46] yields (8.16) for the absorption strength.

8.3.6 RPA Equations in Connection with DFT

Up to now, we have worked as if we were given a microscopic Hamiltonian $\hat{H}$ consisting of a one-body and a two-body part. In density functional theory (see Chap. 6), the one-body part (kinetic energy and external field) is still given as an operator, but the potential energy is packed into a functional of the local density $E_{\mathrm{pot}}[\rho(\mathbf{r})]$ where E_{pot} combines the Hartree part and exchange-correlation functional of (6.5). The question is thus how to compute a double commutator in terms of an energy functional $E[\rho(\mathbf{r})]$.

We consider double commutators $\langle\Phi_0|[\hat{Q},[\hat{H},\hat{P}]]|\Phi_0\rangle$ for arbitrary one-body operators $\hat{Q}$ and $\hat{P}$. First we argue that it suffices to compute the double commutator for Hermitian one-body operators. An arbitrary operator $\hat{Q}$ can be decomposed into Hermitian and anti-Hermitian part and thus be represented by two Hermitian operators $\hat{A}_R$ and $\hat{A}_I$ as $\hat{Q} = \hat{A}_R + \mathrm{i}\hat{A}_I$. Inserting this representation into the double commutator yields the simpler task to evaluate four double commutators all composed of Hermitian one-body operators, i.e., finally to evaluate

$$\langle\Phi_0|[\hat{A},[\hat{H},\hat{B}]]|\Phi_0\rangle , \quad \hat{A}^\dagger = \hat{A} , \quad \hat{B}^\dagger = \hat{B} , \quad \hat{A},\hat{B} \in \{1ph\} .$$

The proper evaluation of such double commutators is achieved by working out the linearization of the time-dependent DFT. The result is

$$\langle\Phi_0|[\hat{A},[\hat{H},\hat{B}]]|\Phi_0\rangle = \langle\Phi_0|[\hat{A},[\hat{h},\hat{B}]]|\Phi_0\rangle + \int d^3r d^3r' \rho_A(\mathbf{r}) \frac{\partial^2 E_{\text{pot}}}{\partial\rho(\mathbf{r})\partial\rho(\mathbf{r}')}\rho_B(\mathbf{r}'), \tag{8.17}$$

$$\rho_A(\mathbf{r}) = \sum_\alpha \left[(\hat{A}\varphi_\alpha)^\dagger(\mathbf{r})\varphi_\alpha(\mathbf{r}) - \varphi_\alpha^\dagger(\mathbf{r})(\hat{A}\varphi_\alpha)(\mathbf{r})\right] , \tag{8.18}$$

where ρ_B is built analogous to ρ_A and $\hat{h}$ is the mean-field Hamiltonian of the DFT scheme (Sect. 6.1). Let us briefly sketch how this result could be derived. Two key ingredients for the reasoning are as follows: (i) the Thouless theorem which states that a unitary transformation $|\Phi'\rangle = e^{i\hat{A}}|\Phi\rangle$ generated by an arbitrary Hermitian one-body operator $\hat{A}$ transforms a Slater state $|\Phi\rangle$ into a new state which is again a Slater state [92] and (ii) the energy-density functional which defines a unique total energy $E = E[|\Phi\rangle]$ for any given Slater state $|\Phi\rangle$ or associated density ρ, respectively.

Since $\hat{A}$ and $\hat{B}$ are one-body operators, they can be used to build a unitary transformation (Thouless transformation) to another Slater state $|\Phi(a,b)\rangle$, i.e.,

$$|\Phi(a,b)\rangle = e^{ib\hat{B}}e^{ia\hat{A}}|\Phi_0\rangle , \tag{8.19}$$

where a and b are some real numbers, in fact, very small ones as we will see. Consider the energy expectation value with $|\Phi(a,b)\rangle$ and perform a Taylor expansion about $|\Phi_0\rangle$ up to first order in both a and b (we abbreviate $\langle\Phi_0|\ldots|\Phi_0\rangle = \langle\ldots\rangle$) :

$$\begin{aligned} E(a,b) &= E[|\Phi(a,b)\rangle] \equiv \langle\Phi(a,b)|\hat{H}|\Phi(a,b)\rangle \\ &\approx E_0 + ib\langle[\hat{H},\hat{B}]\rangle + ia\langle[\hat{H},\hat{A}]\rangle + ab\langle[\hat{A},[\hat{H},\hat{B}]]\rangle . \end{aligned}$$

The double commutator can obviously be extracted through a double derivative, yielding the identification

$$\langle\Phi_0|[\hat{A},[\hat{H},\hat{B}]]|\Phi_0\rangle = \partial_a\partial_b\langle\Phi(a,b)|\hat{H}|\Phi(a,b)\rangle\Big|_{a,b=0} \equiv \partial_a\partial_b E(a,b)\Big|_{a,b=0} . \tag{8.20}$$

In DFT the key quantity is the density. We thus need to express the above double derivative through the density or rather variations thereof. For this the effect of the shift (8.19) on the local density has to be evaluated. This unitary transformation transforms the single-particle states as $\varphi_\alpha \longrightarrow \varphi_{\alpha,ab} \approx \left(1 + ia\hat{A} + ib\hat{B} - ab\hat{B}\hat{A}\right)\varphi_\alpha$ where terms $\propto a^2$, b^2 have been dropped because they will not be needed. This yields the local density (again expanded to the same order)

$$\rho_{ab}(\mathbf{r}) = \rho(\mathbf{r}) + ia\rho_A(\mathbf{r}) + ib\rho_B(\mathbf{r}) + ab\rho_{AB}(\mathbf{r}) ,$$

$$\rho_{AB}(\mathbf{r}) = \Re\left\{\sum_\alpha \left[(\hat{A}\varphi_\alpha)^\dagger(\mathbf{r})(\hat{B}\varphi_\alpha)(\mathbf{r}) - \varphi_\alpha^\dagger(\mathbf{r})(\hat{A}\hat{B}\varphi_\alpha)(\mathbf{r})\right]\right\} ,$$

and ρ_A, ρ_B as given in (8.18). We insert the shifted density $\rho_{ab}(\mathbf{r})$ into the energy-density functional $E[\rho]$. Roughly speaking the double derivation may be attained by deriving either both the energy and the density once or the energy twice. This yields two contributions to the double derivative of the energy:

$$\partial_a \partial_b E[\rho_{ab}]\Big|_{a,b=0} = \int d^3r \frac{\partial E}{\partial \rho(\mathbf{r})} \rho_{AB}(\mathbf{r}) + \int d^3r d^3r' \rho_A(\mathbf{r}) \frac{\partial^2 E}{\partial \rho(\mathbf{r}) \partial \rho(\mathbf{r}')} \rho_B(\mathbf{r}') ,$$

and finally the desired result (8.17), where the term with ρ_{AB} was reformulated as a double commutator with the DFT mean-field Hamiltonian $\hat{h}$ (see Sect. 6.1).

The above reasoning assumes an energy-density functional for the total energy as it is used, e.g., in Thomas–Fermi approximation. The Kohn–Sham scheme treats the kinetic energy in terms of wave functions. The final formula remains valid for this case; one merely has to separate kinetic from potential energy and apply the above steps to the energy-density functional for the potential energy. Recombining kinetic and potential part in the mean-field $\hat{h}$ then again leads to the expression (8.17).

8.3.7 RPA in the LMG Model as Example

The LMG model, discussed extensively in Chap. 7, is also very well suited to exemplify the RPA equations. We consider the under-critical regime where the unperturbed state $|\Phi_0\rangle$ is also the HF ground state. That is technically much simpler and already delivers sufficient insight into the structure of RPA. The basic $1ph$ operators in the LMG model are $\hat{a}^\dagger_{+1,\alpha}\hat{a}_{-1,\alpha} \equiv s^{(\alpha)}_+$. The most general $1ph$ excitation would then read

$$\hat{C}^\dagger_m = \sum_\alpha \left\{ x^{(m)}_\alpha \hat{a}^\dagger_{+1,\alpha}\hat{a}_{-1,\alpha} - y^{(m)}_\alpha \hat{a}^\dagger_{-1,\alpha}\hat{a}_{+1,\alpha} \right\} \equiv \sum_\alpha \left\{ x^{(m)}_\alpha s^{(\alpha)}_+ - y^{(m)}_\alpha s^{(\alpha)}_- \right\} . \tag{8.21}$$

The RPA equations as matrix equations then become

$$\begin{pmatrix} \hat{A} - \omega\hat{1} & \hat{B} \\ \hat{B} & \hat{A} + \omega\hat{1} \end{pmatrix} \begin{pmatrix} x \\ y \end{pmatrix} = 0 , \tag{8.22a}$$

$$A_{\alpha\beta} = \langle \Phi_0 | \left[\hat{s}_{-,\alpha}, \left[\hat{H}, \hat{s}_{+,\beta} \right] \right] | \Phi_0 \rangle , \tag{8.22b}$$

$$B_{\alpha\beta} = \langle \Phi_0 | \left[\hat{s}_{+,\alpha}, \left[\hat{H}, \hat{s}_{+,\beta} \right] \right] | \Phi_0 \rangle . \tag{8.22c}$$

We now need to evaluate the RPA matrices $\hat{A}$ and $\hat{B}$. This amounts to computing the double commutators with $\hat{J}_0$ and $\hat{J}^2_x$. The basic commutators are given by the spin algebra (7.3) leading to

$$\left[\hat{J}_0, \hat{s}_{\pm,\alpha} \right] = \pm \hat{s}_{\pm,\alpha} , \quad \left[\hat{J}_x, \hat{s}_{\pm,\alpha} \right] = \pm \hat{s}_{0,\alpha} .$$

Furthermore, we recall the properties of the ground state $\hat{s}_{-,\alpha}|\Phi_0\rangle = 0$, $\hat{s}_{0,\alpha}|\Phi_0\rangle = -|\Phi_0\rangle$, $\langle\Phi_0|\hat{s}_{+,\alpha} = 0$, and $\langle\Phi_0|\hat{s}_{0,\alpha} = -\langle\Phi_0|$. This yields the expectation values

$$\begin{aligned}
\langle\Phi_0|\big[\hat{s}_{-,\alpha},\big[\hat{J}_0,\hat{s}_{+,\beta}\big]\big]|\Phi_0\rangle &= -2\delta_{\alpha\beta}\langle\Phi_0|\hat{s}_{0,\alpha}|\Phi_0\rangle = \delta_{\alpha\beta}\,,\\
\langle\Phi_0|\big[\hat{s}_{+,\alpha},\big[\hat{J}_0,\hat{s}_{+,\beta}\big]\big]|\Phi_0\rangle &= 0\,,\\
\langle\Phi_0|\big[\hat{s}_{-,\alpha},\big[\hat{J}_x^2,\hat{s}_{+,\beta}\big]\big]|\Phi_0\rangle &= -\langle\Phi_0|\big[\hat{s}_{-,\alpha},\big\{\hat{J}_x,\hat{s}_{0,\beta}\big\}\big]|\Phi_0\rangle\\
&= \langle\Phi_0|\big\{\hat{s}_{0,\alpha},\hat{s}_{0,\beta}\big\} - \delta_{\alpha\beta}\big\{\hat{J}_x,\hat{s}_{-,\alpha}\big\}|\Phi_0\rangle\\
&= 1-\delta_{\alpha\beta}\,,
\end{aligned}$$

$$\begin{aligned}
\langle\Phi_0|\big[\hat{s}_{+,\alpha},\big[\hat{J}_x^2,\hat{s}_{+,\beta}\big]\big]|\Phi_0\rangle &= -\langle\Phi_0|\big[\hat{s}_{+,\alpha},\big\{\hat{J}_x,\hat{s}_{0,\beta}\big\}\big]|\Phi_0\rangle\\
&= \langle\Phi_0| - \big\{\hat{s}_{0,\alpha},\hat{s}_{0,\beta}\big\} + \delta_{\alpha\beta}\big\{\hat{J}_x,\hat{s}_{+,\alpha}\big\}|\Phi_0\rangle\\
&= -1+\delta_{\alpha\beta}\,.
\end{aligned}$$

Combining these results yields the RPA matrices as

$$A_{\alpha\beta} = \delta_{\alpha\beta}\,(\varepsilon - G) + G\,, \quad B_{\alpha\beta} = -G(1-\delta_{\alpha\beta})\,. \tag{8.23}$$

The RPA equations in full detail are

$$\left(\begin{array}{ccccc|ccccc}
\varepsilon-\omega & G & G & \ldots & G & 0 & -G & & \ldots & -G\\
G & \varepsilon-\omega & G & \ldots & G & -G & 0 & & \ldots & -G\\
\vdots & \vdots & \vdots & \vdots & \vdots & \vdots & & \vdots & & \vdots\\
G & G & G & \ldots & \varepsilon-\omega & -G & -G & & \ldots & 0\\
\hline
0 & -G & & \ldots & -G & \varepsilon+\omega & G & G & \ldots & G\\
-G & 0 & & \ldots & -G & G & \varepsilon+\omega & G & \ldots & G\\
\vdots & \vdots & \vdots & \vdots & \vdots & \vdots & & \vdots & & \vdots\\
-G & -G & & \ldots & 0 & G & G & G & \ldots & \varepsilon+\omega
\end{array}\right)
\left(\begin{array}{c} x_1\\ x_2\\ \vdots\\ x_N\\ \hline y_1\\ y_2\\ \vdots\\ y_N \end{array}\right) = 0.$$

This is a combination of a diagonal matrix and a quite homogeneous matrix having the same entry everywhere. The equation is invariant under a shift on a ring $\alpha \longrightarrow \operatorname{mod}(\alpha, N)+1$. This means that it has cyclic $Z(N)$ symmetry (see Sect. 4.3.1). The eigenvectors will thus have the form

$$x_\alpha^{(m)} = x^{(m)}\exp\left(im\alpha\frac{2\pi}{N}\right), \quad y_\alpha^{(m)} = y^{(m)}\exp\left(im\alpha\frac{2\pi}{N}\right),$$

with $m = 0, \ldots, N-1$ and for all $m > 0$ the relation $\sum_\alpha x_\alpha^{(m)} = 0 = \sum_\alpha y_\alpha^{(m)}$ holds so that the contribution of the equidistributed matrices G vanishes. There remains a 2×2 equation for the α-independent coefficients

$$m > 0\ :\quad \begin{array}{rl} (\varepsilon - \omega - G)\, x^{(m)} + & G\ y^{(m)} = 0 \\ G\ x^{(m)} + (\varepsilon + \omega - G)\ y^{(m)} = 0 \end{array},$$

having the positive energy solution $\omega_{m>0} = \epsilon\sqrt{1 - \frac{2G}{\epsilon}}$. The case $m = 0$, on the other hand, collects all contributions from the G accumulating them to a large net result. Summing over the α, a 2×2 problem remains.

$$\begin{array}{rl} \left(\varepsilon(1+\frac{\chi}{2}) - \omega\right) x^{(0)} - & \varepsilon\frac{\chi}{2}\ y^{(0)} = 0 \\ -\varepsilon\frac{\chi}{2}\ x^{(0)} + \left(\varepsilon(1+\frac{\chi}{2}) + \omega\right) y^{(0)} = 0 \end{array}, \quad \chi = \frac{2G(N-1)}{\epsilon},$$

together with the normalization (8.14), which here means $N((x^{(0)})^2 - (y^{(0)})^2) = 1$. Note that here we use the same reduced coupling strength as in the ground-state studies, see (7.19). The resulting homogeneous linear equation for x and y has two solutions. The one with positive energy is $\omega_0 = \varepsilon\sqrt{1+\chi}$. Altogether, we obtain the spectrum

$$\begin{aligned} \omega_0 = \varepsilon\sqrt{1+\chi} \quad &, \ x_\alpha^{(0)} = \frac{1 + \frac{\chi}{2} + \omega_0}{\sqrt{2\omega_0(1 + \frac{\chi}{2} + \omega_0)}}, \\ & \ \ y_\alpha^{(0)} = \frac{\frac{\chi}{2}}{\sqrt{2\omega_0(1 + \frac{\chi}{2} + \omega_0)}}, \\ \omega_{m>0} = \varepsilon\sqrt{1 + \tfrac{\chi}{N-1}} &, \ x_\alpha^{(m>0)} \propto \exp\left(im\alpha\tfrac{2\pi}{N}\right), \\ & \ \ y_\alpha^{(m>0)} \propto \tfrac{G}{\epsilon - \omega_{m>0}} \exp\left(im\alpha\tfrac{2\pi}{N}\right), \end{aligned} \tag{8.24}$$

where we did not care to normalize the eigenvectors for $m > 0$ because that will not be needed in the following. There is one solution, $m = 0$, which gathers the whole effect of the two-body interaction and undergoes a dramatic shift by a factor $\sqrt{1+\chi}$ while the remaining $N-1$ solutions remain almost unchanged near the unperturbed energy ε (remember that the effective interaction strength χ is reduced by a factor $1/(N-1)$ which makes it a negligible effect for large particle numbers). Particularly intriguing is the case of attractive interaction $G < 0$. There is obviously a critical point at $1 + \chi = 0$, i.e., $\chi = -1$, where the lowest oscillation energy becomes imaginary, so that these solutions turn to exponential growth rather than oscillations and the ground-state $|\Phi_0\rangle$ becomes unstable (recall that when $\chi \to -1$, $\omega_0 \to 0$, $x_\alpha^{(0)} \to \infty$, and $y_\alpha^{(0)} \to -\infty$). That is precisely what we observed in Sect. 7.6.3 for the HF ground state. We thus see that the RPA solution signals the phase transition by a mode becoming unstable. The region before onset of instability is associated with a very soft mode having low energy and large amplitude, thus already violating the presupposition of small amplitude which was assumed for the derivation of RPA.

The spectrum (8.24) simulates a further important effect, the reshuffling of transition strength into one collective peak, see the discussion in Sect. 8.1. Let us take as observable the operator $\hat{Q} \equiv \hat{J}_x$. It represents a typical collective operator which acts equally strongly in all $1ph$ states, similar to a multipole operator. The $1ph$ transitions all have equal strength

$$\left|\langle\Phi_0|[\hat{Q},\hat{s}_{+,\alpha}]|\Phi_0\rangle\right|^2 = 1\,.$$

The transition strength for the RPA states becomes, inserting the eigenmode (8.21) with the solutions (8.24),

$$\left|\langle\Phi_0|[\hat{Q},\hat{C}^\dagger_m]|\Phi_0\rangle\right|^2 = \left|\sum_\alpha \left(x^{(m)}_\alpha + y^{(m)}_\alpha\right)\right|^2 = \begin{cases} \frac{N}{\omega_0} & m = 0 \\ 0 & m > 0 \end{cases}. \tag{8.25}$$

The one mode $m = 0$ with the strong shift accumulates all collective transition strength, leaving not the slightest bit for the remaining modes $m > 0$. The LMG model thus achieves a perfect collectivity because it has an exactly degenerate $1ph$ spectrum and exactly equal coupling among all $1ph$ states. This is the mechanism for the collective shift we had seen in Fig. 8.1. The collection of all strength into exactly one resonance is not perfectly realized there. In reality there are often somewhat degenerate $1ph$ spectra and slightly fluctuating coupling matrix elements, which makes the accumulation of strength in one collective mode incomplete. Still, the typical resonance modes in realistic many-fermion systems often exhaust about 90% of the total strength, which shows that the collective coupling mechanism modeled here is a very relevant process.

The total strength (8.25) accumulated in the $m = 0$ mode varies with coupling strength as $\omega_0^{-1} = (\varepsilon\sqrt{1+\chi})^{-1}$ and thus differs from the total strength of uncoupled states which goes as N/ε. It is interesting to note that the photo-absorption strength (8.16) is an energy-weighted sum. We have only one mode contribution here and thus find for the photo-absorption strength $\mathcal{S}_{J_x} = N$, which is independent of the coupling strength and has the same value already for the uncoupled states. The recoupling by RPA has not changed the total $\mathcal{S}_{J_x}$, but rearranged it from dispersion over many $1ph$ states to appear only in one collective resonance state. This observation motivates having a closer look at sum rules, which is done in Sect. 8.4.

8.3.8 A Practical Example: GDR in the Nucleus ^{238}U

One realistic example was given for the cluster case in Fig. 8.1. We complement that here by an example from nuclei. Figure 8.2 shows the isovector dipole strength distribution in ^{238}U (the term "isovector" means the dipole oscillations of neutrons

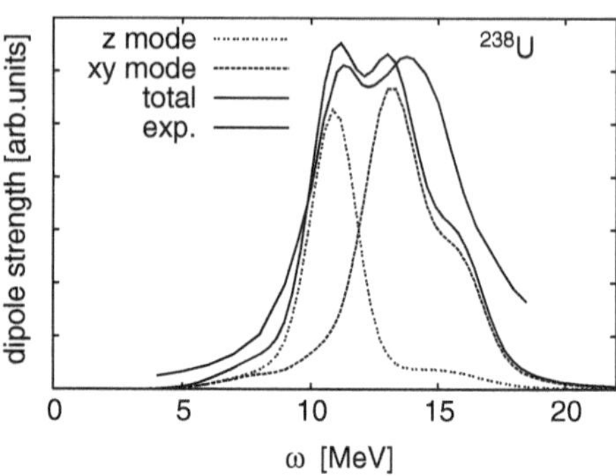

Fig. 8.2 The isovector dipole strength distribution in ^{238}U as computed by RPA with a Skyrme energy functional (SkI3). The contributions from modes in z and x–y direction are provided separately. The experimental strength is shown for comparison. Adapted from [68]

against protons). Similar as in the cluster example, the detailed RPA spectra were folded by some width to simulate broadening effects not included in RPA. The strength distribution, being large and concentrated in a narrow energy range, has resonance character. It is the well-known Giant Dipole Resonance (GDR). We have chosen a well-deformed nucleus (axis ratio $\eta \approx 0.4$) to demonstrate the effect of deformation on the resonance. The calculations allow separating the modes in the three principal directions. The z-mode (elongated axis) has substantially lower resonance energy than the x–y-modes (compressed axes). The total strength still shows the associated double-peak structure, which complies nicely with the experimental data. This example demonstrates how excitation spectra provide crucial information on the underlying ground-state properties. The deformation splitting, demonstrated here, was a key observable to establish the systematics of cluster deformations [44, 89] (see Chap. 3 and Fig. 3.5). In fact, the collective splitting is much better visible in clusters. Nuclei are usually not so strongly deformed which makes it rather hard to clearly see the splitting above background. The above example stems from a rather well-deformed nucleus.

8.4 Sum Rules and Sum-Rule Approximations

8.4.1 RPA Sum Rules

Key quantities in RPA are the multiple commutators which appear notoriously in several places, see, e.g., (8.11) and (8.15). Commutators with observables also play an important role for simple estimates of global excitation strengths, called sum rules. The detailed strength was given in (8.16). Its computation requires the full RPA solution. Global strengths are integrated into sum rules, which are deduced by integrating the transition strength (8.16) over energy augmented with various powers of energy, and amount to energy-weighted moments of the Q transition elements. These are

$$
\begin{aligned}
M_Q^{(N)} &= \int_0^\infty D\omega\omega^{N-1} S_Q(\omega) \\
&= \sum_m \omega_m^N \left|\langle\Phi_0|[\hat{Q},\hat{C}_m^\dagger]|\Phi_0\rangle\right|^2 \Big(\underbrace{\int D\omega\delta(\omega-\omega_m)}_{=1} + \underbrace{\int D\omega\delta(\omega+\omega_m)}_{=0}\Big) \\
&= \sum_m \omega_m^N \left|\langle\Phi_0|[\hat{Q},\hat{C}_m^\dagger]|\Phi_0\rangle\right|^2 .
\end{aligned} \tag{8.26}
$$

For odd moments N, these sum rules can be expressed as multiple commutators of two $\hat{Q}$ with several $\hat{H}$ as

$$
\begin{aligned}
M_Q^{(N)} &= \sum_m \omega_m^N \left|\langle\Phi_0|[\hat{Q},\hat{C}_m^\dagger]|\Phi_0\rangle\right|^2 \\
&= \frac{1}{2}\langle\Phi_0|[\hat{Q},\underbrace{[\hat{H},[[\ldots,[\hat{H},}_{N\text{times}}\hat{Q}]\ldots]]]]|\Phi_0\rangle .
\end{aligned} \tag{8.27a}
$$

Of particular practical importance is the first and the third moment

$$
M_Q^{(1)} = \sum_m \omega_m \left|\langle\Phi_0|[\hat{Q},\hat{C}_m^\dagger]|\Phi_0\rangle\right|^2 = \frac{1}{2}\langle\Phi_0|[\hat{Q},[\hat{H},\hat{Q}]]|\Phi_0\rangle , \tag{8.27b}
$$

$$
M_Q^{(3)} = \sum_m \omega_m^3 \left|\langle\Phi_0|[\hat{Q},\hat{C}_m^\dagger]|\Phi_0\rangle\right|^2 \tag{8.27c}
$$

$$
= \frac{1}{2}\langle\Phi_0|[\hat{Q},[\hat{H},[\hat{H},[\hat{H},\hat{Q}]]]]|\Phi_0\rangle . \tag{8.27d}
$$

To prove these relations, we expand the $1ph$ part of a given operator into the complete set of $1ph$ operators $\hat{C}_m^\dagger$, $\hat{C}_m$ as

$$
\hat{Q} = \sum_m \left[\hat{C}_m^\dagger\langle\Phi_0|[\hat{C}_m,\hat{Q}]|\Phi_0\rangle - \hat{C}_m\langle\Phi_0|[\hat{C}_m^\dagger,\hat{Q}]|\Phi_0\rangle\right] . \tag{8.28}
$$

The proof for the expansion (8.28) relies first on that the $\hat{C}_m^\dagger$, $\hat{C}_m$ constitute a complete set, so that an expansion is capable of describing any $1ph$ operator. The actual expansion coefficients are derived with the help of the orthonormality conditions (8.11c, 8.11d). We start from the general form

$$
\hat{Q} = \sum_m \left[\hat{C}_m^\dagger\lambda_m - \hat{C}_m\mu_m\right] ,
$$

where the λ_m and μ_m are to be determined; we then apply the commutator with $\hat{C}_m$, which successively gives

$$\langle\Phi_0|[\hat{C}_m,\hat{Q}]|\Phi_0\rangle = \sum_{m'}\Big[\underbrace{\langle\Phi_0|[\hat{C}_m,\hat{C}^\dagger_{m'}]|\Phi_0\rangle}_{=\delta_{mm'}}\lambda_{m'} - \underbrace{\langle\Phi_0|[\hat{C}_m,\hat{C}_{m'}]|\Phi_0\rangle}_{=0}\mu_{m'}\Big] = \lambda_m .$$

In a similar fashion one shows the validity of the other expansion coefficient $\langle\Phi_0|[\hat{C}^\dagger_m,\hat{Q}]|\Phi_0\rangle$. With this expansion at hand, we evaluate the double commutator with the Hamiltonian

$$\begin{aligned}\langle\Phi_0|[\hat{Q},[\hat{H},\hat{Q}]]|\Phi_0\rangle = \sum_{mm'}\Big[&\langle\Phi_0|[\hat{C}_{m'},\hat{Q}]|\Phi_0\rangle\underbrace{\langle\Phi_0|[\hat{C}^\dagger_{m'},[\hat{H},\hat{C}^\dagger_m]]|\Phi_0\rangle}_{=0}\langle\Phi_0|[\hat{C}_m,\hat{Q}]|\Phi_0\rangle\\ &-\langle\Phi_0|[\hat{C}^\dagger_{m'},\hat{Q}]|\Phi_0\rangle\underbrace{\langle\Phi_0|[\hat{C}_{m'},[\hat{H},\hat{C}^\dagger_m]]|\Phi_0\rangle}_{=\omega_m\delta_{mm'}}\langle\Phi_0|[\hat{C}_m,\hat{Q}]|\Phi_0\rangle\\ &-\langle\Phi_0|[\hat{C}_{m'},\hat{Q}]|\Phi_0\rangle\underbrace{\langle\Phi_0|[\hat{C}^\dagger_{m'},[\hat{H},\hat{C}_m]]|\Phi_0\rangle}_{=\omega_m\delta_{mm'}}\langle\Phi_0|[\hat{C}^\dagger_m,\hat{Q}]|\Phi_0\rangle\\ &-\langle\Phi_0|[\hat{C}^\dagger_{m'},\hat{Q}]|\Phi_0\rangle\underbrace{\langle\Phi_0|[\hat{C}_{m'},[\hat{H},\hat{C}_m]]|\Phi_0\rangle}_{=0}\langle\Phi_0|[\hat{C}^\dagger_m,\hat{Q}]|\Phi_0\rangle\Big]\\ = 2\sum_m \omega_m\big|\langle\Phi_0|[\hat{Q},\hat{C}^\dagger_{m'}]|\Phi_0\rangle\big|^2 = 2M_Q^{(1)}.&\end{aligned}$$

The other odd sum rules are evaluated in a similar fashion.

Sum rules are powerful tools for several purposes. They can serve as a cross-check for actual calculations. More importantly, they are an extremely useful ingredient for a simple estimate of resonance frequencies from $M_Q^{(1)}$ and $M_Q^{(3)}$ [15], as will be outlined in the next section.

8.4.2 Sum-Rule Approximation

8.4.2.1 Assumption of Exhausting Resonance

As outlined in the previous section, it is rather easy to compute the first and third moments $M_Q^{(N)}$ of a given observable $\hat{Q}$. One often encounters situations of resonant response in which the dominant fraction of the strength distribution (8.16) is confined to a very narrow frequency window. An example is the Mie plasmon resonance, which appears in all finite metal drops or in quantum dots. Under these conditions, it can be assumed that nearly all strength resides in one resonance mode "$m \longrightarrow$ R," such that

$$M_Q^{(N)} = \sum_m \omega_m^N\,|\langle\Phi_0|[\hat{Q},\hat{C}^\dagger_m]|\Phi_0\rangle|^2 \approx \omega_{\mathrm{R}}^N\left|\langle\Phi_0|[\hat{Q},\hat{C}^\dagger_{\mathrm{R}}]|\Phi_0\rangle\right|^2 .$$

This allows to deduce an estimate for the resonant frequency and corresponding transition strength from two moments. The most easy to evaluate is $M_Q^{(1)}$ and $M_Q^{(3)}$, yielding the estimate

$$\omega_{\mathrm{R}} = \sqrt{\frac{M_Q^{(3)}}{M_Q^{(1)}}} = \sqrt{\frac{\langle \Phi_0 | [\hat{Q}, [\hat{H}, [\hat{H}, [\hat{H}, \hat{Q}]]]] | \Phi_0 \rangle}{\langle \Phi_0 | [\hat{Q}, [\hat{H}, \hat{Q}]] | \Phi_0 \rangle}} . \tag{8.29}$$

This is called a sum-rule estimate for a resonant frequency [15].

8.4.2.2 More Details for a Dipole Resonance

We exemplify the power of the sum-rule approximation for the case of the Mie plasmon in metal clusters. The relevant observable is the dipole moment

$$\hat{Q} \quad \longrightarrow \quad \hat{\mathbf{D}} = \sum_{n=1}^{N} e\mathbf{r}_n . \tag{8.30}$$

It is important to note that this is a local and, of course, a one-body operator. In fact, it is a vector of operators, three different operators in one package. The total Hamiltonian of the system reads (assuming N electrons and N ions)

$$\hat{H} = \hat{T} + \hat{U}_{\mathrm{ion}} + \hat{V}_{\mathrm{el}} , \tag{8.31a}$$

$$\hat{T} = \sum_{n=1}^{N} \frac{\hat{\mathbf{p}}_n^2}{2m} , \tag{8.31b}$$

$$\hat{U}_{\mathrm{ion}} = \sum_{n=1}^{N} U_{\mathrm{ion}}(\mathbf{r}_n) , \quad U_{\mathrm{ion}}(\mathbf{r}) = \int D^3r' \, \frac{e^2}{|\mathbf{r} - \mathbf{r}'|} \rho_{\mathrm{ion}}(\mathbf{r}') , \tag{8.31c}$$

$$\hat{V}_{\mathrm{el}} = \frac{1}{2} \sum_{n \neq n'=1}^{N} \int D^3r \, D^3r' \, \frac{e^2}{|\mathbf{r}_n - \mathbf{r}'_{n'}|} . \tag{8.31d}$$

This Hamiltonian consists of the kinetic energy $\hat{T} = \hat{\mathbf{p}}^2/(2m)$, the external Coulomb potential $\hat{U}_{\mathrm{ion}}$ from the ionic distribution (represented by the density $\rho_{\mathrm{ion}}(\mathbf{r})$), which is a local operator, and the Coulomb interaction between the electrons $\hat{V}_{\mathrm{el}}$, which is a two-body operator and again a local one. This implies the following commutators:

$$\left[T, \hat{D}_j\right] = -\frac{\mathrm{i}e}{m} \underbrace{\sum_{n=1}^{N} \hat{p}_{n,j}}_{\hat{P}_j} , \quad \left[\hat{U}_{\mathrm{ion}}, \hat{D}_j\right] = 0 , \quad \left[\hat{V}_{\mathrm{el}}, \hat{D}_j\right] = 0 ,$$

which together yield

$$[\hat{H}, \hat{D}_j] = -\frac{ie}{m}\hat{P}_j\,, \quad \hat{P}_j = \sum_{n=1}^{N} \hat{p}_{n,j}\,, \tag{8.32}$$

where $\hat{\boldsymbol{P}} = (\hat{P}_x, \hat{P}_y, \hat{P}_z)$ is the total momentum of the electrons. With this we obtain

$$\langle \Phi_0 | [\hat{D}_j, [\hat{H}, \hat{D}_j]] | \Phi_0 \rangle = -\frac{ie^2}{m} \underbrace{\langle \Phi_0 | [\hat{D}_j, \hat{P}_j] | \Phi_0 \rangle}_{=Ni} = N\frac{e^2}{m}.$$

Adding up the results of all three components to the total first moment of $\hat{\mathbf{D}}$ yields

$$M_{\mathbf{D}}^{(1)} = \sum_{j=1}^{3} M_{D_j}^{(1)} = 3N\frac{e^2}{m}\,. \tag{8.33}$$

This is an extremely simple result, the famous Thomas–Reiche–Kuhn sum rule, see, e.g., [92, 36]. It stems exclusively from the kinetic energy. There is no contribution from the potential energies as long as these are described by local operators. The third moment becomes somewhat more involved. We first reduce the quadruple commutator using the elementary commutator (8.32) to

$$\begin{aligned} M_{D_j}^{(3)} &= \langle \Phi_0 | [\hat{D}_j, [\hat{H}, [\hat{H}, [\hat{H}, \hat{D}_j]]]] | \Phi_0 \rangle = \langle \Phi_0 | [[\hat{D}_j, \hat{H}], [\hat{H}, [\hat{H}, \hat{D}_j]]] | \Phi_0 \rangle \\ &= \frac{e^2}{m^2} \langle \Phi_0 | [\hat{P}_j, [\hat{H}, \hat{P}_j]] | \Phi_0 \rangle. \end{aligned}$$

Inserting the Hamiltonian (8.31) into the double commutator, there are two vanishing pieces

$$[\hat{T}, \hat{P}_j] = 0, \quad [\hat{V}_{\text{el}}, \hat{P}_j] = 0.$$

The commutator with the kinetic energy vanishes because $\hat{T}$ consists only of momentum operators which all commute with $\hat{P}_j$. The other commutator vanishes because the two-body interaction is translationally invariant and the total momentum $\hat{P}_j$ is the generator of translations. The remaining non-vanishing piece is the double commutator with the external field. We exploit that the commutator of the momentum with a local function produces the derivative of that function ($[f(\hat{x}), \hat{p}]\varphi(x) = i\hbar(\partial f/\partial x)\varphi(x)$). This yields

$$M^{(3)}_{D_j} = \frac{e^2}{m^2}\langle\Phi_0|[\hat{P}_j, [\hat{U}_{\text{ion}}, \hat{P}_j]]|\Phi_0\rangle = \frac{e^2}{m^2}\langle\Phi_0|[-\mathrm{i}\hat{P}_j, \sum_{n=1}^{N}\nabla_j U_{\text{ion}}(\mathbf{r}_n)]|\Phi_0\rangle$$
$$= \frac{e^2}{m^2}\langle\Phi_0|\sum_{n=1}^{N}\nabla_j^2 U_{\text{ion}}(\mathbf{r}_n)|\Phi_0\rangle = \frac{e^2}{m^2}\int D^3r\, \rho_{\text{el}}(\mathbf{r})\nabla_j^2 U_{\text{ion}}(\mathbf{r}).$$

We sum again over all three space directions and obtain

$$M^{(3)}_{\mathbf{D}} = \sum_{j=1}^{3} M^{(3)}_{D_j} = \frac{e^2}{m^2}\int D^3r\, \rho_{\text{el}}(\mathbf{r})\Delta U_{\text{ion}}(\mathbf{r}).$$

At this stage, a compact expression for the average resonance frequency is reached as

$$\omega_{\text{R}}^2 = \frac{M^{(3)}_Q}{M^{(1)}_Q} = \frac{\int D^3r\, \rho_{\text{el}}(\mathbf{r})\Delta U_{\text{ion}}(\mathbf{r})}{Nm}. \tag{8.34}$$

This estimate is the starting point for applications in a broad variety of systems. It can be further reduced for saturating, self-bound systems like metal clusters, see Sect. 8.4.2.4. It is still applicable in situations with dominating external field like in quantum dots where the external field is given by design. This case will be addressed in the following section.

8.4.2.3 Kohn Theorem

The sum-rule estimate (8.34) shows that the dipole resonance frequency is determined exclusively by the external fields. The fermion cloud as such does not contribute to its own dipole oscillation frequency. The case becomes particularly simple for an electron cloud in a trap. Trap potentials usually are largely harmonic (see Sect. 3.2.4). Let us assume that $U_{\text{ion}} = \frac{m}{2}\omega_{\text{ext}}^2\mathbf{r}^2$. This yields $\Delta U_{\text{ion}} = 3\omega_{\text{ext}}^2$ and finally $\omega_{\text{R}}^2 = \omega_{\text{ext}}^2$, which is the remarkable result that the dipole frequency is just the frequency of the external oscillator field.

The result is not as surprising as it may look at first glance. Self-bound fermion systems which are not confined by any external fields have free translation of the static ground-state HF solution as one solution of the TDHF equations. This feature emerges from momentum conservation guaranteed for a Hamiltonian consisting of kinetic energy and a two-body interaction [92]. The estimate (8.34) is more general, as it is also applicable in connection with DFT, and the example of free translation could also be carried through in connection with density functionals, as it is practiced, e.g., with nuclear TDHF calculations which are, in fact, TDDFT calculations. It is a remarkable feature that conservation of center of mass momentum of a free

Fermion cloud is also correctly reproduced in TDLDA. This yields free translational motion if no external field is present. The considerations can be extended to the case of a purely harmonic external field, for then the mean-field state as a whole oscillates with given oscillator frequency in the external field. That is the content of the *harmonic-potential theorem*, an extended version of the Kohn theorem and often also called *zero-force theorem* [26, 110].

8.4.2.4 Mie Plasmon Frequency in a Metal Drop

For a metal drop, we can go one step further and exploit the fact that U_{ion} is the Coulomb field generated by the ionic background density, see (8.31c). The Poisson equation then provides the simple relation $\Delta U_{\text{ion}} = 4\pi e^2 \rho_{\text{ion}}$, which yields

$$M_{\mathbf{D}}^{(3)} = 12\pi \frac{e^4}{m^2} \int D^3 r \, \rho_{\text{el}} \rho_{\text{ion}} \,, \tag{8.35}$$

already a quite simple result, but one can go a few steps further in producing a robust and even simpler estimate. For the purpose of a simple estimate, the ionic background density ρ_{ion} is approximated roughly by the jellium model (1.1). We furthermore assume that at equilibrium $\rho_{\text{el}} = \rho_{\text{ion}}$. There remains the simple integration over a homogeneous sphere yielding its volume $4\pi R_{\text{jel}}^3/3$. This allows evaluating the third moment as

$$M_{\mathbf{D}}^{(3)} == 12\pi \frac{e^4}{m^2} \frac{4\pi R_{\text{jel}}^3}{3} \left(\frac{3}{4\pi R_{\text{jel}}^3} \right)^2 = \frac{e^4}{m^2} \frac{9}{r_s^3 N} \,.$$

Putting all pieces together, we eventually find the sum-rule estimate for the resonance frequency of the Mie plasmon in a spherical metal cluster

$$\omega_{\text{Mie}} = \sqrt{\frac{M_{\mathbf{D}}^{(3)}}{M_{\mathbf{D}}^{(1)}}} = \sqrt{\frac{e^2}{m r_s^3}} \,. \tag{8.36}$$

This is the famous *Mie plasmon frequency*. It was derived a century ago by a purely classical consideration of dielectric vibrations of a charged sphere [71]. The estimate reproduces the observed resonance frequencies in simple metal clusters (Li, Na, K, Cs, Rb) within better than 10% [19]. It thus serves as an extremely useful guideline for the resonance position.

Note that the sum-rule approximations together with local potentials reduce to simple expressions involving only the local densities, see (8.35), and/or local external potentials, see (8.34). The detailed quantum-mechanical information contained in the single-particle wave functions is not required. This means that these estimates can be equally well be used in connection with semi-classical approaches like the Thomas–Fermi method [89].

8.5 Concluding Remarks

Small-amplitude oscillations are a characteristic feature of any physical system. They provide a complementing view on the ground-state properties as well as a doorway for the exploration of dynamical scenarios, typically the response of a system to an external perturbation. The analysis of such small-amplitude oscillations in a complex system such as a many-fermion one requires some formal development to properly account for the interactions between the constituents of the system. In this chapter we thus have derived the linearized version of TDHF, known as RPA, which provides a basis for the analysis of small-amplitude oscillations in many-fermion systems. The RPA gives access to the strength distribution, which can be directly measured experimentally. The RPA furthermore possesses interesting formal properties which can be cast into simple expressions through the so-called sum rules. The latter quantities even provide simple estimates of dominant response peaks and can thus be used for estimating the response frequency.

Chapter 9
Coherent Two-Body Correlations

The previous chapters dealt predominantly with a mean-field description of many-fermion systems. These are all approximate and leave space for correlations beyond the mean field. Correlations naturally cover a very broad field. We will discuss here three illustrative examples. A global measure for the strength of correlations is deduced from perturbation theory: it is the ratio of the average strength of the residual two-body interaction over the typical excitation energies of the system, the latter being easily quantified by the HOMO–LUMO gap. We have learned in Chap. 3 that fermion systems very much prefer situations with a large HOMO–LUMO gap which keeps correlation effects low. We have also seen that such a favorable situation is not necessarily given from the outset. In particular, highly symmetric configurations tend to produce an open-shell situation where the ground state is degenerate. Think, e.g., of the example given in Fig. 3.4, where most system sizes do not relate to closed shells at spherical mean field ($\delta = 0$ in that figure). Other situations may have a unique ground state, but low excitation energies as is the case, e.g., in the Fermi gas, see Chap. 2. One way out of such unwanted configurations is achieved by the *Jahn–Teller effect*. The system breaks a symmetry and undergoes a deformation until the resulting level shifts have produced a maximal HOMO–LUMO gap, see the example in Fig. 3.4. This Jahn–Teller mechanism, originally discovered for molecules [52], is also found in many other systems [32] like solids, nuclei, clusters and shows up in field theories under the label of spontaneous symmetry breaking and Goldstone modes [113]. A second solution to avoid degenerate ground states consists in spin alignment (Hund's rules) which is the preferred remedy in most atoms [36] and in quantum dots [87]. A third way of resolving the open-shell dilemma is to take the residual interaction into account, thus proceeding to a correlated state. A leading portion of the interaction can often be incorporated by pairing correlations. This will be discussed extensively in Sects. 9.3 and 9.4.

Pairing is an example of strong global correlations. There are often cases where correlations are weak at a global scale but can make, nonetheless, significant effects. We will discuss two examples. The first one is the van der Waals interaction (see Sect. 9.1) which is a long-range effect between atoms and/or molecules. The other is Hooke's atom model (Sect. 9.2.2) which allows to scrutinize short-range spatial correlations.

J.A. Maruhn et al., *Simple Models of Many-Fermion Systems*,
DOI 10.1007/978-3-642-03839-6_9, © Springer-Verlag Berlin Heidelberg 2010

9.1 Van der Waals Interaction

Van der Waals interactions play a key role to understand binding in rare gases in which atomic electronic closure leads to high ionization potentials and correspondingly vanishing chemical activity. Binding between rare-gas atoms is thus very faint and stems from correlations between dipole excitations of each atom. The perturbative approach allows to estimate this effect and provides an estimate of its magnitude.

9.1.1 Perturbative Estimate of Dipole–Dipole Correlations

The starting point is the standard formula for $2ph$ correlations from the Coulomb potential $V_{\rm Coul}$ in second-order perturbation theory

$$\Delta E^{(2)} = - \sum_{pp'hh'} \frac{\left|\langle\varphi_h\varphi_{h'}|V_{\rm Coul}|\varphi_p\varphi_{p'}\rangle\right|^2}{\varepsilon_p - \varepsilon_h + \varepsilon_{p'} - \varepsilon_{h'}}, \tag{9.1}$$

where the expression in the numerator is a two-particle matrix element and the denominator contains the unperturbed $2ph$ excitation energies. This expression is hard to evaluate in general and even diverges in the bulk limit [43]. Substantial simplifications are possible if we consider two systems separated by some distance from each other. The perturbation energy (9.1) then can be reduced to a very simple formula which is not only simple to evaluate but also provides an enlightening schematic view of correlations in general and of the van der Waals interaction in particular.

The basic assumption is that the system can be divided into two subsystems which have no spatial overlap, as is typically the case for two atomic or molecular systems with a relative distance R which is larger than each system's spatial dimensions. Let us henceforth call them system "1" and system "2" and label all relevant quantities with this system number. We are interested in the effect of correlations on the atom–atom interaction at long distances. Thus we consider only that part of the correlation energy (9.1) which has one $1ph$ excitation in "1" and the other in "2", i.e., we associate in (9.1) the pair (ph) to system 1 and $(p'h')$ to system "2". Now the fact can be exploited that the two systems have negligible spatial overlap and that the two systems are far from each other. It is then not necessary to stay strictly in the ph perturbational scheme. One can equally well consider the exact excitation spectrum of each of the separate systems at both sites. We call the ground states $|\psi_0^{(i)}\rangle$, the excited states $|\psi_N^{(i)}\rangle$, and $E_N^{(i)}$ their excitation energies. The $1ph$ state becomes the general excitation $(p_i h_i) \longrightarrow |\psi_N^{(i)}\rangle$ and $\varepsilon_{p_1} - \varepsilon_{h_1} \longrightarrow E_N^{(i)}$. This finally yields the van der Waals energy as

$$\Delta E^{(\mathrm{VdW})} = \sum_{N_1 N_2} \frac{\left|\langle \psi_{N_1}^{(1)} \psi_{N_2}^{(2)} | V_{\mathrm{Coul}} | \psi_0^{(1)} \psi_0^{(2)} \rangle\right|^2}{E_{N_1}^{(1)} + E_{N_2}^{(2)}}, \tag{9.2}$$

where $|\psi_{N_1}^{(1)} \psi_{N_2}^{(2)}\rangle$ stands for the combined states of systems 1 and 2. The situation is sketched in Fig. 9.1. We introduce the center of mass for each system $\mathbf{R}_i$ and express the coordinates as $\mathbf{r}_i = \mathbf{R}_i + \mathbf{x}_i$ where all relevant values of $\mathbf{x}_i$ are small compared to the overall distance R. In this case the standard multipole expansion for the Coulomb potential (Taylor expansion in $\mathbf{x}_1$ and $\mathbf{x}_2$) may be used up to second order, i.e.,

$$\begin{aligned}
\frac{e^2}{|\mathbf{r}_1 - \mathbf{x}_2|} &= \frac{e^2}{|\mathbf{x}_1 - \mathbf{x}_2 - \mathbf{R}|} \\
&= \frac{e^2}{R}\left[1 - 2\frac{\mathbf{e}_R \cdot (\mathbf{x}_1 - \mathbf{x}_2)}{R} - \frac{\mathbf{x}_1^2 + \mathbf{x}_2^2 - 2\mathbf{x}_1 \cdot \mathbf{x}_2}{R^2}\right]^{-1/2} \\
&\approx \underbrace{\frac{e^2}{R}}_{\text{monopole}} + \underbrace{e^2 \frac{\mathbf{e}_R \cdot (\mathbf{x}_1 - \mathbf{x}_2)}{R^2}}_{\text{monopole-dipole}} + \underbrace{e^2 \frac{\mathbf{x}_1 \cdot \mathbf{x}_2 - 3(\mathbf{e}_R \cdot \mathbf{x}_1)(\mathbf{e}_R \cdot \mathbf{x}_2)}{R^3}}_{\text{dipole-dipole}} \\
&\quad + \underbrace{e^2 \frac{\mathbf{x}_1^2 - 3(\mathbf{e}_R \cdot \mathbf{x}_1)^2 + \mathbf{x}_2^2 - 3(\mathbf{e}_R \cdot \mathbf{x}_2)^2}{2R^3}}_{\text{monopole-quadrupole}} .
\end{aligned}$$

The monopole operator e^2/R here is just a constant. The monopole transition elements then vanish due to orthogonality, i.e., $\langle \psi_{N_i}^{(i)} | \psi_0^{(i)} \rangle = 0$. There remains only the dipole–dipole coupling. This allows rewriting

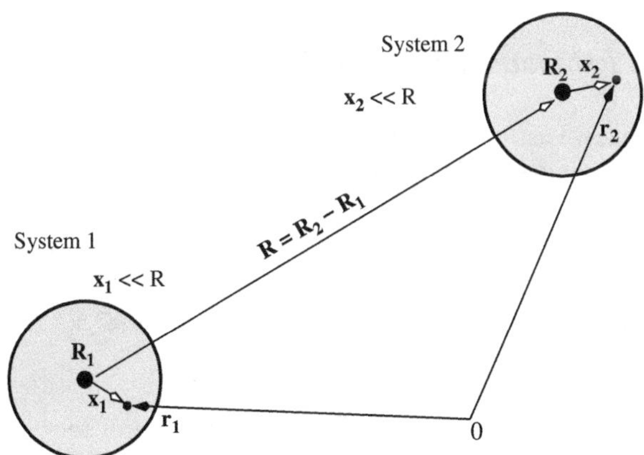

Fig. 9.1 The typical setup for deriving the van der Waals energy. The *grey spheres* indicate the extension of systems 1 and 2

$$\langle\psi_{N_1}^{(1)}\psi_{N_2}^{(2)}|V_{\rm Coul}|\psi_0^{(1)}\psi_0^{(2)}\rangle$$
$$\approx e^2\langle\psi_{N_1}^{(1)}\psi_{N_2}^{(2)}|\frac{\mathbf{x}_1\cdot\mathbf{x}_2-3(\mathbf{x}_1\cdot\mathbf{e}_R)(\mathbf{e}_R\cdot\mathbf{x}_2)}{|\mathbf{R}_1-\mathbf{R}_2|^3}|\psi_0^{(1)}\psi_0^{(2)}\rangle$$
$$=\frac{\langle\psi_{N_1}^{(1)}|\mathbf{x}_1|\psi_0^{(1)}\rangle\cdot\langle\psi_{N_2}^{(2)}|\mathbf{x}_2|\psi_0^{(2)}\rangle-3(\langle\psi_{N_1}^{(1)}|\mathbf{x}_1|\psi_0^{(1)}\rangle\cdot\mathbf{e}_R)(\mathbf{e}_R\cdot\langle\psi_{N_2}^{(2)}|\mathbf{x}_2|\psi_0^{(2)}\rangle)}{|\mathbf{R}_1-\mathbf{R}_2|^3}\,.$$

Thus we obtain, introducing the dipole operator $\hat{\mathbf{D}}_i$ of each system,

$$\Delta E^{(\rm VdW)}=\frac{-e^4}{|\mathbf{R}_1-\mathbf{R}_2|^6}\sum_{N_1N_2}\frac{\left|\mathbf{D}_{N_10}^{(1)}\cdot\mathbf{D}_{N_20}^{(2)}-3\mathbf{D}_{N_10}^{(1)}\cdot\mathbf{e}_R\mathbf{e}_R\cdot\mathbf{D}_{N_20}^{(2)}\right|^2}{E_{N_2}+E_{N_1}}\,,$$
$$\mathbf{D}_{N_i0}^{(i)}=\langle\psi_{N_i}|\hat{\mathbf{D}}_i|\psi_0\rangle_i\quad,\quad\hat{\mathbf{D}}_i=\mathbf{r}_i-\mathbf{R}_i\,,\qquad(9.3)$$

where the index i on the brackets of the matrix element $\mathbf{D}_{N_i0}^{(i)}$ indicates that this is an expectation value in the Hilbert space of system i.

The energy (9.3) depends on the distance of the two systems. It is negative as a second-order perturbative correction should be. It increases in value with decreasing distance, thus representing an attractive potential between the two systems, the well-known van der Waals potential. Its typical behavior $\propto|\mathbf{R}_1-\mathbf{R}_2|^{-6}$ emerges from the squared dipole–dipole interaction. It is interesting to note that the van der Waals contribution does not require the systems to have a non-vanishing ground-state dipole momentum $\langle\psi_0^{(i)}|\hat{\mathbf{D}}_i|\psi_0^{(i)}\rangle$. What is activated here are virtual dipole excitations driven by the perturbation. The "generation" of the two dipoles is weighted with $|\mathbf{R}_1-\mathbf{R}_2|^{-3}$ and the "interaction" between them explores another $|\mathbf{R}_1-\mathbf{R}_2|^{-3}$ which together yield the final $|\mathbf{R}_1-\mathbf{R}_2|^{-6}$ behavior.

9.1.2 Simple Estimate in Terms of Polarizability

For the following, we assume two spherically symmetric systems. It is then advantageous to recouple into spherical representation as

$$\hat{\mathbf{D}}^{(1)}\cdot\hat{\mathbf{D}}^{(2)}-3\hat{\mathbf{D}}^{(1)}\cdot\mathbf{e}_z\mathbf{e}_z\cdot\hat{\mathbf{D}}^{(2)}=\hat{D}_+^{(1)}\hat{D}_-^{(2)}+\hat{D}_-^{(1)}\hat{D}_+^{(2)}-2\hat{D}_z^{(1)}\hat{D}_z^{(2)}\,,$$
$$\hat{D}_\pm^{(i)}=\frac{\hat{D}_x^{(i)}\pm\mathrm{i}\hat{D}_y^{(i)}}{\sqrt{2}}\quad,\quad\hat{D}_m^{(i)}=\sqrt{\frac{4\pi}{3}}r_iY_{1m}(\Omega_i)\,,\qquad(9.4)$$

where we identify $\hat{D}_z=\hat{D}_0$ and where Ω_i is the solid angle in system i. The ground state $|\Psi_0\rangle$ has angular momentum zero and the excitations N_i are classified according to angular momentum. Thus the $\hat{D}_+^{(i)}$, $\hat{D}_-^{(i)}$, and $\hat{D}_0^{(i)}$ all couple to different excited states and the van der Waals energy becomes

$$\begin{aligned}\Delta E^{(\mathrm{VdW})} &= -e^4 \sum_{N_1 N_2} \frac{\sum_{m=-1,0,1}(1+3\delta_{m0}) \left|(\mathbf{D}_m^{(1)})_{N_1 0}\cdot(\mathbf{D}_{-m}^{(2)})_{N_2 0}\right|^2}{(E_{N_2}+E_{N_1})|\mathbf{R}_1-\mathbf{R}_2|^6} \\ &= \frac{-6e^4}{|\mathbf{R}_1-\mathbf{R}_2|^6} \sum_{N_1 N_2} \frac{\left|(\mathbf{D}_z^{(1)})_{N_1 0}\cdot(\mathbf{D}_z^{(2)})_{N_2 0}\right|^2}{E_{N_2}+E_{N_1}},\end{aligned}$$

where we have exploited the relation $|\langle\psi_N|\hat{D}_z|\psi_0\rangle|^2 = |\langle\psi_N|\hat{D}_+|\psi_0\rangle|^2 = |\langle\psi_N|\hat{D}_- |\psi_0\rangle|^2$ which, again, follows from spherical symmetry.

The dipole polarizability α_D of a system as computed by perturbation theory for spherically symmetric systems is [24]

$$\alpha_D = 2\sum_N \frac{|\langle\psi_N|\hat{D}_z|\psi_0\rangle|^2}{E_N}. \tag{9.5}$$

This expression for α_D has some similarity with the expression for the van der Waals energy. There is still a difference in the denominator where the van der Waals energy combines the two energies E_{N_2} and E_{N_1}. We specialize to symmetric cases where systems 1 and 2 belong to the same sort of atom having the same excitation spectrum and assume that the dipole response can be represented by one dominant dipole excitation N_D. This assumption is well justified for simple metal atoms (H, Li, Na, K, Cs, Rb) and for rare gases (He, Ne, Ar, Kr, Xe). Reduction to one relevant excitation yields a simplified polarizability and van der Waals energy

$$\begin{aligned}\alpha_D &= 2e^2 \frac{|\langle\psi_{N_D}|\hat{D}_z|\psi_0\rangle|^2}{E_{N_D}}, \\ \Delta E^{(\mathrm{VdW})} &= \frac{-6e^4}{|\mathbf{R}_1-\mathbf{R}_2|^6} \frac{\left|\langle\psi_{N_D}|\hat{D}_z^{(1)}|\psi_0\rangle\right|^2 \left|\langle\psi_{N_D}|\hat{D}_z^{(2)}|\psi_0\rangle\right|^2}{2E_{N_D}}.\end{aligned}$$

This can finally be resolved to

$$\Delta E^{(\mathrm{VdW})} = -\frac{3}{4}\frac{\alpha_D^2 E_{N_D}}{|\mathbf{R}_1-\mathbf{R}_2|^6}, \tag{9.6}$$

which allows very simple order-of-magnitude estimates for several dimers. Figure 9.2 compares the van der Waals energies for several dimers. It shows the pure van der Waals energy for the H_2 dimer and the full atom–atom interaction potentials for the He_2 and Ar_2 dimer. The latter is known as the Lennard–Jones potential [3] and its long-range, attractive part is just the van der Waals interaction (9.6). The atomic dipole polarizability for each system is also indicated. The relation between α_D and the van der Waals attraction is obvious. It is to be noted that the van der Waals potential is responsible for all the molecular attraction only in the case of the

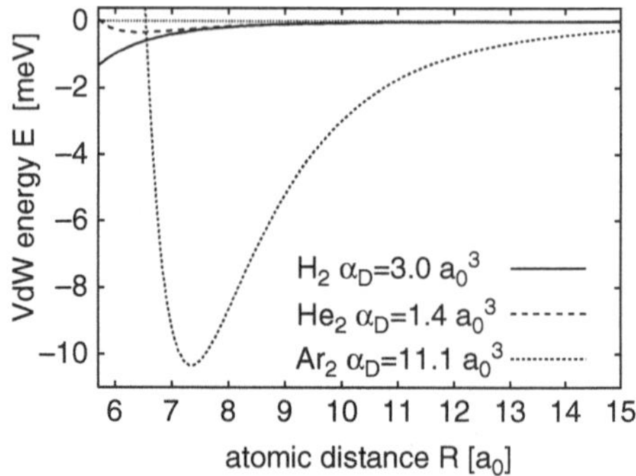

Fig. 9.2 The Born–Oppenheimer (BO) energy curves for several rare-gas dimer molecules as indicated scaled to have zero energy in the asymptotics of two separate atoms. Only the van der Waals energy is shown for H_2 and the full interaction potential for the two rare-gas dimers, He_2 and Ar_2

rare gases where all other molecular binding mechanisms do not contribute. The van der Waals potential is at best a small correction in all other molecules dominated by covalent, ionic, or metallic binding (see Chap. 4). For the H_2 dimer, e.g., the covalent binding overrules van der Waals by orders of magnitude and is thus not included in the figure.

9.2 Electron Correlations in Small Systems

9.2.1 Empirical Findings

Modern experimental techniques such as collisions of highly charged and fast ions on atoms or intense and short laser pulses combined with elaborate coincidence measurements have allowed to obtain information on spatial electron correlations in atoms, for a review see [28]. An example was given in Fig. 1.15. This motivates us to present a model study of spatial correlations adapted to a two-electron system. It deals with a widely used, analytically solvable model called *Hooke's atom.*

9.2.2 Hooke's Atom

Hooke's atom is a model for two electrons in an external field having the Hamiltonian

$$\hat{H} = -\frac{\hbar^2}{2m}\left(\varDelta_{\mathbf{r}_1} + \varDelta_{\mathbf{r}_2}\right) + \frac{c}{2}\left(r_1^2 + r_2^2\right) + \frac{e^2}{|\mathbf{r}_1 - \mathbf{r}_2|}. \tag{9.7}$$

It is the Hamiltonian (5.27) of the He atom, but with the external Coulomb fields $\propto e^2/r_i$ replaced by harmonic potentials $\propto r_i^2$. The basic mechanisms, binding through external field versus electron–electron repulsion, remain the same, but the harmonic external field allows to exactly separate relative from center of mass coordinates, which is the key to simplicity. At first glance this modeling appears like wishful thinking, as it leads to a solvable model. Yet Hooke's atom also has a touch of reality. It corresponds to a quantum dot with two electrons [23]. The harmonic confining field is a reality for quantum dots.

Hooke's atom model is used for a great variety of studies. It is a nice testing ground for developments in DFT (see, e.g., [2]). Such investigations, however, go deeply into the subtleties of DFT and thus beyond the scope of this book. Another useful aspect of the model is that it provides a correlated wave function in closed form, which allows analyzing correlation effects in coordinate or momentum space. It is this second line of application which we will follow here.

9.2.2.1 The Analytic Solution for the Ground State

We introduce center of mass and relative coordinates via

$$\mathbf{R}=\frac{1}{2}(\mathbf{r}_1+\mathbf{r}_2)\,,\quad \mathbf{r}=\mathbf{r}_1-\mathbf{r}_2 \quad\longleftrightarrow\quad \mathbf{r}_{1,2}=\mathbf{R}\pm\frac{1}{2}\mathbf{r}\,, \tag{9.8}$$

use the well-known separation of the kinetic energy in these coordinates, and obtain the Hamiltonian in the form

$$\hat{H}=-\frac{\hbar^2}{4m}\Delta_{\mathbf{R}}+cR^2-\frac{\hbar^2}{m}\Delta_{\mathbf{r}}+\frac{c}{4}r^2+\frac{e^2}{r}, \tag{9.9}$$

where $R=|\mathbf{R}|$ and $r=|\mathbf{r}|$ as usual.

The equation to be solved is the two-electron Schrödinger equation

$$\hat{H}\Psi(x_1,x_2)=E\Psi(x_1,x_2), \tag{9.10}$$

with x_1 and x_2 referring to the combination of spin and spatial coordinates. We are interested in the ground-state solution, for which the spin is in the antisymmetric state and the spatial wave function remains symmetric. As the model Hamiltonian (9.9) separates into relative and c.m. motion, we can separate the spatial ground-state wave function similarly and obtain

$$\Psi(x_1,x_2)=\tilde{\Psi}(\mathbf{R})\psi(\mathbf{r})\frac{1}{\sqrt{2}}\left(\chi^{(1)}_{+\frac{1}{2}}\chi^{(2)}_{-\frac{1}{2}}-\chi^{(1)}_{-\frac{1}{2}}\chi^{(2)}_{+\frac{1}{2}}\right), \tag{9.11a}$$

$$\left(-\frac{\hbar^2}{4m}\Delta_{\mathbf{R}}+cR^2\right)\tilde{\Psi}=E_{\mathrm{cm}}\tilde{\Psi}\,, \tag{9.11b}$$

$$\left(-\frac{\hbar^2}{m}\Delta_{\mathbf{r}}+\frac{c}{4}r^2+\frac{e^2}{r}\right)\psi=E_{\mathrm{rel}}\psi\,. \tag{9.11c}$$

The center of mass equation (9.11b) is a pure oscillator problem with the standard solution [24]

$$\tilde{\Psi}(R) = \left(\frac{\Lambda}{\pi}\right)^{3/4} \exp\left(-\frac{\Lambda}{2}R^2\right), \quad \Lambda = \frac{2\sqrt{mc}}{\hbar}. \tag{9.12}$$

The asymptotic behavior for $r \longrightarrow \infty$ of (9.11c) for the relative wave function is also governed by a harmonic potential. Thus we make the ansatz

$$\psi = f(r)\exp\left(-\frac{\lambda}{2}r^2\right), \quad \lambda = \frac{\sqrt{mc}}{2\hbar},$$

where $f(r)$ is in general a series in r. The equation determining $f(r)$ becomes

$$-\frac{1}{r}\partial_r^2(rf) - 2\lambda r\partial_r f + \frac{me^2}{\hbar^2 r}f = \underbrace{\left(\frac{2mE_{\text{rel}}}{\hbar^2} - \frac{\hbar}{2}\sqrt{\frac{c}{m}}\right)}_{\tilde{E}} f.$$

Inserting the power series ansatz $f = \sum_n c_n r^n$ yields a recurrence relation for the coefficients as

$$c_{n+1} = \frac{me^2}{\hbar^2(n+1)(n+2)}c_n + \frac{\tilde{E} + 2\lambda(n-1)}{(n+1)(n+2)}c_{n-1},$$

which may be iterated to

$$c_0 = 1, \quad c_1 = \frac{me^2}{2\hbar^2}, \quad c_2 = +\left(\frac{me^2}{2\hbar^2}\right)^2\frac{1}{3} + \frac{\tilde{E}}{6},$$
$$c_3 = +\left(\frac{me^2}{2\hbar^2}\right)^3\frac{1}{18} + \frac{me^2}{2\hbar^2}\frac{\tilde{E}}{9} + \frac{me^2}{2\hbar^2}\frac{2\lambda}{12}.$$

The choice of c_0 is arbitrary and selected for convenience. The aim is to terminate the series as early as possible. The coefficient c_1 is fixed by the interaction strength. We try to arrange $c_2 = 0 = c_3$ which then would guarantee the disappearance of all further terms. The condition $c_2 = 0$ determines $\tilde{E} = -2(me^2/(2\hbar^2))^2$, which is inserted into condition $c_3 = 0$. This yields

$$\lambda = -\left(\frac{me^2}{2\hbar^2}\right)^2\frac{1}{3} + \left(\frac{me^2}{2\hbar^2}\right)^2\frac{4}{3} = \left(\frac{me^2}{2\hbar^2}\right)^2. \tag{9.13}$$

It turns out that the series terminates for one certain choice of c or λ, respectively. We take the liberty to choose that parameter according to (9.13) and so obtain an exactly solvable model for two electrons interacting via Coulomb repulsion.

Altogether, the model and its solution are

$$\hat{H} = -\frac{\hbar^2}{2m}\left(\Delta_{\mathbf{r}_1} + \Delta_{\mathbf{r}_2}\right) + 2\frac{\hbar^2}{2m}\lambda^2\left(r_1^2 + r_2^2\right) + \frac{e^2}{|\mathbf{r}_1 - \mathbf{r}_2|},$$

$$\Psi \propto \left(1 + \sqrt{\lambda}r\right)\mathrm{e}^{-\lambda(r_1^2+r_2^2)}\frac{1}{\sqrt{2}}\left(\chi^{(1)}_{+\frac{1}{2}}\chi^{(2)}_{-\frac{1}{2}} - \chi^{(1)}_{-\frac{1}{2}}\chi^{(2)}_{+\frac{1}{2}}\right),$$

with λ given by (9.13). The correlation is contained in the factor $1 + \sqrt{\lambda}r = 1 + \sqrt{\lambda}|\mathbf{r}_1 - \mathbf{r}_2|$, having a non-analytic point, a cusp, at $\mathbf{r}_1 \longrightarrow \mathbf{r}_2$, which is obviously related to the singularity of the repulsive electron–electron interaction at that point of contact. The result should be compared with a HF solution which, however, cannot be worked out analytically. But there are some qualitative features which can safely be concluded. The one-body density is regular at the origin and starts out as $1 + cr_i^2$. The same property is acquired by the self-consistent mean field. This, in turn, yields single-particle wave functions which also start with $1 + c'r_i^2$. The behavior at small relative distances is then also a smooth function starting also as $1 + c''(\mathbf{r}_1 - \mathbf{r}_2)^2$, quite different from the non-analytic behavior of the correlated wave function Ψ.

The further discrimination of correlation effects versus mean-field properties can be carried forth analytically to a large extent but becomes quite involved. We close the example with a few qualitative remarks. The most prominent global measure is, of course, the correlation energy, i.e., the energy gain when proceeding from a Hartree–Fock (HF) calculation to a fully correlated wave function. This has traditionally been studied intensively for the He atom, and it is found that correlations in the energy are rather weak (compare the exact energy with the HF result in Table 6.1) and can be well described by perturbative corrections [53]. A more specific and still global measure for correlations is the entanglement L which can be deduced from the one-body density matrix as

$$L = \mathrm{tr}\{\hat{\varrho}_1\} - \mathrm{tr}\{\hat{\varrho}_1^2\}, \quad \varrho_1(x, x') = \int dx_2 \Psi^*(x, x_2)\Psi(x', x_2).$$

This measure exploits the fact that the one-body density from a pure mean-field state is a projector, i.e., $\hat{\varrho}_1^2 = \hat{\varrho}_1$, in which case we would have $L = 0$. Correlated states produce a $\hat{\varrho}_1$ with eigenvalues $0 \leq \varrho_n \leq 1$. The trace is the sum of eigenvalues and $\varrho_n^2 \leq \varrho_n$. Thus we find $L \geq 0$ and the limit $L = 0$ indicates an uncorrelated state. The entanglement has been evaluated for Hooke's atom in [23]. Considering a quantum dot of width $\lambda = 10$ nm and CdS as carrier material, one finds $L \approx 0.005$ which is an extremely small value, again indicating weak correlations at a global scale. On the other hand, there is a strong, even qualitative, correlation effect when looking at spatial structures with high resolution. The correlated wave function at short relative distance has a kink $\propto \lambda|\mathbf{r}_1 - \mathbf{r}_2|$. This is a non-analytic point which cannot be reproduced by any perturbation series. A similar kink is to be expected for the correlated wave function of the He atom, and there are model wave functions which explicitly take that into account as, e.g., the Hylleraas ansatz [103]. The question remains how one could access these spatial details experimentally. This is done indirectly through measuring the momentum distributions [28]. A study of momentum-space effects in Hooke's atom compared to a realistic He atom is found in [76], again analytically solvable but cumbersome. The effects are subtle, though, and would require high resolution to be seen.

9.3 Pairing Correlations in a Simple Model

9.3.1 Evidence for Pairing Correlations

As discussed in several places in this book, there often arise situations where the mean-field description produces a high-level density near the Fermi energy which, in turn, means a high density of unperturbed excitations with low energy. We know from quantum-mechanical perturbation theory that small energy differences between unperturbed states can make perturbations look relatively large, so that a full treatment of them in the subspace near degeneracy becomes compulsory [24]. In the many-body context this means that the residual interaction can cause large effects. It adds a substantial amount to binding and re-establishes the gap in the excitation spectrum. Probably the most prominent member of this class of mechanisms is the pairing gap appearing in a great variety of many-fermion systems. In this section the basic mechanisms of pairing correlations will be discussed in terms of simple models.

A well-known example of pairing is superconductivity in metals [91, 79, 3]. The traditionally most prominent example for pairing in finite systems is atomic nuclei where the pairing gap is identified from the odd–even staggering of binding energies and, less directly, from superfluidity in nuclear rotations [17, 18, 92, 42]. More recent achievements are the observation of superfluidity in liquid ^{3}He (bulk as well as finite droplets) [41, 111] and in fermion clouds in atomic traps [14]. Table 9.1 gives an order-of-magnitude overview of physical systems where pairing plays a role. The finite members of that table are fermionic atoms in a trap, droplets of ^{3}He, and nuclei. The mechanisms for pairing are very similar in all systems such that the schematic models outlined below have a broad range of applications. Note that the

Table 9.1 A selection of systems where pairing plays a role with their Fermi energy ϵ_{F} and critical temperature T_c. The pairing gap is typically twice T_c. These physical quantities indicate orders of magnitude

System	Fermion	ϵ_{F}	T_c	Evidence
Atoms trap	Whole atom	nK	nK	Superfluidity
Liquid ^{3}He	^{3}He atom	2.7 K	2.5 mK	Superfluidity
Bulk Al	Electron	11.7 eV	1.2 K	Superconductivity
Bulk Nb	Electron	5.3 eV	9.3 K	Superconductivity
Nuclei	Nucleons	≈35 MeV	0.5–1 MeV	Odd–even staggering
Neutron star	Neutron	≈ 60 MeV	∼ 1 MeV	Superfluidity

famous high-temperature superconductors were not included in the discussion nor in Table 9.1 because their pairing mechanisms are not yet fully understood [72] and, up to now, they are all bulk material.

9.3.2 A Schematic Description of the Pairing Interaction

The starting point is a case where the pure mean-field description in terms of single-fermion states produces an open-shell configuration with zero or very small $1ph$ excitation energies, so that the residual interaction can become important. The full Hamiltonian has to be reconsidered. We write it in the form

$$\hat{H} = \sum_{\alpha} \varepsilon_\alpha \hat{a}^\dagger_\alpha \hat{a}_\alpha + \sum_{\alpha\beta\gamma\delta} V^{(\text{res})}_{\alpha\beta\delta\gamma} \hat{a}^\dagger_\alpha \hat{a}^\dagger_\beta \hat{a}_\gamma \hat{a}_\delta , \tag{9.14}$$

where the optimized mean field already went into the single-particle states α and energies ε_α and where the residual interaction $V^{(\text{res})}$ embraces everything not accounted for by the mean field (see Sect. 5). The $1ph$–$1ph$ residual interaction is dealt with in the RPA, see Chap. 8. Much more important for the open-shell situation is the part of the residual interaction which describes two-fermion encounters. We point that out by collecting the pair operators in brackets

$$\hat{V}^{(\text{res})} \longrightarrow \sum_{\alpha\beta\gamma\delta} V^{(\text{res})}_{\alpha\beta\delta\gamma} \left(\hat{a}^\dagger_\alpha \hat{a}^\dagger_\beta\right) \left(\hat{a}_\gamma \hat{a}_\delta\right) .$$

It is now found that the pp residual interaction is strongest between pairs of time-reversal conjugate states. In bulk systems, these are the partners of opposite momentum and spin, i.e., $\varphi_{\mathbf{k}\uparrow} \longleftrightarrow \varphi_{-\mathbf{k}\downarrow}$. In finite systems, we associate the partners in Kramers-degenerate doublets (see Appendix A.6), e.g., for axially symmetric systems the states $\varphi_{nm,1/2} \longleftrightarrow \varphi_{n-m,-1/2}$ where n is the combination of principal quantum numbers, m the z-component of orbital angular momentum, and $\pm\frac{1}{2}$ labels the spin (for the quantum numbers, see the discussion of the cylindrical oscillator in A.1.2). The strong coupling between Kramers pairs is due to a short range of the interaction, which favors coupling between states with the same spatial probability distribution.

To stay independent of a system's symmetry, we will henceforth indicate these conjugate states by the general labeling $+\alpha \longleftrightarrow -\alpha$. The most relevant part of the residual interaction then becomes

$$\hat{V}^{(\text{res})} \longrightarrow \hat{V}^{(\text{pair})} = \sum_{\alpha\beta} V^{(\text{pair})}_{\alpha\beta} \left(\hat{a}^\dagger_\alpha \hat{a}^\dagger_{-\alpha}\right) \left(\hat{a}_{-\beta} \hat{a}_\beta\right) , \quad V^{(\text{pair})}_{\alpha\beta} = V^{(\text{res})}_{\alpha-\alpha\beta-\beta} .$$

A further piece of experience is that the matrix elements $V^{(\text{pair})}_{\alpha\beta}$ have a maximum near the Fermi energy, i.e., for $\varepsilon_\alpha, \varepsilon_\beta \approx \epsilon_{\text{F}}$ and decrease quickly when moving away from ϵ_{F}. There remains a narrow zone ("active" zone Ω) where pairing is

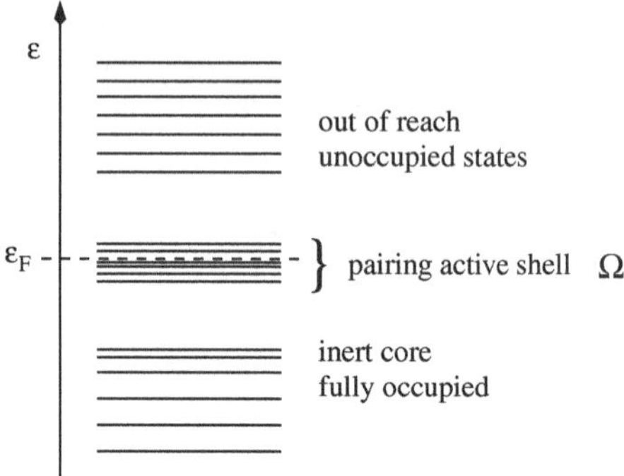

Fig. 9.3 Sketch of a typical level sequence for a many-fermion system with the pairing active shell near the Fermi surface $\epsilon_{\rm F}$

active as sketched in Fig. 9.3. This suggests simplifying the residual interaction to a schematic pairing interaction, thus dealing with

$$\hat{H} = \sum_\alpha \varepsilon_\alpha \hat{a}^\dagger_\alpha \hat{a}_\alpha - G \underbrace{\sum_\alpha \hat{a}^\dagger_{+\alpha} \hat{a}^\dagger_{-\alpha}}_{\hat{A}^\dagger} \underbrace{\sum_\beta \hat{a}_{-\beta} \hat{a}_{+\beta}}_{\hat{A}} . \tag{9.15}$$

The single-particle spectrum needs further modeling if one aims at analytic solutions. We will discuss the seniority model which approximates the active space Ω by one degenerate shell, in Sect. 9.3 and the Fermi gas model as introduced in Sect. 2.2.

It is important to note that the pair creator $\hat{A}^\dagger$ and the annihilator $\hat{A}$ together with a third operator $\hat{S}_0 = [\hat{A}^\dagger, \hat{A}]$ form a quasispin algebra such that the analytic solution can reuse many of the useful relations established for the quasispin model in Sect. 7.2.

9.3.3 Seniority Model and Quasispin Algebra

The seniority model starts from the model Hamiltonian (9.15) and simplifies it further by assuming that all single-particle energies in the pairing active shell (see Fig. 9.3) are practically degenerate, i.e., $\varepsilon_\alpha = \varepsilon =$ constant. We thus can ignore the single-particle term and remain with the seniority Hamiltonian

$$\hat{H} = \hat{V}_{\rm P} = -G\hat{A}^\dagger \hat{A} , \quad \hat{A}^\dagger = \sum_{\alpha=1}^{\Omega} \hat{a}^\dagger_{+\alpha} \hat{a}^\dagger_{-\alpha} . \tag{9.16}$$

The degenerate shell covers 2Ω single-particle states. The particle number thus can range from $N = 0$ (empty shell) up to $N = 2\Omega$ (totally filled shell). We are now going to develop the spectrum for this seniority Hamiltonian. Thereby we will confine ourselves to the case of even particle numbers N with the limiting cases $N = 0$ and $N = 2\Omega$.

First, we realize that the operators $\hat{A}^\dagger$ and $\hat{A}$ together with the particle number operator $\hat{N}$ form a quasispin algebra. We can identify

$$\hat{J}_+ = \hat{A}^\dagger = \sum_{\alpha=1}^{\Omega} \hat{a}_\alpha^\dagger \hat{a}_{-\alpha}^\dagger ,$$
$$\hat{J}_- = \hat{A} = \sum_{\alpha=1}^{\Omega} \hat{a}_{-\alpha} \hat{a}_\alpha ,$$
$$\hat{J}_0 = \tfrac{1}{2} \sum_{\alpha=1}^{\Omega} \left(\hat{a}_\alpha^\dagger \hat{a}_\alpha + \hat{a}_{-\alpha}^\dagger \hat{a}_{-\alpha} - 1 \right) = \tfrac{1}{2}\left(\hat{N} - \Omega \right) . \tag{9.17}$$

It is straightforward to show that these operators $\hat{J}_\pm$ and $\hat{J}_0$ fulfill the quasispin algebra (7.3). Thus all useful consequences of angular momentum algebra can be taken over from Chap. 7.

Note that the total quasispin can be decomposed as

$$\hat{J}^2 = \hat{J}_+ \hat{J}_- - \hat{J}_0 + \hat{J}_0^2 ,$$

which yields the pairing potential in the form

$$\hat{V}_\mathrm{P} = -G\left(\hat{J}^2 - \hat{J}_0^2 + \hat{J}_0 \right) . \tag{9.18}$$

In this way it can be expressed completely in terms of the diagonal quasispin operators. The problem is thus essentially solved, at least formally. The eigenvalues of $\hat{J}_0$ are simply $j_0 = \frac{1}{2}\left(N - \Omega\right)$, where N is the actual particle number. This eigenvalue is fixed for given N. The different classes of solutions for given N are characterized by the principal quantum number j for spectrum of $\hat{J}^2$. The allowed values for J are the integers or half-integers $j = |j_0|, |j_0| + 1, \ldots, \frac{1}{2}\Omega$ (note that the sequence proceeds in steps of 1). This yields the spectrum for an N-particle system

$$E_j^{(N)} = -G\left[j(j+1) - \tfrac{1}{4}(N-\Omega)^2 + \tfrac{1}{2}(N-\Omega) \right] , \tag{9.19}$$
$$j = \frac{|N-\Omega|}{2}, \ldots, \frac{\Omega}{2} - 1, \frac{\Omega}{2} .$$

It is obvious that the ground state is the state with maximum quasispin $j = \frac{1}{2}\Omega$ and the highest excited state the one with lowest quasispin $j = \frac{1}{2}|\left(N - \Omega\right)|$. It is to be noted that the spin quantum numbers j and j_0 yield only part of the classifying quantum numbers, fortunately just the part which suffices to compute the spectrum. We do not need to specify all other quantum numbers but want to find out the degeneracy of a given energy level in the spectrum (9.19). To this end, it has to be examined in more detail how the states are built.

The starting point is the state $|0\rangle$ associated with an empty shell, i.e., $N = 0$. It is a non-degenerate state with the properties

$$N = 0 \quad \longleftrightarrow \quad |0\rangle \ : \quad j_0 = -\frac{\Omega}{2}, \quad j = |j_0| = \frac{\Omega}{2},$$

much in analogy to the ground state in the LMG model, see Sect. 7.3. The next stage along even particle numbers is $N = 2$. The value of $N = 2$ fixes $j_0 = \frac{1}{2}(2-\Omega)$ and in turn the two possible values of j, namely $j = |j_0| = |\frac{1}{2}(2-\Omega)| = \frac{1}{2}\Omega - 1$ and $j = j_0 + 1 = \frac{1}{2}\Omega$, the largest acceptable value of j. There are $\Omega(2\Omega - 1)$ ways (binomial of 2Ω over 2) to build an $N = 2$ state $\hat{a}^\dagger_\alpha \hat{a}^\dagger_\beta|0\rangle$. The ground state is built from the coherent superposition of pairs

$$\hat{A}^\dagger|0\rangle = \hat{J}_+|0\rangle \quad \Longrightarrow \quad j = \frac{\Omega}{2}, \quad j_0 = -\frac{\Omega}{2} + 1 = \frac{2-\Omega}{2}.$$

This ground state is non-degenerate because $\hat{A}^\dagger = \hat{J}_+$ is unique. There remain the combinations of $\hat{a}^\dagger_\alpha \hat{a}^\dagger_\beta$ orthogonal to $\hat{A}^\dagger$. We denote them by $\hat{B}^{(2)}_i|0\rangle$, $i = 1, \ldots, \Omega(2\Omega-1)-1$ and orthogonality implies $\langle 0|\hat{A}\hat{B}^{(2)}_i|0\rangle = 0$. By the rules of angular momentum coupling, all $\hat{B}^{(2)}_i$ states have to belong to the only alternative $j = \frac{1}{2}\Omega - 1$ which complies perfectly well with the degeneracy of $\Omega(2\Omega-1)-1$. The energies of the quasispin states are given by the general expression (9.19) using the specific values of j defined above for $N = 2$. Altogether, the spectrum becomes

state	quasispin		energy	degeneracy
$\hat{A}^\dagger|0\rangle$	$j = \frac{\Omega}{2}$	:	$-G\Omega$	$\nu = 1$
$\hat{B}^{(2)}_i|0\rangle$	$j = \frac{\Omega}{2} - 1$	:	0	$\nu = \binom{2\Omega}{2} - 1$

The ground state gathers all binding energy from the seniority Hamiltonian and the many other states, generated by the $\hat{B}^{(2)}_i$, remain at the unperturbed value. For $N = 4$, the total number of quasispin states equals the binomial of 2Ω over 4. They are distributed over ground-state $(\hat{A}^\dagger)^2|0\rangle$, again unique with maximum $j = \frac{1}{2}\Omega$, a group of states $\hat{A}^\dagger\hat{B}^{(2)}_i|0\rangle$ with $j = \frac{1}{2}\Omega$, and a group constructed from new generators $\hat{B}^{(4)}_i$. This can be continued to general N. The degeneracies of the various states are attained in a recursive way by counting the number of excitations. The degeneracies sum up to the total number of states for a given N, namely, the binomial of 2Ω over N.

state	quasispin		energy	degeneracy
$(\hat{A}^\dagger)^{N/2}\|0\rangle$	$j=\frac{\Omega}{2}$	:	$E^{(N)}_{\Omega/2}$	$\nu=1$
$(\hat{A}^\dagger)^{N/2-1}\hat{B}^{(2)}_i\|0\rangle$	$j=\frac{\Omega}{2}-1$	:	$E^{(N)}_{\Omega/2-1}$	$\nu=\binom{2\Omega}{2}-1$
$(\hat{A}^\dagger)^{N/2-2}\hat{B}^{(4)}_i\|0\rangle$	$j=\frac{\Omega}{2}-2$	:	$E^{(N)}_{\Omega/2-2}$	$\nu=\binom{2\Omega}{4}-\binom{2\Omega}{2}$
.	.		.	.
.	.		.	.
.	.		.	.
$\hat{B}^{(N)}_i\|0\rangle$	$j=\frac{\Omega-N}{2}$	:	$E^{(N)}_{(\Omega-N)/2}$	$\nu=\binom{2\Omega}{N}-\binom{2\Omega}{N-2}$

up to $N \leq \Omega$. That series holds up to $N \leq \Omega$. The number of states shrinks for larger N as the binomials shrink. The extreme of a totally filled shell again has only one unique state. The branch $N > \Omega$ can be coped with by starting from the filled shell $N = 2\Omega$ and depleting it successively, thus considering the holes as quasi-particles. We will not detail that here.

So far, we have characterized the states by j_0, which is equivalent to particle number N, and by the total quasispin j. The latter quantum number is often expressed in terms of a more physical quantity, the *seniority* s, which is the number of particles not generated in pairs by $\hat{A}^\dagger$. The ground-state $(\hat{A}^\dagger)^{N/2}|0\rangle$ is fully paired and has $s = 0$. The next excitation has $s = 2$ and so forth. Thus we have in general the relation between j and s as $s = \Omega - 2j$.

Figure 9.4 shows the spectrum of the seniority model for the case of $\Omega = 6$ (the trivial cases of empty or filled shell are omitted). The fully paired ground state (seniority $s = 0$) is always well separated in energy from the many other states. The energy difference between the ground state ($s = 0$ or $j = \Omega/2$) and the multitude of first excited states ($s = 2$ or $j = \Omega/2 - 1$) is always $\Delta E = G\Omega$, as indicated in the figure. This is the well-known pairing gap, which separates the unique, fully paired ground states from the many other states in a system.

9.3.4 Odd N and Odd–Even Staggering

Figure 9.4 has shown one important observable consequence of pairing, namely, the gap in the excitation spectrum. It leads to effects as superfluidity and superconductivity in very large systems or bulk. Another important effect, which is particularly well visible in small systems, is the *odd–even staggering* of binding energies. Odd particle number N means that there is at least one non-paired particle which will thus be "kept apart" from the particles taking part in pairing. Let us assume that the odd particle resides in state β. Then both states β and $-\beta$ will not contribute

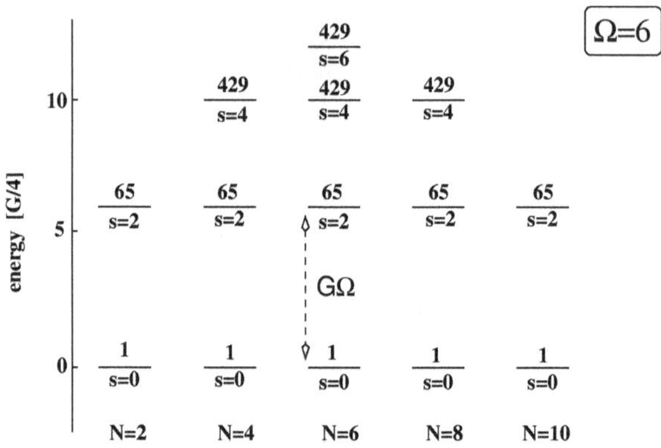

Fig. 9.4 The spectrum of the seniority model for $\Omega = 6$. The states are labeled in terms of particle number N and seniority $s = \Omega - 2j$. The number *above a line* indicates the degeneracy ν. The spectra are scaled to place all ground states at energy zero

to the $\sum_\alpha$ in $\hat{A}^\dagger$ and $\hat{A}$, so that odd systems are consistently less well bound. The energy difference to even systems is directly related to the pairing gap as we will see shortly.

The ground state of an odd system contains as many pairs as possible and one unpaired particle remains, i.e.,

$$|\Phi_{N,\beta}\rangle = (\hat{A}^\dagger)^{(N-1)/2}\hat{a}^\dagger_\beta|0\rangle \quad \longleftrightarrow \quad E_{\text{odd}}(N) = E^{(N-1)}_{\Omega/2-1}\,, \tag{9.20}$$

with $E^{(N-1)}_{\Omega/2-1}$ as given in (9.19). Note that the ground state is not unique but can be built with any choice for the odd-particle state β. The energy is found very quickly. The one state $\hat{a}^\dagger_\beta$ effectively removes the pair $\hat{a}^\dagger_\beta\hat{a}^\dagger_{-\beta}$ from the sum in the operator $\hat{A}^\dagger$ and similarly in $\hat{A}$. What remains is a seniority model for $N-1$ particles in a shell of $\Omega - 2$ states, which produces precisely the energy as given in (9.20).

The ground-state energy of an even system is $E_{\text{even}}(N) = E^{(N)}_{\Omega/2}$. The staggering between better-bound even systems and less well-bound odd ones can be characterized by the second difference of energies $\Delta^{(3)}_N$. With trivial algebra we find the three-point formula

$$\begin{aligned}\Delta^{(3)}_N &= \frac{1}{2}\left[E_{\text{even}}(N+1) - 2E_{\text{odd}}(N) + E_{\text{even}}(N-1)\right] \\ &= \frac{1}{2}\left[E^{(N+1)}_{\Omega/2} - 2E^{(N-1)}_{\Omega/2-1} + E^{(N-1)}_{\Omega/2}\right] = \frac{G\Omega}{2} \equiv \Delta\,.\end{aligned} \tag{9.21}$$

This shows the relation between the odd–even staggering quantified in $\Delta^{(3)}_N$ and the gap Δ in the excitation spectrum. In fact, the odd-even staggering is a key observable to assess pairing gaps in nuclei. Figure 9.5 shows $\Delta^{(3)}_N \equiv \Delta$ for a great variety of nuclei. It is to be noted that two gaps can be discussed in nuclei, for neutrons and for protons. Here we show the neutron gap deduced from varying neutron number

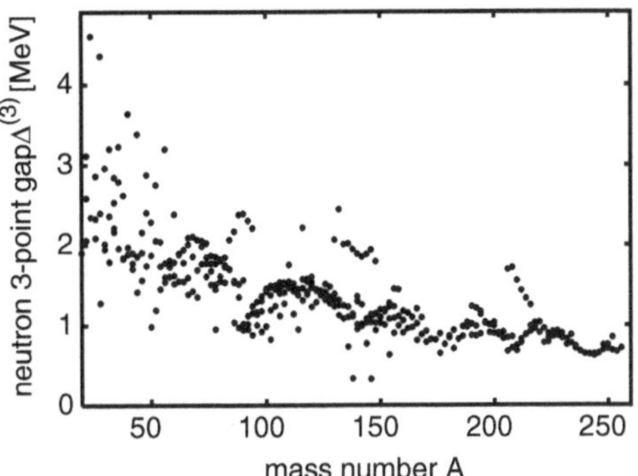

Fig. 9.5 The odd–even staggering (9.21) for the neutron gap in nuclei for a broad variety of sizes drawn versus total nucleon number A. The vicinity of magic neutron shells ($N = 8, 20, 28, 50, 82, 126$) has been excluded to avoid artifacts from shell effects

by ± 1 while keeping the proton number fixed. The figure shows that the nuclear pairing gap has only a weak dependence on particle number A. We will come back to this feature at the end of this chapter.

9.4 Pairing and the BCS Model

9.4.1 The BCS State

The seniority and quasispin models illustrate the action of the pairing force and the interplay of fermion and boson-like behavior quite well, but are severely restricted by the assumption of a partially filled single shell of states with a quite special form for the pairing potential. In a more general situation, the pairing potential will not treat all levels on the same footing. But Kramers degeneracy, related to time-reversal invariance [70] (see Appendix A.6), ensures the existence of pairs of degenerate, mutually time-reversed conjugate states. It turns out that these are still strongly coupled pairwise by the pairing force. An example is continuous systems where $-\alpha$ stands for the time-reversed state of α having opposite momentum and spin. Another example is spherically symmetric systems where $-\alpha$ stands just for the state which has the z-component of angular momentum reversed as compared to α. In the following α will be used to represent all quantum numbers of the single-particle states; the only important property we need is that there are always two states α and $-\alpha$ related to each other by time reversal and coupled preferentially by the pairing force.

For the Hamiltonian, we take up the reasoning in Sect. 9.3.1 leading to the form (9.15) which we write here as

$$\hat{H} = \sum_{\alpha} \varepsilon_{\alpha}^{0} \, \hat{a}_{\alpha}^{\dagger} \hat{a}_{\alpha} - G \sum_{\alpha\alpha'>0} \hat{a}_{\alpha}^{\dagger} \hat{a}_{-\alpha}^{\dagger} \hat{a}_{-\alpha'} \hat{a}_{\alpha'} \,, \tag{9.22}$$

where we now allow the ε_α^0 to be non-degenerate. The eigenstates in this case cannot be constructed analytically, but there is an approximate solution based on the *BCS state* named after Bardeen, Cooper, and Schrieffer [7]:

$$|\text{BCS}\rangle = \prod_{\alpha>0}^{\infty} \left(u_\alpha + v_\alpha \hat{a}_\alpha^\dagger \hat{a}_{-\alpha}^\dagger\right)|0\rangle \,. \tag{9.23}$$

In this state, each pair of single-particle levels $(\alpha, -\alpha)$ is occupied with a probability $|v_\alpha|^2$ and remains empty with probability $|u_\alpha|^2$. The parameters u_α and v_α will be determined through a variational principle. We will assume that they are real numbers; this will prove sufficiently general for time-independent problems. It is clear that the state contains contributions with different, but always even, particle numbers.

We first summarize a few key expectation values of the BCS state:

$$1 \stackrel{!}{=} \langle\text{BCS}|\text{BCS}\rangle = \prod_{\alpha>0}^{\infty} \left(u_\alpha^2 + v_\alpha^2\right), \tag{9.24a}$$

$$N \stackrel{!}{=} \langle\text{BCS}|\hat{N}|\text{BCS}\rangle = 2\sum_{\alpha>0} v_\alpha^2 \,, \tag{9.24b}$$

$$E = \langle\text{BCS}|\hat{H}|\text{BCS}\rangle = 2\sum_{\alpha>0} \varepsilon_\alpha^0 v_\alpha^2 - G\Big(\sum_{\alpha>0} u_\alpha v_\alpha\Big)^2 - G\sum_{\alpha>0} v_\alpha^4 \,. \tag{9.24c}$$

In order to meet the normalization condition (9.24a), it is most convenient to impose normalization on each pair of states $(\alpha, -\alpha)$, i.e.,

$$u_\alpha^2 + v_\alpha^2 \stackrel{!}{=} 1 \quad \longleftrightarrow \quad u_\alpha^2 = 1 - v_\alpha^2 \,. \tag{9.24d}$$

This ensures $\langle\text{BCS}|\text{BCS}\rangle = \prod_{\alpha>0}^{\infty}\left(u_\alpha^2 + v_\alpha^2\right) = \prod_{\alpha>0}^{\infty} 1 = 1$. We are now going to prove these relations.

A few basic properties of the vacuum are needed:

$$\hat{a}_\alpha|0\rangle = 0 = \langle 0|\hat{a}_\alpha^\dagger \quad \text{and} \quad \langle 0|\hat{a}_{-\alpha}\hat{a}_\alpha\hat{a}_\alpha^\dagger\hat{a}_{-\alpha}^\dagger|0\rangle = 1 \,.$$

The terms $\left(u_\alpha + v_\alpha \hat{a}_\alpha^\dagger \hat{a}_{-\alpha}^\dagger\right)$ in the BCS ansatz (9.23) all commute with each other so that the vacuum expectation value factorizes and only the products of $\alpha = \alpha'$ terms finally appear. This leads for the norm to

$$
\begin{aligned}
\langle \mathrm{BCS}|\mathrm{BCS}\rangle &= \langle 0|\prod_{\alpha>0}^{\infty}\left(u_\alpha + v_\alpha \hat{a}_{-\alpha}\hat{a}_\alpha\right)\prod_{\alpha'>0}^{\infty}\left(u_{\alpha'} + v_{\alpha'}\hat{a}^\dagger_{\alpha'}\hat{a}^\dagger_{-\alpha'}\right)|0\rangle \\
&= \prod_{\alpha>0}^{\infty}\langle 0|\left(u_\alpha + v_\alpha \hat{a}_{-\alpha}\hat{a}_\alpha\right)\left(u_\alpha + v_\alpha\hat{a}^\dagger_\alpha\hat{a}^\dagger_{-\alpha}\right)|0\rangle \\
&= \prod_{\alpha>0}^{\infty}\underbrace{\langle 0|u_\alpha^2 + u_\alpha v_\alpha\left(\hat{a}^\dagger_\alpha\hat{a}^\dagger_{-\alpha} + \hat{a}_{-\alpha}\hat{a}_\alpha\right) + v_\alpha^2\,\hat{a}_{-\alpha}\hat{a}_\alpha\hat{a}^\dagger_\alpha\hat{a}^\dagger_{-\alpha}|0\rangle}_{u_\alpha^2+v_\alpha^2}\,.
\end{aligned}
$$

In the following it will be assumed that the normalization condition (9.24a) is fulfilled such that the operator-free factors (in an average over the BCS state) become $\langle 0|\left(u_\alpha + v_\alpha\hat{a}_{-\alpha}\hat{a}_\alpha\right)\left(u_\alpha + v_\alpha\hat{a}^\dagger_\alpha\hat{a}^\dagger_{-\alpha}\right)|0\rangle = 1$. We proceed to the expectation value of the particle number operator $\hat{N} = \sum_{\alpha>0}\left(\hat{a}^\dagger_\alpha\hat{a}_\alpha + \hat{a}^\dagger_{-\alpha}\hat{a}_{-\alpha}\right)$ and first compute a single term $\hat{a}^\dagger_\alpha\hat{a}_\alpha$. The commutator relation $[\hat{a}^\dagger_\alpha\hat{a}_\alpha, \hat{a}^\dagger_\alpha\hat{a}^\dagger_{-\alpha}] = \hat{a}^\dagger_\alpha\hat{a}^\dagger_{-\alpha}$ will be used to move the operator $\hat{a}^\dagger_\alpha\hat{a}_\alpha$ to the right and then apply $\hat{a}^\dagger_\alpha\hat{a}_\alpha|0\rangle = 0$. This leads to

$$
\begin{aligned}
\langle \mathrm{BCS}|\hat{a}^\dagger_\alpha\hat{a}_\alpha|\mathrm{BCS}\rangle &= \langle 0|\left(u_\alpha + v_\alpha\hat{a}_{-\alpha}\hat{a}_\alpha\right)\hat{a}^\dagger_\alpha\hat{a}_\alpha\left(u_\alpha + v_\alpha\hat{a}^\dagger_\alpha\hat{a}^\dagger_{-\alpha}\right)|0\rangle \\
&= \langle 0|\left(u_\alpha + v_\alpha\hat{a}_{-\alpha}\hat{a}_\alpha\right)v_\alpha\hat{a}^\dagger_\alpha\hat{a}^\dagger_{-\alpha}|0\rangle = v_\alpha^2\,,
\end{aligned}
$$

and similarly $\langle \mathrm{BCS}|\hat{a}^\dagger_{-\alpha}\hat{a}_{-\alpha}|\mathrm{BCS}\rangle = v_\alpha^2$. This together summed over all α yields (9.24b). It also immediately provides the first term in the energy (9.24c). The interaction term in the energy requires two separate calculations. First, we consider the case $\alpha \neq \alpha'$:

$$
\begin{aligned}
\langle \mathrm{BCS}|\hat{a}^\dagger_\alpha\hat{a}^\dagger_{-\alpha}\hat{a}_{-\alpha'}\hat{a}_{\alpha'}|\mathrm{BCS}\rangle &= \langle 0|\left(u_\alpha + v_\alpha\hat{a}_{-\alpha}\hat{a}_\alpha\right)\hat{a}^\dagger_\alpha\hat{a}^\dagger_{-\alpha}\left(u_\alpha + v_\alpha\hat{a}^\dagger_\alpha\hat{a}^\dagger_{-\alpha}\right)|0\rangle \\
&\quad \langle 0|\left(u_{\alpha'} + v_{\alpha'}\hat{a}_{-\alpha'}\hat{a}_{\alpha'}\right)\hat{a}_{-\alpha'}\hat{a}_{\alpha'}\left(u_{\alpha'} + v_{\alpha'}\hat{a}^\dagger_{\alpha'}\hat{a}^\dagger_{-\alpha'}\right)|0\rangle \\
&= (u_\alpha v_\alpha)(u_{\alpha'} v_{\alpha'})\,,
\end{aligned}
$$

where we used $\langle 0|\hat{a}^\dagger_\alpha\hat{a}^\dagger_{-\alpha}\hat{a}^\dagger_\alpha\hat{a}^\dagger_{-\alpha}|0\rangle = 0$ and $\langle 0|\hat{a}_{-\alpha'}\hat{a}_{\alpha'}\hat{a}_{-\alpha'}\hat{a}_{\alpha'}|0\rangle = 0$, which can be deduced from the Pauli principle in terms of Fermi operators, $\hat{a}^\dagger_\alpha\hat{a}^\dagger_\alpha|0\rangle = 0$ and $\hat{a}_\alpha\hat{a}_\alpha|0\rangle = 0$. Second, we go for $\alpha = \alpha'$ which reads

$$
\begin{aligned}
\langle \mathrm{BCS}|\hat{a}^\dagger_\alpha\hat{a}^\dagger_{-\alpha}\hat{a}_{-\alpha}\hat{a}_\alpha|\mathrm{BCS}\rangle &= \langle 0|\left(u_\alpha + v_\alpha\hat{a}_{-\alpha}\hat{a}_\alpha\right)\hat{a}^\dagger_\alpha\hat{a}^\dagger_{-\alpha}\hat{a}_{-\alpha}\hat{a}_\alpha\left(u_\alpha + v_\alpha\hat{a}^\dagger_\alpha\hat{a}^\dagger_{-\alpha}\right)|0\rangle \\
&= v_\alpha^2\langle 0|\hat{a}_{-\alpha}\hat{a}_\alpha\hat{a}^\dagger_\alpha\hat{a}^\dagger_{-\alpha}\hat{a}_{-\alpha}\hat{a}_\alpha\hat{a}^\dagger_\alpha\hat{a}^\dagger_{-\alpha}|0\rangle = v_\alpha^2\,.
\end{aligned}
$$

Realizing that $\sum_{\alpha\neq\alpha'}(u_\alpha v_\alpha)(u_{\alpha'}v_{\alpha'}) = \sum_{\alpha\alpha'}u_\alpha v_\alpha u_{\alpha'}v_{\alpha'} - \sum_\alpha u_\alpha^2 v_\alpha^2$ and using (9.24d), we come to the expression for the G terms in the energy (9.24c).

It may be mentioned in passing that the variance of particle number can be evaluated in a similar fashion and becomes

$$
\Delta^2 N = \langle \mathrm{BCS}|(\hat{N} - N)^2|\mathrm{BCS}\rangle = 4\sum_{\alpha>0}u_\alpha^2 v_\alpha^2 = 4\sum_{\alpha>0}(1 - v_\alpha^2)v_\alpha^2\,.
$$

This shows that the BCS ansatz sacrifices a clean particle number. The uncertainty in the particle number is caused just by those single-particle states that are fractionally occupied, i.e., for which $0 < v_\alpha^2 < 1$. Allowing only values of 0 or 1 for the occupation probabilities would mean reverting to the pure single-particle model with well-defined particle number. The uncertainty in the particle number, while strictly speaking incorrect, may not be important as long as $\Delta N \ll N$. This has to be checked in the practical results.

9.4.2 The Gap Equation

We are now going to determine the optimal occupation amplitudes v_α by minimizing the total energy. As the trial wave function does not conserve the particle number, the desired expectation value has to be achieved through a constraint with a Lagrange multiplier. This leads to the variational problem

$$\delta\langle \mathrm{BCS}|\hat{H} - \lambda\hat{N}|\mathrm{BCS}\rangle = 0\,,$$

where the variation is with respect to the occupation probabilities u_α and v_α. Note that u_α depends on v_α through (9.24d).

For the detailed evaluation, we insert the expectation values (9.24b) and (9.24c) and reformulate them all in terms of v_α inserting the relation $u_\alpha = \sqrt{1-v_\alpha^2}$ both ways several times. The variational equation thus becomes

$$\begin{aligned} 0 &= \frac{\partial}{\partial v_\alpha}\left[2\sum_{\alpha'>0}(\varepsilon_{\alpha'}^0 - \lambda)\, v_{\alpha'}^2 - G\Big(\sum_{\alpha'>0}\sqrt{1-v_{\alpha'}^2}\,v_{\alpha'}\Big)^2 - G\sum_{\alpha'>0} v_{\alpha'}^4\right] \\ &= 4(\varepsilon_\alpha^0 - \lambda)v_\alpha - 2G\left(u_\alpha - \frac{v_\alpha^2}{u_\alpha}\right)\sum_{\alpha'>0} u_{\alpha'}v_{\alpha'} - 4Gv_\alpha^3 \\ &= u_\alpha^{-1}\Big[4\underbrace{(\varepsilon_\alpha^0 - \lambda - Gv_\alpha^2)}_{\epsilon_\alpha}u_\alpha v_\alpha - 2\left(u_\alpha^2 - v_\alpha^2\right)\underbrace{G\sum_{\alpha'>0} u_{\alpha'}v_{\alpha'}}_{\Delta}\Big], \end{aligned}$$

where we have used as abbreviation an effective single-particle energy ε_α and the pairing gap Δ. The latter will turn out to be a key quantity describing pairing properties. The variational equations thus take the compact form

$$2\varepsilon_\alpha v_\alpha u_\alpha + \Delta\left(v_\alpha^2 - u_\alpha^2\right) = 0\,. \tag{9.25}$$

Squaring this equation allows replacing u_α^2 by $1 - v_\alpha^2$, and it may then be solved for the latter:

$$v_\alpha^2 = \frac{1}{2}\left(1 \pm \sqrt{1 - \frac{\Delta^2}{\varepsilon_\alpha^2 + \Delta^2}}\right) = \frac{1}{2}\left(1 \pm \frac{\varepsilon_\alpha}{\sqrt{\varepsilon_\alpha^2 + \Delta^2}}\right).$$

The ambiguous sign results from the fourth-order equation appearing during the calculation; taking the square root of ε_α^2 contributed no further ambiguity. The correct sign can be selected by noting that for very large single-particle energies $\varepsilon_\alpha \to \infty$ the occupation probabilities must go to zero; this requires the negative sign.

The final result is thus

$$v_\alpha^2 = \frac{1}{2}\left(1 - \frac{\varepsilon_\alpha}{\sqrt{\varepsilon_\alpha^2 + \Delta^2}}\right), \quad u_\alpha^2 = \frac{1}{2}\left(1 + \frac{\varepsilon_\alpha}{\sqrt{\varepsilon_\alpha^2 + \Delta^2}}\right).$$

If we assume that λ and Δ have been determined, the behavior of these expressions is easily seen. For $\varepsilon_\alpha = 0$, i.e., when $\varepsilon_\alpha^0 - Gv_\alpha^2 = \lambda$, both u_α^2 and v_α^2 are equal to $\frac{1}{2}$. For large negative values of ε_α we will have $u_\alpha^2 \approx 0$ and $v_\alpha^2 \approx 1$ and the reverse is true for large positive values. The width of the transition is governed by Δ. This behavior is illustrated in Fig. 9.6. We note that λ obviously plays the role of a generalized Fermi energy.

The unknown parameters λ and Δ can now be determined by inserting the explicit forms for u_α and v_α into the variational equations (9.25) with $\Delta = G\sum_\alpha u_\alpha v_\alpha$ and into the particle-number condition (9.24b). This yields

$$\frac{2}{G} = \sum_{\alpha>0} \frac{1}{\sqrt{\varepsilon_\alpha^2 + \Delta^2}}, \tag{9.26a}$$

$$N = \sum_{\alpha>0} 2v_\alpha^2 = \sum_{\alpha>0}\left(1 - \frac{\varepsilon_\alpha}{\sqrt{\varepsilon_\alpha^2 + \Delta^2}}\right), \tag{9.26b}$$

$$\varepsilon_\alpha = \varepsilon_\alpha^{(0)} - Gv_\alpha^2 - \lambda \approx \varepsilon_\alpha^{(0)} - \lambda, \tag{9.26c}$$

where the first is the famous *gap equation*. The coupled equations can be solved iteratively using the known values of G and the single-particle energies ε_α^0. The other parameter λ then follows from simultaneously fulfilling the condition for the total particle number (9.26b). To do this the term $-Gv_\alpha^2$ in the definition of the

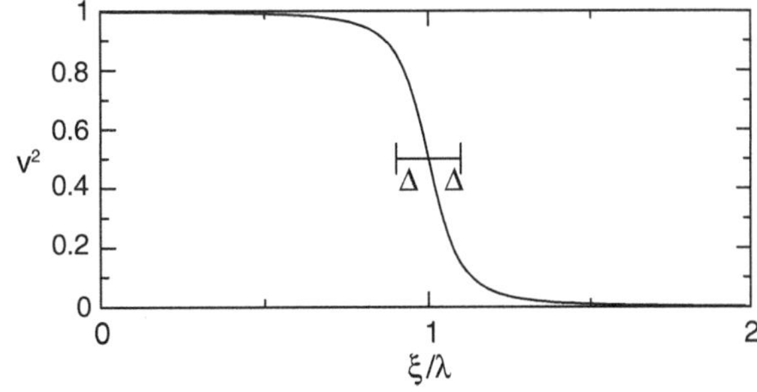

Fig. 9.6 The BCS occupation numbers in dependence on the single-particle energies (scaled in terms of the Fermi energy λ). The size of the gap parameter is indicated by the *horizontal line*

ε_α of (9.26a) has to be neglected. This is usually done with the argument that it corresponds only to a renormalization of the single-particle energies.

The gap equation, (9.26a), always has the trivial solution $\Delta = 0$. For sufficient pairing strength, there is also the non-trivial paired solution with $\Delta > 0$ and lower total energy. The switch from a non-pairing to a pairing regime is sort of a *phase transition* somewhat similar to the one discussed in connection with the quasispin model, see Sect. 7.7.

9.4.3 BCS for Continuous Spectra

The gap equation (9.26a) can be solved analytically for the case of a continuous spectrum. This is a good approximation if the single-particle energies have a spacing small compared to the gap Δ. In this case we can replace the sum by an integral and obtain the gap equation together with the particle number condition (9.26b) as

$$\frac{2}{G} = \int_{\varepsilon_1^0-\lambda}^{\varepsilon_2^0-\lambda} d\varepsilon\, \mathcal{D}(\varepsilon)\frac{1}{\sqrt{\varepsilon^2+\Delta^2}}\,, \quad N = \int_{\varepsilon_1^0-\lambda}^{\varepsilon_2^0-\lambda} d\varepsilon\, \mathcal{D}(\varepsilon)\left(1-\frac{\varepsilon}{\sqrt{\varepsilon^2+\Delta^2}}\right),$$

where $\mathcal{D}(\varepsilon)$ denotes the density of states (see Sect. 1.2.4.3 and Chap. 2). The limits of integration, ε_1 and ε_2, are required when using a simple constant pairing matrix element G. More realistic interactions $V_{\alpha\beta}^{(\mathrm{pair})}$ cover the whole space of states, but fall off quickly with increasing energy difference $|\varepsilon_\alpha - \varepsilon_\beta|$. The simple model here replaces that by a finite energy interval with constant G inside. The proper choice of the bounds depends on the system. In nuclear physics, e.g., it typically covers a major shell $\varepsilon_2^0 - \varepsilon_1^0 \sim 5 - 10\,\mathrm{MeV}$. Assuming a constant density of states $\mathcal{D}(\varepsilon) = \overline{\mathcal{D}}$ allows to develop a fully analytic solution of these coupled equations which, however, is rather lengthy to derive. We here consider for simplicity the limit of small G, and subsequently small gap $\Delta \ll |\varepsilon_i^0 - \lambda|$ for $i = 1, 2$. Concerning signs, it is important to note that $\varepsilon_1^0 < \lambda < \varepsilon_2^0$.

The two basic equations can be integrated for constant density $\overline{\mathcal{D}}$ yielding

$$\begin{aligned}
\frac{2}{G\overline{\mathcal{D}}} &= \log\left(\frac{\sqrt{(\varepsilon_2^0-\lambda)^2+\Delta^2}+\varepsilon_2^0-\lambda}{\sqrt{(\varepsilon_1^0-\lambda)^2+\Delta^2}-(\lambda-\varepsilon_1^0)}\right) \\
&= \log\left(\frac{\varepsilon_2^0-\lambda}{\lambda-\varepsilon_1^0}\right) + \log\left(\sqrt{1+\frac{\Delta^2}{(\varepsilon_2^0-\lambda)^2}}+1\right) - \log\left(\sqrt{1+\frac{\Delta^2}{(\varepsilon_1^0-\lambda)^2}}-1\right) \\
&\approx \log\left(\frac{\varepsilon_2^0-\lambda}{\lambda-\varepsilon_1^0}\right) + 2\log(2) - 2\log\left(\frac{\Delta}{\lambda-\varepsilon_1^0}\right),
\end{aligned}$$

$$\frac{N}{\overline{\mathcal{D}}} = \varepsilon_2^0 - \varepsilon_1^0 - \sqrt{(\varepsilon_2^0-\lambda)^2+\Delta^2} + \sqrt{(\varepsilon_1^0-\lambda)^2+\Delta^2}$$
$$\approx \quad \varepsilon_2^0 - \varepsilon_1^0 - (\varepsilon_2^0-\lambda) + (\lambda-\varepsilon_1^0) = 2(\lambda-\varepsilon_1^0)\,,$$

where we have already invoked the limit of small gap Δ. We resolve the second equation for $\lambda - \varepsilon_1^0$ and insert that into the first one. This yields

$$\log\left(\frac{2\Delta\overline{\mathcal{D}}}{N}\right) = \frac{1}{2}\log\left(\frac{\Omega}{N}-1\right) - \frac{1}{G\overline{\mathcal{D}}} \qquad \Omega = 2\overline{\mathcal{D}}(\varepsilon_2-\varepsilon_1).$$

The factor $2\overline{\mathcal{D}}(\varepsilon_2^0 - \varepsilon_1^0)$ has a simple physical interpretation. $\overline{\mathcal{D}}$ is the number of states per energy interval. Thus $2\overline{\mathcal{D}}(\varepsilon_2-\varepsilon_1) = \Omega$ is the number of states embraced in the active energy interval between ε_1 and ε_2. It corresponds to the parameter Ω of the seniority model (see Sect. 9.3).

Thus we finally obtain the estimate

$$\Delta = (\varepsilon_2^0 - \varepsilon_1^0)\sqrt{\frac{N}{2\Omega}\left(1-\frac{N}{2\Omega}\right)}\exp\left(\frac{1}{\overline{\mathcal{D}}G}\right) \sim \frac{\varepsilon_2^0-\varepsilon_1^0}{2}\exp\left(-\frac{1}{\overline{\mathcal{D}}G}\right). \tag{9.27}$$

For the last step of simplification, we have exploited that a mid-shell situation is associated typically with $N/(2\Omega) \approx 1/2$. The very interesting point to be seen from this formula is that the gap parameter depends in a quite unexpected way on G when the pairing strength goes to zero. Although the gap goes to zero with G, it cannot be expanded into a power series. This is again an indication that there is a phase transition between the BCS and the HF states, not a gradual transition.

This continuous modeling of the pairing state complements the simple seniority model as outlined in Sect. 9.3. It serves as a reliable estimate in finite systems and becomes a realistic description for larger systems with high-level density. Let us briefly discuss the orders of magnitude in nuclear pairing. As pairing between protons and neutrons is usually negligible, one considers each species, protons and neutrons, separately. The typical density of states is $\overline{\mathcal{D}} \sim A/26\,\text{MeV}^{-1}$ (see Sect. 1.2.4.3) and the pairing strength $G \sim 20\,\text{MeV}/A$. Thus we have the remarkable result that the exponential factor in the pairing gap is $\exp\left(-1/\overline{\mathcal{D}}G\right) \approx 0.5$ independent of the size of the nucleus A. There remains the pre-factor. It depends only weakly on particle number A and the energy span of the active zone is typically $\varepsilon_2^0 - \varepsilon_1^0 \sim 5 - 10$ MeV yielding together $\Delta \sim 0.7 - 1.4$ MeV which complies nicely with the empirical findings in Fig. 9.5. Note that the exponential factor is independent of A while the active interval $\varepsilon_2^0 - \varepsilon_1^0$ depends weakly on A as Δ does.

Another quite typical example is also fermion gases in a trap (see Sect. 1.1.8). Their pairing usually relies on the continuous description [14, 38]. The Fermi energy and the active zone around it ranges typically in the nK (see Table 9.1). The appealing feature of traps is that the average interaction strength can be tuned almost

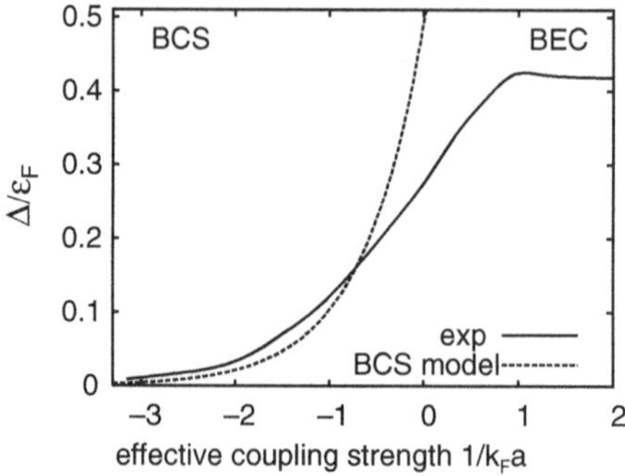

Fig. 9.7 Pairing gap (critical temperature T_c respectively) in units of Fermi energy as function of inverse dimensionless interaction strength as computed in [47]. The BCS estimate is shown for comparison (*dotted line*)

arbitrarily, thus allowing the exploration of the regimes from breakdown of pairing deep into very strong pairing (the latter appearing in the unitary limit of infinite scattering length a, see Sect. 6.5.2). One usually identifies a dimensionless coupling strength as $\overline{\mathcal{D}}G \propto k_{\mathrm{F}}a$ where k_{F} is the Fermi momentum of the gas and a is the scattering length. In case of attractive interactions, one is in the BCS regime of pairing and can derive for the gap [38]

$$\Delta = \frac{8\varepsilon_{\mathrm{F}}}{\mathrm{e}2} \exp\left(-\frac{\pi}{2k_{\mathrm{F}}|a|}\right), \tag{9.28}$$

which is exactly the trend of (9.27), but now with a detailed prediction for the prefactor. Figure 9.7 shows the trend of the pairing gap with interaction strength. The data are taken from a detailed many-body calculation in [47] yielding the critical temperature T_c and rescaled to a pairing gap with the BCS relation $\Delta = 1.76\,T_c$ [3]. The result is compared with the BCS trend (9.28) in the pairing regime of negative interaction strength. The predictions of the rather simple BCS description are fairly relevant. The regime of positive interaction strength is associated with the Bose–Einstein Condensate (BEC). Here the atoms form dimers which then behave like bosons and reach the Bose condensate in the limit of low temperatures. This is the famous BCS-BEC crossover [14, 38], somewhat outside the scope of this book.

9.5 Concluding Remarks

In this chapter, we have discussed correlation effects in terms of three very different examples. The first example was the van der Waals interaction between atoms in a molecule which is deduced from a perturbative correction. It is nearly negligible where other binding mechanisms (covalent, ionic) prevail, but it becomes the leading long-range contribution for rare gases where all other mechanisms fade away. The second example was Hooke's atom, a solvable model for two-body correlations in coordinate space. Here we have seen how a singularity of the two-body interaction strongly modifies the wave function at short distances in the relative coordinate while the overall effect on global observables remains small. The third example dealt with pairing correlations, first in a simple solvable model and then in BCS approximation. The seniority model makes it very clear how a unique ground state with large gap to first excitation is produced out of a fully degenerate unperturbed spectrum. The gap is in some sense a collective effect cumulating small contributions from each single pair to a large energy. The more realistic case of non-degenerate single-particle spectra is handled in the BCS approximation. Particularly the approximation of continuous spectra allows a closed solution which provides good order-of-magnitude estimates for finite systems and is often even used as quantitative description, e.g., in the case of fermion gases in atomic traps.

Chapter 10
Conclusions

The physics of finite fermion systems covers an impressive variety of situations. Fermion systems exist at all scales of distances and energies in the universe from inside atomic nuclei to compact stars. The topics covered in this book thus span a bunch of different scenarios. The theoretical description of finite fermion systems is usually involved for many reasons. The number of constituents is often large, which naturally leads to some complexity. The difficulties are strongly enhanced due to the fermionic nature of the constituents, which, to a large extent, prevents the use of classical approaches. Finally one is forced to handle possibly singular interactions and/or interactions spanning a large range of energies. These various constraints make the description of finite fermion systems a difficult task which requires well-developed theoretical tools.

The aim of this book is to exemplify the essentials of these involved theories in terms of simple approaches and so to provide some insight into the basic mechanisms at work in many-fermion systems. As a side product it turns out that these simple models usually have a large range of applications and allow to describe on a common basis seemingly very different physical systems at very different energy and size scales. This aspect is all the more true, the simpler the model under consideration, and it was thus one of our goals to try to keep most developments and applications at an analytic or quasi-analytic level. It also simply reflects the physical fact that the systems under consideration share common properties, once their specific scales are properly re-scaled to generic ones.

Altogether, when looking back at the various developments we have discussed, we can tentatively make a few general remarks, to some extent draw conclusions, and also open perspectives for more detailed investigations.

- *Old recipes for new physics* As illustrated in the course of the book, old simple models like the Fermi gas or the harmonic oscillator model provide a remarkable richness in terms of applications even in the most recently emerging domains such as quantum dots or fermion traps. This is a quite welcome feature in many respects. First it proves that basic features (Pauli principle, shell effects) as exemplified in these simple approaches are extremely robust and remain a key ingredient to sort our understanding of various physical systems. Of course, in quantum dots or fermion traps the harmonic oscillator dominates by construction, but even

J.A. Maruhn et al., *Simple Models of Many-Fermion Systems*,
DOI 10.1007/978-3-642-03839-6_10, © Springer-Verlag Berlin Heidelberg 2010

in self-bound systems such as nuclei, or more recently metal clusters, the harmonic oscillator picture provides a remarkably successful first-order approach. A second aspect concerns the pedagogical interest of such simple models. Having at hand so beautiful examples of applications render simple approaches especially attractive for a newcomer in quantum mechanics. It is thus quite instructive to be able to address such fascinating complex systems as nuclei, metal clusters, quantum dots, or fermion traps with so few technicalities.

- *A simple account of complexity* Simplicity beyond any doubt is a key issue in this book. But such simplicity does not necessarily exclude complexity. As we have tried to show at many places for various systems, the complexity can often, to a large extent, be lightened by a properly tuned simple model. This strategy would be somewhat limited if the building of such approaches led to highly specific models. Clearly this is not the case, and we have seen that simple models most often have a strong generic character, allowing to address on the same footing several a priori very different physical systems. The notion of simplicity also requires some further comment. We have tried all over the text to exemplify basic mechanisms in terms of analytic solutions. This usually implies a loss of detail but a gain of understanding, as a compact formula often allows a quick grasp of physics and relevant parameters. The case of finite fermion systems, nevertheless, usually implies painful calculations and analytical solutions are thus rather rare, whence the particular relevance of the few existing models.
- *Complexity beyond simplicity* While simple models offer remarkable clues to the description of complex systems, they do not exhaust all our understanding of such systems. We have in many places throughout the book identified some limitations of our approaches. Our line of progression consisting in including interactions and correlations at a deeper and deeper level is illustrative in that respect. While the simplest account of interactions may allow rather simple treatments (at the price of detail), including interactions and/or correlations at a more refined degree usually leads to more and more complex modeling. This progressively reduces the chances to find analytical treatments. Still, the stepwise inclusion of correlations helps to disentangle their influence and so provides a very telling illustration of how more and more subtle effects affect and modify our understanding of such fascinating systems as finite fermion systems.

Appendix A
Quantum-Mechanical Background

A.1 Spherical Single-Particle Wave Functions

A.1.1 General Form

In nuclear, atomic, and cluster physics the case of spherically symmetric systems has always been attractive for getting an understanding of the underlying phenomena because of its simplicity both in analytic and in computational applications. This appendix summarizes some properties of single-particle wave functions in spherically symmetric potentials.

The single-particle Hamiltonian for this case can be written in spherical coordinates as

$$\hat{h} = -\frac{\hbar^2}{2m}\nabla^2 + V(r) = -\frac{\hbar^2}{2m}\frac{1}{r^2}\frac{\partial}{\partial r}\left(r^2\frac{\partial}{\partial r}\right) + \frac{\hat{L}^2}{2mr^2} + V(r).$$

An important consequence of spherical symmetry is that the stationary wave functions can be chosen also as eigenfunctions of the angular momentum operators $\hat{L}^2$ and $\hat{L}_z$. As is known from elementary quantum mechanics the corresponding eigenfunctions are the spherical harmonics and the single-particle wave function can be decomposed as

$$\varphi_{nlm}(\mathbf{r}) = f_n(r)Y_{lm}(\Omega),$$

with n indicating the quantum numbers associated with the radial wave function. It fulfills the eigenvalue relations

$$\begin{aligned} \hat{h}\varphi_{nlm} &= \varepsilon_{nl}\varphi_{nlm} \\ \hat{L}^2\varphi_{nlm} &= \hbar^2 l(l+1)\varphi_{nlm} \\ \hat{L}_z\varphi_{nlm} &= \hbar m\varphi_{nlm}. \end{aligned}$$

Note that the single-particle energies must be degenerated with respect to the quantum number m.

J.A. Maruhn et al., *Simple Models of Many-Fermion Systems*,
DOI 10.1007/978-3-642-03839-6, © Springer-Verlag Berlin Heidelberg 2010

The radial wave functions then are solutions of the ordinary differential equation

$$\left[-\frac{\hbar^2}{2m}\frac{1}{r^2}\frac{\mathrm{d}}{\mathrm{d}r}\left(r^2\frac{\mathrm{d}}{\mathrm{d}r}\right)+\frac{\hbar^2 l(l+1)}{2mr^2}+V(r)\right]f_n(r)=\varepsilon_{nl}f_n(r).$$

In the Hamiltonian as written down above spin does not play a role. The eigenstates including the spin degree of freedom therefore simply have to be multiplied by a spinor,

$$\varphi_{nlms}(x)=f_n(r)Y_{lm}(\Omega)\chi_s.$$

In this case spin practically only influences the theories by allowing double occupation of these single-particle states. It becomes a more active participant in the theory only if it is present in the Hamiltonian, for example, via a spin–orbit coupling containing the operator $\hat{\mathbf{L}}\cdot\hat{\mathbf{s}}$.

A.1.2 Harmonic Oscillator

The most important set of spherical wave functions is the one for the harmonic oscillator potential. For $V(r)=\frac{1}{2}m\omega^2 r^2$ the complete wave functions are given by

$$\begin{aligned}\psi_{nlm}(r,\Omega)=&\sqrt{\frac{2^{n+l+2}}{n!(2n+2l+1)!!\sqrt{\pi}x_0^3}}\\&\times\frac{r^l}{x_0^l}L_n^{l+1/2}(r^2/x_0^2)\,\exp^{-r^2/2x_0^2}\,Y_{lm}(\Omega).\end{aligned}$$

with $x_0=\sqrt{\hbar/m\omega}$. Here $n=1,2,\ldots$ and, as usual, $l=0,1,\ldots$ and $m=-l,-l+1,\ldots+l$.

The symbol $L_n^m(x)$ stands for the generalized Laguerre polynomial (for some properties see Appendix B.2). The eigenenergies are determined by the principal quantum number $N=2(n-1)+l$ as

$$E_N=\hbar\omega\left(N+\tfrac{3}{2}\right)\tag{A.1}$$

The computational effectiveness of these functions is due to the fact that many integrals involving them can be calculated analytically. In addition, many recursion relations found in the mathematical handbooks make the calculation of derivatives and specific values quite efficient.

An alternative treatment of the spherical harmonic oscillator can be done in Cartesian coordinates. The Hamiltonian is decomposed into three parts corresponding to the three coordinate directions

$$\hat{h} = -\frac{\hbar^2}{2m}\frac{\mathrm{d}^2}{\mathrm{d}x^2} + \frac{1}{2}m\omega^2 x^2 - \frac{\hbar^2}{2m}\frac{\mathrm{d}^2}{\mathrm{d}y^2} + \frac{1}{2}m\omega^2 y^2 - \frac{\hbar^2}{2m}\frac{\mathrm{d}^2}{\mathrm{d}z^2} + \frac{1}{2}m\omega^2 z^2,$$

so that the eigenfunction becomes a product of three eigenfunctions of the 1D harmonic oscillator:

$$\varphi_{n_x n_y n_z} = \varphi_{n_x}(x)\varphi_{n_y}(y)\varphi_{n_z}(z)$$

with

$$\varphi_n(x) = \frac{1}{\sqrt{2^n n! \sqrt{\pi} x_0}} H_n(x/x_0) \exp(-\tfrac{1}{2}x/x_0^2)$$

and the H_n the Hermite polynomials (for properties again see Appendix B.2). The eigenenergy in this case is

$$\varepsilon_{n_x n_y n_z} = \hbar\omega\left(n_x + n_y + n_z + \tfrac{3}{2}\right).$$

The advantage of this formulation is that for the Hermite polynomials even more useful properties are known than for the Laguerre ones. The drawback is that the functions are no longer angular momentum eigenstates.

In *cylindrical coordinates* (ρ, z) the eigenfunctions of the harmonic oscillator can also be constructed in a simple way. If the z-axis is the symmetry axis, we still have a quantum number m from $\hat{L}_z$, and in the z-direction there is a simple 1D harmonic oscillator with quantum number n_z, while for ρ things are a bit more complicated. Its excitation is given by $n_\rho = 0, 1, \ldots$. The complete solution is

$$\psi_{n_z n_\rho \mu}(z, \rho, \phi) = N \exp\left[-\tfrac{1}{2}k^2(z^2 + \rho^2)\right] \mathrm{H}_{n_z}(kz)\, \rho^{|\mu|}\, L_{n_\rho}^{|\mu|}(k\rho^2)\, \mathrm{e}^{\mathrm{i}\mu\phi}$$

with N an unspecified normalization constant and $k = m\omega/\hbar$. The energy of the levels is given by

$$E = \hbar\omega\left(n_z + 2n_\rho + |\mu| + \tfrac{3}{2}\right) \quad .$$

Note that the number of "quanta" n_ρ in the ρ direction counts twice in the energy formula because it contains two oscillator directions and that the angular-momentum projection contributes to the energy because of the centrifugal potential.

Of course the degeneracy of the levels is the same independent of the coordinate system used, and the principal quantum number N can be split up in three ways:

$$N = n_x + n_y + n_z = n_z + 2n_\rho + |\mu| = 2n + l.$$

The degeneracy can be computed most easily from the Cartesian form. The different possibilities of choosing $0 \le n_x, n_y, n_z \le N$ with $n_x + n_y + n_z = N$ fixed is given by choosing a value for n_x and then one for n_y from the remaining choices, which determines n_z completely, this we get

$$\mathcal{N} = \sum_{n_x=0}^{N} \sum_{n_y=0}^{N-n_x} 1 = \sum_{n_x=0}^{N} (N - n_x + 1) = \tfrac{1}{2}(N+1)(N+2). \qquad \text{(A.2)}$$

The various coordinate systems are useful in different situations; for example, the spherical basis makes the spin–orbit coupling diagonal, while non-spherical but still axially symmetric systems are often treated more simply in the cylindrical basis.

A.1.3 The Hydrogen Atom

In the hydrogen atom the potential is $V(r) = -e^2/r$. This problem can be solved analytically only in spherical coordinates. The solution of the radial part leads to a radial quantum number $n_r = 0, 1, 2, \ldots$, which is, however, replaced by the *principal quantum number* $n = n_r + l + 1$, since the energy eigenvalues are found to depend only on n. The familiar angular momentum quantum numbers are, of course, also present, so that the set of quantum numbers becomes:

$$n = 1, 2, 3, \ldots \quad l = 0, 1, \ldots, n-1, \quad m = -l, -l+1, \ldots, +l$$

The energy eigenvalues are given by

$$E_N = -\frac{e^4 m_e}{2\hbar^2}\frac{1}{n^2} = -\frac{\text{Ry}}{n^2},$$

defining the *Rydberg constant* $\text{Ry} \approx 13.6\,\text{eV}$. The normalized eigenfunctions are

$$\varphi(r, \theta, \phi) = \sqrt{\left(\frac{2}{n a_0}\right)^3 \frac{(n-l-1)!}{2n(n+l)!}}\, \mathrm{e}^{-r/2}\, r^l\, L_{n-l-1}^{2l+1}(r)\, Y_{lm}(\theta, \phi).$$

They again contain the generalized Laguerre polynomials (see Appendix B.2).

The states discussed here are all bound. For positive energies there is a continuous spectrum with distorted plane waves as eigenfunctions, which will not be used in this book.

A.1.4 The Spherical Square Well (Box)

The radial wavefunctions for the spherical box potential (as given in Table 3.1) are determined by

Table A.1 *Left*: The zeroes of the spherical Bessel functions $j_{nl}(x)$. *Right*: The lowest states in the spectrum of the spherical box potential with quantum numbers n and l, momentum k_{nl}, and single-particle energy ε_{nl}. The last column shows the multiplicity of the states including spin

Zeroes of $j_{nl}(x)$				
$l\backslash n$	1	2	3	4
0	3.142	6.283	9.425	12.566
1	4.493	7.725	10.904	14.066
2	5.764	9.095	12.323	
3	6.988	10.417	13.698	
4	8.183	11.705		
5	9.356	12.967		
6	10.513	14.207		
7	11.657			
8	12.791			
9	13.916			

Spectrum of spherical box				
n	l	k_{nl} $[R^{-1}]$	ε_{nl} $[\frac{\hbar^2}{2m}]$	$2(2l+1)$
1	0	3.142	9.872	2
1	1	4.493	20.187	6
1	2	5.764	33.224	10
2	0	6.283	39.476	2
1	3	6.988	48.832	14
2	1	7.725	59.676	6
1	4	8.183	66.961	18
2	2	9.095	82.719	10
1	5	9.356	87.535	22
3	0	9.425	88.831	2

$$\left[-\frac{\hbar^2}{2m}\frac{1}{r^2}\frac{\partial}{\partial r}\left(r^2\frac{\partial}{\partial r}\right)+\frac{\hbar^2 l(l+1)}{2mr^2}\right] f_{nl}(r) = \varepsilon_{nl} f_{nl}(r), \tag{A.3a}$$

$$f_{nl}(R) = 0. \tag{A.3b}$$

These are free spherical waves, except for the boundary condition (A.3b). Solutions are the spherical Bessel functions

$$f_{nl}(r) = j_{nl}(k_{nl}r), \qquad \varepsilon_{nl} = \frac{\hbar^2}{2m}k_{nl}^2, \tag{A.3c}$$

where the k_{nl} are determined such that the boundary condition (A.3b) is fulfilled. The zeroes of the j_{nl} can only be found numerically. They are given for small l and n on the left part of Table A.1. The resulting spectrum is shown on the right part.

A.2 Angular Momentum

A.2.1 Angular Momentum Algebra

In this and the following section a few facts about angular momenta and their coupling are presented. For anything going beyond the elementary treatment given here, the reader is directed to one of the many excellent textbooks on the field, e.g., [31] or [108].

The basis for all of angular momentum theory lies in the commutation relations between the components,

$$[\hat{J}_x, \hat{J}_y] = \mathrm{i}\hbar\hat{J}_z, \qquad [\hat{J}_y, \hat{J}_z] = \mathrm{i}\hbar\hat{J}_x, \qquad [\hat{J}_z, \hat{J}_x] = \mathrm{i}\hbar\hat{J}_y, \tag{A.4}$$

which define the *angular momentum algebra*. The algebra is independent of whether $\hat{\mathbf{J}}$ is replaced by the orbital angular momentum operator $\hat{\mathbf{L}} = -\mathrm{i}\hbar\mathbf{r} \times \nabla$ or the spin operator $\hat{\mathbf{s}} = \frac{1}{2}\hbar\boldsymbol{\sigma}$.

As should be familiar, an immediate consequence is that the square of the angular momentum operator

$$\hat{\mathbf{J}}^2 = \hat{J}_x^2 + \hat{J}_y^2 + \hat{J}_z^2$$

commutes with all of the components and can therefore be diagonalized together with one of them, for which conventionally $\hat{J}_z$ is chosen. In a spherically symmetric system, for which $[\hat{\mathbf{J}}, \hat{H}] = 0$ holds, we can therefore select eigenstates $|\alpha J M\rangle$ fulfilling

$$\begin{aligned} \hat{H}|\alpha J M\rangle &= E_\alpha|\alpha J M\rangle, \\ \hat{\mathbf{J}}^2|\alpha J M\rangle &= \hbar^2 J(J+1)|\alpha J M\rangle, \\ \hat{J}_z|\alpha J M\rangle &= \hbar M|\alpha J M\rangle. \end{aligned} \tag{A.5}$$

Here α summarizes all non-angular momentum quantum numbers. The derivation of the eigenvalues $J = 0, \frac{1}{2}, 1, \frac{3}{2}, \ldots$ and $M = -J, -J+1, \ldots, +J$ can be found in quantum-mechanics textbooks [70, 24].

An especially useful alternative set of operators is given by $\hat{J}_z$ together with the combinations

$$\hat{J}_+ = \hat{J}_x + \mathrm{i}\hat{J}_y, \qquad \hat{J}_- = \hat{J}_x - \mathrm{i}\hat{J}_y. \tag{A.6}$$

Their commutation relations are

$$[\hat{\mathbf{J}}^2, \hat{J}_\pm] = 0, \qquad [\hat{J}_z, \hat{J}_\pm] = \pm\hat{J}_\pm.$$

These operators have the simple effect of raising or lowering the angular-momentum projection by one:

$$\hat{J}_\pm|J M\rangle = \sqrt{(J \mp M)(J \pm M + 1)}|J, M \pm 1\rangle.$$

The matrix element contained in the equation corresponds to a specific choice of relative phases of the angular momentum eigenstates according to *Condon and Shortley*.

It is also worth keeping in mind that rotation of wave functions can be done using angular momentum operators. A rotation of a state $|\Psi\rangle$ around the x-axis through an angle φ, e.g., can be written as

$$|\Psi'\rangle = \exp\left(-\frac{\mathrm{i}}{\hbar}\hat{J}_x\varphi\right)|\Psi\rangle,$$

and $\hat{J}_x$ therefore describes *infinitesimal rotations*.

A.2.2 Angular Momentum Coupling

A new problem arises when there is more than one angular momentum in a system. This can occur in two ways: a single particle can have internal angular momentum, i.e., spin, in addition to its orbital angular momentum, or there may be more than one particle, each contributing its own orbital angular momentum and spin.

Let us first examine the case of several particles. As long as they all are affected only by a spherical external potential, they are completely independent of each other and their angular momenta are seperately conserved. As soon as they interact with each other, however, the rotational symmetry is fulfilled only for *rotating all the particles together*, which is described by the total angular momentum operator. For two particles it is given by

$$\hat{\mathbf{J}} = \hat{\mathbf{J}}_1 + \hat{\mathbf{J}}_2.$$

Calculating the commutation relations for the components of total angular momentum reveals that they again fulfill the angular momentum algebra as defined in (A.4). This is very important, because *the sum of any number of angular momenta behaves the same as an elementary angular momentum operator* — the eigenvalues and eigenstates of the total angular momentum of a complete atom or nucleus just follow the same rule (A.5) as that of an elementary particle.

Rotational symmetry assures us that $[\hat{\mathbf{J}}, \hat{H}] = 0$, so that eigenstates of the Hamiltonian can also be constructed to be eigenstates of the operators $\hat{\mathbf{J}}^2$ and $\hat{J}_z$,

$$\begin{aligned}
\hat{H}|\alpha, J, M\rangle &= E_\alpha|\alpha, J, M\rangle \\
\hat{\mathbf{J}}^2|\alpha, J, M\rangle &= \hbar^2 J(J+1)|\alpha, J, M\rangle \\
\hat{J}_z|\alpha, J, M\rangle &= \hbar M|\alpha, J, M\rangle.
\end{aligned}$$

For the spin and orbital angular momentum of a single particle things are quite analogous, since as soon as the two are coupled, rotational invariance will be valid only for rotations involving both orbital and internal rotation, so again the total angular momentum operator, in this case $\hat{\mathbf{J}} = \hat{\mathbf{L}} + \hat{\mathbf{s}}$, will be a conserved quantity and produce good quantum numbers. Things are thus completely analogous to the coupling of two different particles.

A.2.3 Coupled and Uncoupled Basis

In constructing states of good angular momentum, normally one has to start with the eigenstates of the individual angular momenta. Thus we have eigenstates $|\alpha_1 j_1 m_1\rangle$ for particle 1 fulfilling

$$\hat{\mathbf{J}}_1^2|\alpha_1 j_1 m_1\rangle = \hbar^2 j_1(j_1+1)|\alpha_1 j_1 m_1\rangle, \qquad \hat{J}_{1z}|\alpha_1 j_1 m_1\rangle = \hbar m_1|\alpha_1 j_1 m_1\rangle,$$

where α_1 summarizes all other quantum numbers of particle 1. Similarly for particle 2 the same properties hold with the index replaced by 2. A set of basis states, the *uncoupled basis*, can then be constructed from the direct products,

$$|\alpha_1 j_1 m_1\rangle|\alpha_2 j_2 m_2\rangle,$$

and they are eigenstates of all four operators.

To make the transition to a coupled basis we have to investigate which operators can still be diagonal together with the total angular operators $\hat{\mathbf{J}}^2$ and $\hat{J}_z$. Since $\hat{\mathbf{J}}_1$ and $\hat{\mathbf{J}}_2$ refer to different degrees of freedom and thus commute trivially, we get $[\hat{\mathbf{J}}_{1,2}, \hat{\mathbf{J}}^2] = 0$ and $[\hat{\mathbf{J}}_{1,2}, \hat{J}_z] = 0$, while $\hat{\mathbf{J}}_{1z}$ and $\hat{\mathbf{J}}_{2z}$ do not commute with the total angular momentum operators. A complete set of commuting operators is thus given by

$$\hat{\mathbf{J}}^2, \quad \hat{J}_z, \quad ,\hat{\mathbf{J}}_1^2, \quad \hat{\mathbf{J}}_2^2.$$

The corresponding quantum numbers are J, M, j_1, and j_2 and lead to the *coupled basis*

$$|JM, j_1 j_2\rangle.$$

The linear transformation between the coupled and uncoupled bases is given by the so-called *Clebsch–Gordan coefficients*, but since they are not used in this book we refer the reader to the angular momentum theory textbooks mentioned above.

The physical meaning of the angular momentum coupling as the addition of two vectors implies some restrictions on the quantum numbers involved. First, because $\hat{J}_z = \hat{J}_{1z} + \hat{J}_{2z}$, the corresponding quantum numbers must also add up: $M = m_1 + m_2$.

Second, there is the *triangle rule* concerning the lengths of the vectors: this can visualized geometrically in that the length of the sum must lie between the difference and the sum of lengths, corresponding to limiting possible angles between the vectors. This is expressed as

$$|j_1 - j_2| \le J \le j_1 + j_2.$$

A.2.4 Spin Coupling

An important special case is the coupling of the spins of two spin 1/2 particles. According to the triangle rule, the resulting total spin S is either 0 or 1. The explicit formulas are for $S = 0$

$$|S=0, M=0\rangle = \frac{1}{\sqrt{2}}\left(\chi^{(1)}_{+\frac{1}{2}}\chi^{(2)}_{-\frac{1}{2}} - \chi^{(1)}_{-\frac{1}{2}}\chi^{(2)}_{+\frac{1}{2}}\right),$$

where $\chi^{(1,2)}_{\pm\frac{1}{2}}$ are the basis spinors of the two particles. This wave function is *anti-symmetric* under exchange of the two particles. On the other hand, for $S = 1$ the wave functions are symmetric:

$$\begin{aligned}
|S=1, M=+1\rangle &= \chi^{(1)}_{+\frac{1}{2}}\chi^{(2)}_{+\frac{1}{2}},\\
|S=1, M=0\rangle &= \frac{1}{\sqrt{2}}\left(\chi^{(1)}_{+\frac{1}{2}}\chi^{(2)}_{-\frac{1}{2}} + \chi^{(1)}_{-\frac{1}{2}}\chi^{(2)}_{+\frac{1}{2}}\right),\\
|S=1, M=-1\rangle &= \chi^{(1)}_{-\frac{1}{2}}\chi^{(2)}_{-\frac{1}{2}}.
\end{aligned}$$

A.2.5 Matrix Element of $\hat{\mathbf{L}} \cdot \hat{\mathbf{s}}$

In some simple cases properties of the coupled state can be calculated without explicit construction of the pertinent eigenstates. We consider the case of an orbital angular momentum $\hat{\mathbf{L}}$ coupled with spin $\hat{\mathbf{s}}$ to a resulting $\hat{\mathbf{J}}$. The coupled basic states are $|\alpha JM, l\tfrac{1}{2}\rangle$.

Now in many physical systems there is a *spin–orbit interaction* term in the Hamiltonian proportional to $\hat{\mathbf{L}} \cdot \hat{\mathbf{s}}$. The matrix element of this in the coupled states can be calculated by noting

$$\hat{\mathbf{J}}^2 = (\hat{\mathbf{L}} + \hat{\mathbf{s}})^2 = \hat{\mathbf{L}}^2 + \hat{\mathbf{s}}^2 + 2\hat{\mathbf{L}} \cdot \hat{\mathbf{s}},$$

so that

$$\hat{\mathbf{L}} \cdot \hat{\mathbf{s}} = \tfrac{1}{2}(\hat{\mathbf{J}}^2 - \hat{\mathbf{L}}^2 - \hat{\mathbf{s}}^2).$$

The operators on the right-hand side are all diagonal in the coupled basis and substituting the eigenvalues yields the diagonal matrix elements

$$\langle\alpha JM, l\tfrac{1}{2}|\hat{\mathbf{L}} \cdot \hat{\mathbf{s}}|\alpha JM, l\tfrac{1}{2}\rangle = \frac{\hbar^2}{2}\left(J(J+1) - l(l+1) - \tfrac{3}{4}\right)$$

with all other matrix elements zero.

A.3 Independent-Particle Wave Functions

The simplest kind of many-particle system, which serves as the basis for most of the more complex theories, is one of identical non-interacting particles: an *independent-particle model*. In this case the Hamiltonian $\hat{H}$ of the N-particle system is just the sum of the identical single-particle Hamiltonians $\hat{h}$ for the individual particles,

$$\hat{H}(x_1, x_2, \ldots x_N) = \sum_{k=1}^{N} \hat{h}(x_k). \tag{A.7}$$

Here we have introduced the super-vector $x = (\mathbf{r}, \nu)$ which stands for both the configuration space coordinate $\mathbf{r}$ and the spin projection $\nu = \pm\frac{1}{2}$.

Clearly for this Hamiltonian an eigenfunction for the many-body case can be obtained simply as a product of single-particle eigenfunctions: if we have eigenstates fulfilling

$$\hat{h}(x)\varphi_\alpha(x) = \varepsilon_\alpha \phi_\alpha(x), \tag{A.8}$$

the many-body wave function

$$\Psi(x_1, \ldots x_N) = \prod_{k=1}^{N} \phi_{\alpha_k}(x_k) \tag{A.9}$$

corresponds to a state with the particles in the orbitals labeled α_k, $k = 1, \ldots N$, and fulfilling

$$\hat{H}\Psi = E\Psi, \qquad \text{with} \quad E = \sum_{k=1}^{N} \varepsilon_k. \tag{A.10}$$

In general there will be an infinite number of single-particle eigenstates of *hop* forming a complete basis, and the many-particle wave function is determined by the set of *occupied states* chosen from them. We will always assume that the single-particle states are orthonormal,

$$\int \mathrm{d}^3x \, \varphi_\alpha^*(x)\varphi_\beta(x) = \delta_{\alpha\beta}.$$

Here the "integration" over x is meant to include the integration over configuration space as well the scalar product of the spinors. If the single-particle wave functions are orthonormal, then the many-body one is also normalized.

As it stands, however, the product wave function is not correct, since we are dealing with Fermions and the many-body wave function must be antisymmetric under exchange of two-particle coordinates. This can be achieved by using a *Slater determinant*

$$\Psi(x_1, \dots x_N) = \frac{1}{\sqrt{N!}} \det \begin{pmatrix} \varphi_{\alpha_1}(x_1) & \varphi_{\alpha_2}(x_1) & \cdots & \varphi_{\alpha_N}(x_1) \\ \varphi_{\alpha_1}(x_2) & \varphi_{\alpha_2}(x_2) & \cdots & \varphi_{\alpha_N}(x_2) \\ \vdots & \vdots & \ddots & \vdots \\ \varphi_{\alpha_1}(x_N) & \varphi_{\alpha_2}(x_N) & \cdots & \varphi_{\alpha_N}(x_N) \end{pmatrix} . \tag{A.11}$$

The normalization factor takes into account the fact that when multiplying out the overlap of the determinant with itself, the $N!$ contributions with the same permutation of coordinates in both wave function products yield a contribution of 1, while the other products vanish because of the orthonormality.

Since a determinant with two or more identical columns vanishes, we get the *Pauli principle* as an immediate consequence: all occupied single-particle states must be distinct.

Using Slater determinants in calculations quickly becomes extremely cumbersome. The way to a simpler description of the many-particle product states lies in the recognition that the physical information characterizing the state is only the set of occupied states: In (A.11), the particle coordinates are introduced but then immediately made irrelevant by the antisymmetrization. We are thus led to a formulation of the many-body states in an *occupation number representation.*

For a few derivations later we will need an explicit form of the Slater determinant which is easier to manipulate. This is given by

$$\Psi(x_1, x_2, \dots, x_N) = \frac{1}{\sqrt{N!}} \sum_{\pi} (-1)^{\pi} \prod_{k=1}^{N} \varphi_{\alpha_{k_\pi}}(x_k),$$

for the case of N fermions, where π is a permutation of the indices $1, \dots, N$ and $(-1)^{\pi}$ is its sign, i.e., $+1$ for even and -1 for odd permutations. The permutation changes the index i into i_π. The sum is over all $N!$ permutations of the N indices.

A Slater determinant is a highly restricted state of the many-body system. For independent-particle models, Slater determinants are the correct eigenstates of the Hamiltonian, but once interactions are introduced, the true many-body wave function can be expanded into a sum over a large number of Slater determinants, all with the same particle number, but with all possible selections of occupied states. The number of such Slater determinants increases extremely rapidly with particle number and single-particle basis size.

A.4 Algebra of Fermion Operators

A.4.1 Fock Space

In the occupation number representation each single-particle basis state with label α has an associated occupation number n_α which for fermions can be only zero or one. For N particles there must be N occupied states, so that $N = \sum_{\alpha=1}^{\infty} n_\alpha$. The many-body states are denoted by ket vectors as $|n_1, n_2, \ldots\rangle$ and the Slater determinant of (A.11) would correspond to $n_{\alpha_k} = 1$ for $k = 1, \ldots N$ and all other occupation numbers equal to zero.

States of different particle number N can be combined in what is called *Fock space*. Mathematically speaking, Fock space is the direct sum of the Hilbert spaces for fixed particle numbers, so that a state can contain components with different N; *direct sum* essentially means that states from the different Hilbert spaces can be added up, but the overlap between states of different N is defined to be zero — in conventional Hilbert space it would not make sense to calculate such an overlap.

We now introduce operators that allow changing the particle number. The *creation operator* $\hat{a}_\alpha^\dagger$ inserts one particle into state α, while the *annihilation operator* $\hat{a}$ deletes it. Their basic action is thus, depending on whether the single-particle state is already filled or not,

$$
\begin{aligned}
\hat{a}_\alpha^\dagger |n_1, n_2, \ldots n_\alpha = 0, \ldots\rangle &= |n_1, n_2, \ldots n_\alpha = 1, \ldots\rangle, \\
\hat{a}_\alpha |n_1, n_2, \ldots n_\alpha = 1, \ldots\rangle &= |n_1, n_2, \ldots n_\alpha = 0, \ldots\rangle, \\
\hat{a}_\alpha^\dagger |n_1, n_2, \ldots n_\alpha = 1, \ldots\rangle &= 0, \\
\hat{a}_\alpha |n_1, n_2, \ldots n_\alpha = 0, \ldots\rangle &= 0.
\end{aligned}
$$

The third and fourth conditions reflect the Pauli principle and the impossibility to delete a nonexisting particle, respectively. These conditions define the behavior of the operators almost completely, we note especially that it is impossible to create or destroy two particles in the same state, so that $\hat{a}_\alpha^\dagger \hat{a}_\alpha^\dagger = 0$ and $\hat{a}_\alpha \hat{a}_\alpha = 0$, while applying a combination of both on a state with occupation n_α yields

$$
\begin{aligned}
\hat{a}_\alpha^\dagger \hat{a}_\alpha |n_1, n_1, \ldots n_\alpha, \ldots\rangle &= n_\alpha |n_1, n_1, \ldots n_\alpha, \ldots\rangle \\
\hat{a}_\alpha \hat{a}_\alpha^\dagger |n_1, n_1, \ldots n_\alpha, \ldots\rangle &= (1 - n_\alpha) |n_1, n_1, \ldots n_\alpha, \ldots\rangle,
\end{aligned}
$$

making it apparent that a *particle-number operator* can be defined as

$$
\hat{N}_\alpha = \hat{a}_\alpha^\dagger \hat{a}_\alpha
$$

that counts how many particles (0 or 1) are in state α.

These properties of the operators are represented by the diagonal ($\alpha = \beta$) versions of the abstract *anticommutation relations*

$$
\{\hat{a}_\alpha^\dagger, \hat{a}_\beta^\dagger\} = 0, \qquad \{\hat{a}_\alpha, \hat{a}_\beta\} = 0, \qquad \{\hat{a}_\alpha, \hat{a}_\beta^\dagger\} = \delta_{\alpha\beta}, \tag{A.12}
$$

where the anticommutator of two operators is defined as $\{\hat{A}, \hat{B}\} = \hat{A}\hat{B} + \hat{B}\hat{A}$. It should be noted as a consequence that

$$\hat{a}^\dagger_\alpha \hat{a}^\dagger_\beta = -\hat{a}^\dagger_\beta \hat{a}^\dagger_\alpha. \qquad \text{(A.13)}$$

The nondiagonal versions of these relations imply that creating or annihilating particles in two different single-particle states in order $\alpha \rightarrow \beta$ or $\beta \rightarrow \alpha$ produces the same many-particle wave function, but with opposite sign. This expresses the antisymmetry for fermions.

We thus have set up a *second-quantization* formalism for fermions characterized by the anticommutation rules A.12. Many-particle states can now be produced out of the *vacuum* $|0\rangle$, which is simply the state with all $n_\alpha = 0$, by the application of creation operators. Thus, the state corresponding to the Slater determinant A.11 can be written as

$$\hat{a}^\dagger_{\alpha_1} \hat{a}^\dagger_{\alpha_2} \ldots \hat{a}^\dagger_{\alpha_N} |0\rangle. \qquad \text{(A.14)}$$

The final sign depends on the ordering of the operators, but that is usually no problem. A convention is, e.g., that the order of the creation operators be the same as the order of the states in the columns of the Slater determinant. We then have a one-to-one correspondence between the Slater determinants in configuration space and the Fock-space states as defined in (A.14).

A.4.2 Operators in Fock Space

For the use of the formalism in actual calculations it is necessary to also transform operators into the second-quantization formalism. To that end it is sufficient to construct an operator that yields the same matrix elements between the Slater determinants as between the corresponding Fock states. The form of such an operator will depend on its type, the two most useful being

- **one-body operators** depending only on the coordinates of one particle, but for identical particles summed over identical contributions for all particles. An example is the kinetic energy

$$\hat{T} = \sum_{k=1}^{N} \hat{t}_k = \sum_{k=1}^{N} \frac{-\hbar^2}{2m} \nabla_k^2.$$

A general one-body operator will have the form

$$\hat{F} = \sum_{k=1}^{N} \hat{f}(x_k).$$

- **two-body operators** depending on the coordinates of two particles. A typical example is the interaction potential given by

$$\hat{V}(x_1, \ldots x_N) = \tfrac{1}{2} \sum_{j,k=1}^{N} \hat{v}(x_j, x_k).$$

The matrix elements of these expressions have to be evaluated between two Slater determinants and an operator in second-quantization notation must be constructed that produces identical matrix elements. We take two Slater determinants

$$\Psi(x_1, \ldots, x_N) = \frac{1}{\sqrt{N!}} \sum_{\pi} (-1)^{\pi} \prod_{j=1}^{N} \varphi_{\alpha_{j\pi}}(x_j),$$

$$\Psi'(x_1, \ldots, x_N) = \frac{1}{\sqrt{N!}} \sum_{\pi'} (-1)^{\pi'} \prod_{j'=1}^{N} \varphi'_{j'_{\pi'}}(x_{j'}).$$

with the indices $\alpha'_{j'}$ and α_j describe the different choices of N occupied single-particle wave functions from a complete orthonormal set $\varphi_k(x)$, $k = 1, \ldots, \infty$. The matrix element becomes

$$\langle \Psi | \hat{f} | \Psi' \rangle = \sum_{k=1}^{N} \frac{1}{N!} \int \mathrm{d}^3 x_1 \cdots \int \mathrm{d}^3 x_N$$
$$\sum_{\pi\pi'} (-1)^{\pi+\pi'} \Big(\prod_{j=1}^{N} \varphi^*_{\alpha_{j\pi}}(x_j) \Big) \, \hat{f}(x_k) \Big(\prod_{j'=1}^{N} \varphi_{\alpha'_{j'_{\pi'}}}(x_{j'}) \Big).$$

From the products we can form single-particle integrals by taking the single-particle wave functions with identical arguments together. For the special case $j = j' = k$ the matrix element

$$\int \mathrm{d}^3 x_k \, \varphi^*_{\alpha_{k\pi}}(x_k) \, \hat{f}(x_k) \, \varphi_{\alpha'_{k_{\pi'}}}(x_k) = f_{\alpha_{k\pi} \alpha'_{k_{\pi'}}}, \tag{A.15}$$

results, whereas the others simply reduce to

$$\int \mathrm{d}^3 x_j \, \varphi^*_{\alpha_{j\pi}}(x_j) \, \varphi_{\alpha'_{j'_{\pi'}}}(x_j) = \delta_{\alpha_{j\pi} \alpha'_{j'_{\pi'}}}.$$

The many-body matrix element becomes

$$\langle \Psi | \hat{f} | \Psi' \rangle = \sum_{k} \frac{1}{N!} \sum_{\pi\pi'} (-1)^{\pi+\pi'} f_{k_\pi k_{\pi'}} \prod_{\substack{j,j'=1 \\ j \neq k, j' \neq k}}^{N} \delta_{j_\pi j'_{\pi'}}.$$

For this matrix element not to vanish, the Kronecker symbols require that the same states must be occupied in Ψ as in Ψ' with a single exception. This implies that a single-particle operator changes the state of a single particle only.

Now if at most one single-particle state is different between the two occupations, for a given permutation π there is only one π' that makes all Kronecker symbols nonzero; so the sum over π' can be dropped by choosing π' correctly for each π. How is the total sign to be determined? As π and π' are the permutations needed to bring the states numbered by i and i' from their original ordering into the same order (with the indices k_π and $k_{\pi'}$ also in the same position), the factor $\sigma = (-1)^{\pi+\pi'}$ tells whether an even or odd permutation is needed to transform these original orderings into each other and it does not depend on π or π'. The sum over permutations π then effectively runs only over the various ways to number the $N-1$ states occupied in both Ψ and Ψ'.

The matrix element is totally independent of the permutation. It only contains the factor σ, and the single-particle matrix element $f_{k_\pi k_{\pi'}}$ always has the same indices: those of the two single-particle states which differ between Ψ and Ψ'; let us simply call them j and j'. The matrix element is now

$$\langle\Psi|\,\hat{f}\,|\Psi'\rangle = \sum_{k=1}^{N} \frac{1}{N!}\,\sigma\, f_{jj'} \sum_{\pi,\,j \text{ fixed}} 1 = \frac{1}{N!}\,\sigma\, N\, f_{jj'}\,(N-1)! = \sigma\, f_{jj'}.$$

What should the equivalent operator in second quantization look like? It must remove one particle from state j' and put it into the state j while not doing anything to the other states, the resulting matrix element being $f_{jj'}$. Since this must operate for all combinations of j and j' we are led to

$$\hat{F} = \sum_{jj'} f_{jj'}\,\hat{a}_j^\dagger \hat{a}_{j'}.$$

It has to be checked whether the signs are correct. The Fock-space states are

$$|\Psi\rangle = \hat{a}_{\alpha_1}^\dagger \cdots \hat{a}_{\alpha_N}^\dagger |0\rangle \quad , \quad |\Psi'\rangle = \hat{a}_{\alpha'_1}^\dagger \cdots \hat{a}_{\alpha'_N}^\dagger |0\rangle,$$

and the matrix element is

$$\langle\Psi|\hat{F}|\Psi'\rangle = \sum_{jj'} f_{jj'}\langle 0|\hat{a}_{\alpha_N} \cdots \hat{a}_{\alpha_1}\hat{a}_j^\dagger \hat{a}_{j'}\hat{a}_{\alpha'_1}^\dagger \cdots \hat{a}_{\alpha'_N}^\dagger|0\rangle.$$

It is clear that again the set of indices α must denote the same states as α', except that j replaces j'. Permute the i' in such a way that they are in the same order as the i (and with j' at the same place as j), and this will yield the same sign factor σ as defined above for the Slater determinants. Permuting the operator combination $\hat{a}_j^\dagger \hat{a}_{j'}$ in front of $\hat{a}_{j'}^\dagger$ does not change the sign, and we get

$$\langle \Psi | \hat{F} | \Psi' \rangle = \sum_{jj'} f_{jj'} \langle 0 | \hat{a}_{\alpha_N} \cdots \hat{a}_{\alpha_1} \hat{a}^\dagger_{\alpha_1} \cdots \hat{a}^\dagger_j \hat{a}_{j'} \hat{a}^\dagger_{j'} \cdots \hat{a}^\dagger_{\alpha_N} | 0 \rangle .$$

In this expression the operator combinations $\hat{a}_i^\dagger \hat{a}_{i'}$ all yield factors of 1, and the final result is

$$\langle \Psi | \hat{F} | \Psi' \rangle = \sigma \; f_{jj'}$$

in agreement with the result calculated in the Slater-determinant formulation.

The rule for transcribing a single-particle operator into second-quantized form is thus

$$\hat{F} = \sum_{jj'} f_{jj'} \hat{a}^\dagger_j \hat{a}_{j'} \text{ with } f_{jj'} = \langle \varphi_j | \hat{f} | \varphi_{j'} \rangle .$$

The analogous result for **two-body operators** such as the potential energy

$$\hat{V} = \tfrac{1}{2} \sum_{k \neq k'} \hat{v}(x_k, x_{k'})$$

can be obtained in a similar way. Each individual term in the sum can change the states of two particles now, so that a product of two creation and two annihilation operators is needed. The calculation can be done in a similar way as above, but is of course more complex. We skip the cumbersome details and give the final form for the second-quantized operator:

$$\hat{V} = \tfrac{1}{2} \sum_{ijkl} v_{ijkl} \, \hat{a}^\dagger_i \hat{a}^\dagger_j \hat{a}_l \hat{a}_k \quad , \tag{A.16}$$

with the two-particle matrix element defined by

$$v_{ijkl} = \int \mathrm{d}^3x \int \mathrm{d}^3x' \; \varphi_i^*(x) \, \varphi_j^*(x') \, v(x, x') \, \varphi_k(x) \, \varphi_l(x') .$$

Note the opposite ordering of the last two indices in the matrix element as compared to the operators in (A.16).

In many calculations the evaluation of the matrix elements leads to an antisymmetric combination, which is therefore given a special abbreviation:

$$\bar{v}_{ijkl} = v_{ijkl} - v_{ijlk}.$$

Note also the symmetry of the matrix element under the interchange of the two pairs of single-particle wave functions:

$$v_{ijkl} = v_{jilk}.$$

Two more general properties of these second-quantized expressions are worth mentioning: they conserve particle number, since both contain the same number of creation as annihilation operators in each term, and they do not explicitly depend on the number of particles anymore: this is instead supplied by the Fock-space states.

A.4.3 Field Operators

The creation and annihilation operators discussed up to now have affected particles in specific states. It is sometimes useful to construct operators for particles at a specific location in space by defining

$$\hat{\psi}^\dagger(\mathbf{r}) = \sum_k \hat{a}_k^\dagger \varphi_k^*(\mathbf{r}), \qquad \hat{\psi}(\mathbf{r}) = \sum_k \hat{a}_k \varphi_k(\mathbf{r}). \tag{A.17}$$

These *field operators* fulfill the anticommutation relations

$$\begin{aligned}
\{\hat{\psi}^\dagger(\mathbf{r}), \hat{\psi}^\dagger(')\} &= 0, \\
\{\hat{\psi}(\mathbf{r}), \hat{\psi}(\mathbf{r}')\} &= 0, \\
\{\hat{\psi}(\mathbf{r}), \hat{\psi}^\dagger(\mathbf{r}')\} &= \delta(\mathbf{r} - \mathbf{r}').
\end{aligned}$$

The last relation can be probed using the completeness of the single-particle basis.

These definitions also show the origin of the name "second quantization": the wave functions themselves become field operators.

A.5 Hierarchy of Density Operators

A.5.1 The One-Body Density

From the many-body wave function it is relatively straightforward to calculate the probability to find one particle at the location $\mathbf{r}$; this probability is normalized to one and has to be multiplied by the total number of particles N to obtain the density. Since we are not interested in the positions of the other particles, they have to be integrated over and the result is

$$\varrho(\mathbf{r}) = N \int \Psi^*(\mathbf{r}, \mathbf{r}_2, \ldots \mathbf{r}_N)\Psi(\mathbf{r}, \mathbf{r}_2, \ldots \mathbf{r}_N)\, \mathrm{d}^3 r_2 \ldots \mathrm{d}^3 r_N. \tag{A.18}$$

If Ψ is a Slater determinant built out of the single-particle states ϕ_k, $k = 1, \ldots N$, this can be reduced to

$$\varrho(\mathbf{r}) = \sum_{k=1}^{N} |\varphi_k(\mathbf{r})|^2$$

by using the orthonormality of the single-particle wave functions. It can also be formulated as the matrix element of an operator

$$\hat{\varrho}(\mathbf{r}) = \sum_{j=1}^{N} \delta(\mathbf{r} - \mathbf{r}_j),$$

so that

$$\varrho(\mathbf{r}) = \langle \Psi | \hat{\varrho} | \Psi \rangle.$$

A useful generalization of this concept is the nondiagonal density

$$\varrho(\mathbf{r}, \mathbf{r}') = N \int \Psi^*(\mathbf{r}', \mathbf{r}_2, \ldots \mathbf{r}_N)\Psi(\mathbf{r}, \mathbf{r}_2, \ldots \mathbf{r}_N)\, \mathrm{d}^3 r_2 \ldots \mathrm{d}^3 r_N. \tag{A.19}$$

This concept can be carried over to second quantization. The density can be obtained by finding the number of particles in a single-particle state, multiplying it by the corresponding density, and summing over all states,

$$\varrho(\mathbf{r}) = \sum_{j=1}^{\infty} \varphi_j^*(\mathbf{r})\varphi_j(\mathbf{r}) \langle \Psi | \hat{a}_j^\dagger \hat{a}_j | \Psi \rangle = \langle \Psi | \hat{\psi}^\dagger(\mathbf{r}) \hat{\psi}(\mathbf{r}) | \Psi \rangle,$$

where the result was also expressed through field operators.

For the nondiagonal density we get similarly

$$\varrho(\mathbf{r}, \mathbf{r}') = \sum_{j,k=1}^{\infty} \varphi_j^*(\mathbf{r}')\varphi_k(\mathbf{r}) \langle \Psi | \hat{a}_j^\dagger \hat{a}_k | \Psi \rangle = \langle \Psi | \hat{\psi}^\dagger(\mathbf{r}') \hat{\psi}(\mathbf{r}) | \Psi \rangle. \tag{A.20}$$

This also motivates the definition of a nondiagonal density operator

$$\hat{\varrho}(\mathbf{r}, \mathbf{r}') = \hat{\psi}^\dagger(\mathbf{r}') \hat{\psi}(\mathbf{r}).$$

A.5.2 The One-Body Density Matrix

We are thus naturally led to introduce the operator combination just discussed as the Fock-space version of the nondiagonal one-body density. Given a many-particle state $|\Psi\rangle$, the *one-body density matrix* is defined as

$$\varrho_{kl} = \langle\Psi|\hat{a}_l^\dagger \hat{a}_k|\Psi\rangle, \tag{A.21}$$

where k and l run over the single-particle basic states. $|\Psi\rangle$ need not be a simple Slater determinant built out of these states but can be a general many-body wave function, a superposition of such Slater determinants. Note that the one-body density matrix depends both on the state $|\Psi\rangle$ and on the single-particle basis defining the operators $\hat{a}_k^\dagger$ and $\hat{a}_l$. It is customary to use the shorter term "density matrix" for the one-body density matrix if no confusion with other types of density matrix is possible.

The following elementary properties of the density matrix are easily derived:

- ϱ_{kl} is hermitian:

$$\varrho_{lk} = \langle\Psi|\hat{a}_k^\dagger \hat{a}_l|\Psi\rangle = \langle\Psi|(\hat{a}_l^\dagger \hat{a}_k)^\dagger|\Psi\rangle = \varrho_{kl}^*.$$

- Expectation values of single-particle operators such as

$$\hat{t} = \sum_{kl} t_{kl}\,\hat{a}_k^\dagger \hat{a}_l$$

can be calculated via

$$\langle\Psi|\hat{t}|\Psi\rangle = \sum_{kl} t_{kl}\langle\Psi|\hat{a}_k^\dagger \hat{a}_l|\Psi\rangle = \sum_{kl} t_{kl}\,\varrho_{lk},$$

which can be rewritten using matrix trace notation:

$$\langle\Psi|\hat{t}|\Psi\rangle = \mathrm{Tr}\{\hat{\varrho}\hat{t}\}.$$

This also explains why in the definition (A.21) the order of indices is chosen oppositely on both sides.

Here t on the right-hand side stands for the matrix t_{kl} representing the operator $\hat{t}$.[1]

- If the state $|\Psi\rangle$ is a simple Slater determinant, the form of the density matrix is quite restricted. We first regard the case that $|\Psi\rangle$ is built out of the same single-particle states as those contained in the single-particle basis defining the density matrix. Then we must have

$$\varrho_{kl} = \begin{cases} \delta_{kl} & \text{for } k \text{ and } l \text{ occupied in } |\Psi\rangle \\ 0 & \text{otherwise} \end{cases} .$$

Thus ϱ is diagonal in this case with ones and zeroes on the diagonal depending on whether the corresponding single-particle state is occupied or empty. Furthermore it fulfills the important relation

$$\varrho^2 = \varrho, \tag{A.22}$$

which follows immediately from the special form of the matrix. It is important that this relation continues to hold in the more general case. If the single-particle states that make up $|\Psi\rangle$ are not included in the basis defining ϱ, they may in any case be expanded in those using some unitary matrix U,

$$\hat{\beta}_k = \sum_{k'} U_{kk'} \, \hat{a}_{k'},$$

where $\hat{\beta}_k$ now denotes the second-quantization operator for these states occupied in $|\Psi\rangle$. The density matrix $\tilde{\varrho}$ defined in the basis of the $\hat{\beta}_k$ now is given by

$$\tilde{\varrho} = U \varrho U^\dagger,$$

so that conversely $\varrho = U^\dagger \tilde{\varrho} U$. Since $\tilde{\varrho}$ fulfills (A.22), we get

$$\varrho^2 = U^\dagger \tilde{\varrho} U U^\dagger \tilde{\varrho} U = U^\dagger \tilde{\varrho}^2 U = U^\dagger \tilde{\varrho} U = \varrho.$$

So $\varrho^2 = \varrho$ holds for a Slater determinant $|\Psi\rangle$ no matter what single-particle basis is used for defining ϱ.

This relation for *one-body* density matrices should not be confused with the analogous equation for a general density matrix. A general density matrix fulfilling $\varrho^2 = \varrho$ describes a *pure state*, a totally different implication.

The relation (A.22) shows that ϱ is a projection matrix: it projects any vector describing a superposition of the basis states onto the subspace of occupied ones.

---

[1] The following notation is used: $\hat{t}$ denotes an operator, t the corresponding matrix, and t_{kl} the elements of the matrix.

A.5.3 The Two-Body Density

The one-body density is suited for the description of one-body operators. Since the interactions between particles always depend on at least two-particle coordinates, it is necessary to introduce a two-body density as well. In configuration space it is the simple generalization of (A.19)

$$\varrho(\mathbf{r}_1, \mathbf{r}_2; \mathbf{r}_1', \mathbf{r}_2') = \int \Psi^*(\mathbf{r}_1, \mathbf{r}_2, \mathbf{r}_3, \ldots \mathbf{r}_N)\Psi(\mathbf{r}_1', \mathbf{r}_2', \mathbf{r}_3, \ldots \mathbf{r}_N)\, \mathrm{d}^3r_3 \ldots \mathrm{d}^3r_N. \tag{A.23}$$

(In fact, this process can be carried further to yield higher and higher densities up to the full N-body density $|\Psi(\mathbf{r}_1, \ldots \mathbf{r}_N)|^2$)

We can now define operators and matrices as in the one-body case. In terms of field operators the two-body density is given by

$$\varrho(\mathbf{r}_1, \mathbf{r}_2; \mathbf{r}_1', \mathbf{r}_2') = \langle \Psi | \hat{\psi}^\dagger(\mathbf{r}_1')\hat{\psi}^\dagger(\mathbf{r}_2')\hat{\psi}(\mathbf{r}_2)\hat{\psi}(\mathbf{r}_1) | \Psi \rangle.$$

Going over to creation and annihilation operators and a matrix formulation we get the definition of the *two-body density matrix*

$$\varrho_{jklm} = \langle \Psi | \hat{a}_l^\dagger \hat{a}_m^\dagger \hat{a}_k \hat{a}_j | \Psi \rangle, \tag{A.24}$$

which is, however, not as useful for the purposes of this book. The most important property of the two-body density is that for a Slater-determinant state $|\Psi\rangle$ it can be split up into products of single-particle densities:

$$\varrho(\mathbf{r}_1, \mathbf{r}_2, \mathbf{r}_1', \mathbf{r}_2') = \varrho(\mathbf{r}_1, \mathbf{r}_1')\varrho(\mathbf{r}_2, \mathbf{r}_2') - \varrho(\mathbf{r}_1, \mathbf{r}_2')\varrho(\mathbf{r}_2, \mathbf{r}_1').$$

This result can be shown by using a basis containing the occupied states, evaluating the matrix element in (A.24) as shown in Chap. 5 and going back to coordinate space.

A.6 Time Reversal and Kramers Degeneracy

Time-reversal invariance is usually not treated thoroughly in elementary quantum-mechanics courses, because it is somewhat more complicated than space inversion. For fermionic systems, however, it has important consequences which we will derive here.

Time reversal is defined by its effect on the principal physical quantities position, momentum, angular momentum, spin, and time itself:

$$\mathbf{r} \to \mathbf{r} \quad , \quad \mathbf{p} \to -\mathbf{p} \quad , \quad \mathbf{s} \to -\mathbf{s} \quad , \quad \mathbf{L} \to -\mathbf{L} \quad , \quad t \to -t \quad .$$

Hamiltonians should be time-reversal invariant, so that, for example, a term such as $\mathbf{r} \cdot \mathbf{p}$ cannot be allowed. This leads to the usual property in *microscopic* physics that the evolution of a system and its time-reversed analogue should both be possible.

Time reversal has unusual properties compared with the other symmetries. Denoting the time-reversal operator by $\hat{\mathcal{T}}$, the commutator of position and momentum transforms as

$$\hat{\mathcal{T}}\left[x, p_x\right]\hat{\mathcal{T}}^{-1} = \left[x, -p_x\right] = -\mathrm{i}\hbar,$$

so that we must demand $\hat{\mathcal{T}}\mathrm{i}\hbar\hat{\mathcal{T}}^{-1} = -\mathrm{i}\hbar$! The only way to achieve this is to assume that the time reversal operator includes complex conjugation. This also works for the Schrödinger equation, where the time-reversal invariance of the Hamiltonian demands that $\mathrm{i}\hbar\partial/\partial t$ be invariant, too.

The operator for time reversal, $\hat{\mathcal{T}}$, thus cannot be unitary, because it is not even linear. A linear operator commutes with an arbitrary c number, so that one should have

$$\hat{\mathcal{T}}\mathrm{i}\hbar\hat{\mathcal{T}}^{-1} = \mathrm{i}\hbar\hat{\mathcal{T}}\hat{\mathcal{T}}^{-1} = \mathrm{i}\hbar,$$

whereas we need as the transformation of the constant $\hat{\mathcal{T}}\mathrm{i}\hbar\hat{\mathcal{T}}^{-1} = -\mathrm{i}\hbar$. Operators with the property $\hat{\mathcal{T}}\alpha = \alpha^*\hat{\mathcal{T}}$ for an arbitrary complex number α are called *antilinear operators*.

Special care needs to be taken when constructing the eigenstates of $\hat{\mathcal{T}}$, because many of the familiar operator properties do not hold. Assume that $|A\rangle$ is an eigenstate of $\hat{\mathcal{T}}$ with eigenvalue A. Applying $\hat{\mathcal{T}}^2$ then yields

$$\hat{\mathcal{T}}^2|A\rangle = \hat{\mathcal{T}}A|A\rangle = A^*\hat{\mathcal{T}}|A\rangle = |A|^2|A\rangle.$$

Now the basic definition of time reversal shows that it must satisfy $\hat{\mathcal{T}}^2 = 1$, so that the eigenvalue A must have a magnitude of unity:

$$A = \exp^{\mathrm{i}\phi_A}$$

with some phase angle ϕ_A. This angle actually depends on the choice of phase for the eigenstate, because if $|A\rangle$ is replaced by a new state

$$|A'\rangle = \exp^{\frac{1}{2}\mathrm{i}\phi_A}|A\rangle,$$

the eigenvalue also changes:

$$\hat{\mathcal{T}}|A'\rangle = \hat{\mathcal{T}}\exp^{\frac{1}{2}\mathrm{i}\phi_A}|A\rangle = \exp^{-\frac{1}{2}\mathrm{i}\phi_A}\hat{\mathcal{T}}|A\rangle = \exp^{-\frac{1}{2}\mathrm{i}\phi_A}\exp^{\mathrm{i}\phi_A}|A\rangle = |A'\rangle.$$

So it is possible to make the eigenvalue of $\hat{\mathcal{T}}$ equal to 1 by a change of phase. It is interesting to note that the properties of time reversal are thus intimately linked to the phases of the wave functions, which otherwise play little role in quantum mechanics.

For spherical systems the interplay between angular momentum and time reversal is interesting. Let us look at total angular momentum $\hat{\mathbf{J}} = \hat{\mathbf{L}} + \hat{\mathbf{s}}$, although the arguments are equally valid for spin and orbital angular momenta. $\hat{\mathcal{T}}$ inverts the sign of $\hat{\mathbf{J}}$, so that it commutes with $\hat{J}^2$ but not with $\hat{J}_z$. If time reversal is combined with a rotation $\mathcal{R}$ that inverts the direction of the z-axis, it should commute with both:

$$\left[\mathcal{R}\hat{\mathcal{T}}, \hat{J}^2\right] = 0, \quad \left[\mathcal{R}\hat{\mathcal{T}}, \hat{J}_z\right] = 0,$$

and this operator $\mathcal{R}\hat{\mathcal{T}}$ can then be diagonalized together with $\hat{J}^2$ and $\hat{J}_z$. The possible eigenvalues, by the same argument as for $\hat{\mathcal{T}}$, must have the form $\exp(\mathrm{i}\phi_A)$ and can again be made equal to 1 by a change of phase of the wave function. Denoting any additional quantum numbers by α, we can thus construct a system of eigenfunctions $|\alpha JM\rangle$ such that

$$\mathcal{R}\hat{\mathcal{T}}|\alpha JM\rangle = |\alpha JM\rangle$$

This works with the usual angular momentum eigenstates, but requires a special choice of phase for the basic functions.

The rotation $\mathcal{R}$ is arbitrary except for the condition of inverting the z-axis. It is customary to choose a rotation by an angle π about the y-axis, $\mathcal{R} = R_y(\pi)$.

Let us now examine the action of $(\mathcal{R}\hat{\mathcal{T}})^2$. On the one hand, since its eigenvalue is 1, we have

$$(\mathcal{R}\hat{\mathcal{T}})^2|\alpha JM\rangle = |\alpha JM\rangle.$$

On the other hand, $\mathcal{R}$ commutes with $\hat{\mathcal{T}}$, because the rotation $R_y(\pi)$ can be expressed as $\exp(\mathrm{i}\pi\hat{J}_y)$ and $\hat{\mathcal{T}}$ inverts the signs of both $\hat{J}_y$ and the imaginary factor $\mathrm{i}\pi$. Thus we can also write

$$\left(\mathcal{R}\hat{\mathcal{T}}\right)^2|\alpha JM\rangle = \mathcal{R}^2\hat{\mathcal{T}}^2|\alpha JM\rangle = R_y(2\pi)\hat{\mathcal{T}}^2|\alpha JM\rangle.$$

Now comes the crucial part. The rotation by 2π is the identity for wave functions with integral spin, but produces a -1 for half-integer spins. The result is thus

$$\hat{\mathcal{T}}^2 = \begin{cases} +1 & \text{for integral spin,} \\ -1 & \text{for half-integer spin.} \end{cases}$$

Here we are only interested in the latter case. An eigenvalue of -1 is not possible because $\hat{\mathcal{T}}^2$ is the identity operator and thus has only eigenvalues of $+1$.

The solution for this apparent contradiction is that for fermions there are no eigenvectors of $\hat{T}$, or, in other words, $\hat{T}|A\rangle$ is always linearly independent of $|A\rangle$. As the Hamiltonian is invariant under time reversal and commute with $\hat{T}$, $|A\rangle$ and $\hat{T}|A\rangle$ have the same energy: there are thus two linearly independent but degenerate eigenstates of the Hamiltonian.

This twofold degeneracy of fermionic states is called *Kramers degeneracy*. In the single-particle wave functions, for example, consequences appear in the following way: for a wave function with orbital angular momentum projection M combining with spin projection s to total angular momentum projection Ω, the state produced by time reversal has quantum numbers $(-M, -s, -\Omega)$. In the case of spherical symmetry, this is hidden by the degeneracy of all angular momentum projections (for other quantum numbers identical), but if the system deviates from spherical to cylindrical geometry, Kramers degeneracy implies that there will still be degenerate states with opposite Ω projection.

This general result also applies to many-body states. If the Hamiltonian for a system consisting of an odd number of fermions (and thus having a fractional spin) is invariant under time reversal, its eigenstates will always show twofold degeneracy with the two states being time reversed with respect to each other.

Appendix B
Useful Tables

B.1 Units

We list here basic physical constants and units (data are taken from [102]). Note that the Gaussian system of units for electromagnetic properties is used throughout.

Energy scales:

$$10^{-6}\,\mathrm{eV} = 0.2418\,h\,\mathrm{GHz} = 8.066\times 10^{-3}\frac{hc}{\mathrm{cm}}$$
$$1\,h\,\mathrm{GHz} = 4.136\times 10^{-6}\,\mathrm{eV}\ ;\ 1\frac{hc}{\mathrm{cm}} = 0.1240\times 10^{-3}\,\mathrm{eV}$$

Rydberg constant

$$\mathrm{Ry} = 13.6\,\mathrm{eV}$$

Boltzmann constant:

$$k_\mathrm{B} = 8.6174\,10^{-5}\,\mathrm{eV\,K^{-1}}$$

Timescales:

$$1\,\mathrm{fs} = 10^{-15}\,\mathrm{s} = 1.519\frac{\hbar}{\mathrm{eV}} = 20.66\frac{\hbar}{\mathrm{Ry}}$$
$$\frac{\hbar}{\mathrm{Ry}} = 0.0484\,\mathrm{fs}$$

Scale factors:

$$\hbar c = 1.9731\times 10^{-7}\,\mathrm{eV\,m} = 1973.1\,\mathrm{eV\AA} = 274.12\,\mathrm{Ry\,a_0}$$
$$\frac{\hbar^2}{m_e} = 2\,\mathrm{Ry\,a_0^2} = 7.617\,\mathrm{eV\,\AA^2}$$

Electron mass:

$$m_e c^2 = 510.9\,\text{keV} = 37.57 \times 10^3\,\text{Ry}$$
$$m_e = 0.0156\,\text{eV}\,\text{fs}^2 \text{a}_0^{-2} = 0.5\,\text{Ry}^{-1} \text{a}_0^{-2}$$

Light velocity:

$$c = 5670\,\text{a}_0\,\text{fs}^{-1} = 274.12\,\text{Ry}\,\text{a}_0$$

Gravitational constant:

$$G = 6.67 \times 10^{-11} \text{m}^3/(\text{kg}\,\text{s}^2)$$

Fine-structure constant:

$$\alpha = \frac{e^2}{\hbar c} = 0.007297 = \frac{1}{137.03}$$

Charge:

$$e^2 = 2\,\text{Ry}\,\text{a}_0 = 14.40\,\text{eV}\,\text{Å}$$

Dielectric constant:

$$\epsilon_0 = \frac{1}{4\pi} \equiv \text{Gaussian system of units}$$

Bohr energy:

$$E_B = \frac{e^4 m_e}{2\hbar^2} = 1\,\text{Ry} = \frac{\alpha^2 m_e c^2}{2} = 13.604\,\text{eV}$$

Bohr radius:

$$\text{a}_0 = \frac{\hbar^2}{m_e c^2} = 0.5291\,\text{Å} = 0.05291\,\text{nm} = 0.5291 \times 10^{-10}\,\text{m}$$

Bohr magneton:

$$\mu_\text{B} = \frac{\hbar e}{2m_e} = 5.788\,\text{eV}\,\text{T}^{-1}$$

A few nuclear quantities [29]:

Proton:

$$m_p c^2 = 938.2\,\text{MeV} = 1836.1\,m_e$$
$$\frac{\hbar^2}{m_p} = 41.494\,\text{MeV}\,\text{fm}^2$$
$$\mu_p = 2.7928\,\mu_N$$

Nuclear magneton:

$$\mu_N = \frac{e\hbar}{2m_p c} = 0.3152 * 10^{-11}\frac{\text{eV}}{\text{Gs}} = 0.5446 * 10^{-3}\mu_e$$

Neutron:

$$m_n c^2 = 939.5\,\text{MeV} = 1838.9\,m_e$$

Deuteron:

$$m_d c^2 = 1875.4\,\text{MeV} = 3670.8\,m_e$$
$$\mu_d = 0.8574\,\mu_N$$

B.2 Hermite and Laguerre Polynomials

Othogonal polynomials occur in the solutions of many differential equations in physics. Distinct polynomials arise for different boundary conditions and dimensionalities. Because many useful properties can be derived for them, they are used extensively in analytic calculations. The most important properties are the differential equations they obey, their normalization integrals, and simple recursion relations.

For more information on special functions the classic reference is [118], downloadable from `http://www.math.sfu.ca/cbm/aands/`.

B.2.1 The Hermite Polynomials

These occur most prominently in the solution of the 1D oscillator. The differential equation

$$\frac{\text{d}^2}{\text{d}x^2} f(x) - 2x\frac{\text{d}}{\text{d}x} f(x) + 2n f(x) = 0$$

with $n \geq 0$ an integer (for the harmonic oscillator problem corresponding to the energy) has as solutions the Hermite polynomials $H_n(x)$, which are normalized according to

$$\int_{-\infty}^{\infty} e^{-x^2)} H_{n'}(x) H_n(x)\,dx = 2^n\, n! \sqrt{\pi}\,\delta_{nn'}.$$

The polynmials for different n are related by the recursion relation

$$H_{n+1}(x) = 2x H_n(x) - 2n H_{n-1}(x).$$

Note that this allows to calculate values for all n if $H_0(x)$ is known. Explicitly the lowest Hermite polynomials are

$$\begin{array}{ll} H_0(x) = 1 & H_1(x) = 2x \\ H_2(x) = 4x^2 - 2 & H_3(x) = 8x^3 - 12x \\ H_4(x) = 16x^4 - 48x^2 + 12 & H_5(x) = 32x^5 - 160x^3 + 120x. \end{array}$$

It is worth noting that the polynomials have alternating parity given by $(-1)^n$.

B.2.2 The Laguerre Polynomials

These occur in the solution of the Schrödinger equations for cylindric geometry and therefore contain two quantum numbers: n is again related to the energy in the radial and angular coordinates, and α related to the projection of orbital angular momentum, although α need not be integer. The polynomials with $\alpha = 0$ are simply called Laguerre polynomials, while for the case of $\alpha \neq 0$ the name *generalized Laguerre polynomials* is used.

The differential equation for $f(x) = L_n^{(\alpha)}(x)$ is

$$x \frac{\mathrm{d}^2}{\mathrm{d}x^2} f(x) + (\alpha + 1 - x) \frac{\mathrm{d}}{\mathrm{d}x} f(x) + n f(x) = 0$$

and the orthogonality relation

$$\int_0^{\infty} e^{-x} x^{\alpha} L_{n'}^{(\alpha)}(x) L_n^{(\alpha)}(x)\,dx = \frac{\Gamma(\alpha + n + 1)}{n!} \delta_{nn'}.$$

The recursion relation is

$$(n+1) L_{n+1}^{(\alpha)}(x) = (2n + \alpha + 1 - x) L_n^{(\alpha)}(x) - (n + \alpha) L_{n-1}^{(\alpha)}(x).$$

Note that all of these relations only refer to polynomials with the same α in each case. There are also some working in α:

$$L_n^{(\alpha+1)}(x) = \frac{1}{x}\left[(x-n)L_n^{(\alpha)}(x) + (\alpha+n)L_{n-1}^{(\alpha)}(x)\right],$$
$$L_n^{(\alpha-1)}(x) = L_n^{(\alpha)}(x) - L_{n-1}^{(\alpha)}(x).$$

Special cases are

$$L_0^{(\alpha)}(x) = 1,$$
$$L_1^{(\alpha)}(x) = -x + \alpha + 1,$$
$$L_2^{(\alpha)}(x) = 1/2\,a^2 + 3/2\,a + 1 - xa - 2\,x + 1/2\,x^2.$$

Clearly these become quite complicated quickly but can readily be generated using the recursion relations.

References

1. M. Abramowitz, I.A. Stegun, *Handbook of Mathematical Functions with Formulas, Graphs, and Mathematical Tables*, ninth Dover printing, tenth GPO printing edn. (Dover, New York, 1964)
2. A. Artemyev, E.V. Ludena, V. Karasiev, J. Mol. Struct.: THEOCHEM **580**, 47 (2002)
3. N.W. Ashcroft, N.D. Mermin, *Solid State Physics* (Saunders College, Philadelphia, 1976)
4. P.W. Atkins, *Physical Chemistry* (Oxford University, Oxford, 1977)
5. M. Baldo, C. Maieron, P. Schuck, X. Vinas, Nucl. Phys. A **736**, 241 (2004)
6. R. Balian, C. Bloch, Ann. Phys. **85**, 514 (1974)
7. J. Bardeen, L.N. Cooper, J.R. Schrieffer, Phys. Rev. **108**, 1175 (1957)
8. M. Bender, P.H. Heenen, P.G. Reinhard, Rev. Mod. Phys. **75**, 121 (2003)
9. G.F. Bertsch, *The Practitioner's Shell Model* (North-Holland, Amsterdam, 1972)
10. G.F. Bertsch, R. Broglia, *Oscillations in Finite Quantum Systems* (Cambridge University Press, Cambridge, 1994)
11. G.F. Bertsch, S. Das Gupta, Phys. Rep. **160**, 190 (1988)
12. K. Binder, *Encyclopedia of Mathematics* (Kluwer Academic, Dordrecht, 2001)
13. J.P. Blaizot, G. Ripka, *Quantum Theory of Finite Systems* (MIT Press, Cambridge Massachusetts and London England, 1985)
14. I. Bloch, J. Dalibart, W. Zwerger, Rev. Mod. Phys. **80**, 885 (2008)
15. O. Bohigas, A.M. Lane, J. Martorell, Phys. Rep. **51**, 267 (1979)
16. D. Bohm, D. Pines, Phys. Rev. **92**, 609 (1953)
17. Å. Bohr, B.R. Mottelson, *Struktur der Atomkerne I. Einteilchenbewegung* (Akademie Verlag, Berlin, 1975)
18. Å. Bohr, B.R. Mottelson, *Struktur der Atomkerne II. Kerndeformationen* (Akademie Verlag, Berlin, 1980)
19. M. Brack, Rev. Mod. Phys. **65**, 677 (1993)
20. M. Brack, R.K. Bhaduri, *Semiclassical Physics* (Addison-Wesley, Reading, 1997)
21. D.M. Ceperley, B.J. Alder, Phys. Rev. Lett. **45**, 566 (1980)
22. E. Chabanat, P. Bonche, P. Haensel, J. Meyer, R. Schaeffer, Nucl. Phys. A **627**, 710 (1997)
23. J.P. Coe, A. Sudbery, I. D'Amico, Phys. Rev. B **77**, 205122 (2008)
24. C. Cohen-Tanuudji, B. Diu, F. Laloe, *Quantum Mechanics* (Wiley, New York, 2006)
25. P.M. Dinh, J. Navarro, E. Suraud, *Océans et gouttelettes quantiques* (CNRS Editions, Paris, 2007)
26. J.F. Dobson, Phys. Rev. Lett. **73**, 2244 (1994)
27. R.M. Dreizler, E.K.U. Gross, *Density Functional Theory: An Approach to the Quantum Many-Body Problem* (Springer-Verlag, Berlin, 1990)
28. R. Dörner, V. Mergel, O. Jagutzki, L. Spielberger, J. Ullrich, R. Moshammer, H. Schmidt-Böcking, Phys. Rep. **330**, 95 (2000)
29. G. Eder, *Kernkräfte* (Verlag G. Braun, Karlsruuhe, 1965)

30. A.R. Edmonds, *Angular Momentum in Quantum Mechanics* (Princeton University Press, Princeton, 1960)
31. A.R. Edmonds, *Angular Momentum in Quantum Mechanics* (Princeton University Press, Princeton, 1996)
32. R. Englman, *The Jahn-Teller Effect in Molecules and Crystals* (Wiley, London, 1972)
33. T.E.O. Ericson, W. Weise, *Pions and Nuclei* (Clarendon, Oxford, 1988)
34. E. Fermi, Z. Phys. **48**, 73 (1928)
35. S. Flügge, *Practical Quantum Mechanics* (Springer, Heidelberg, 1974)
36. H.S. Friedrich, *Theoretical Atomic Physics* (Springer, Heidelberg, 2006)
37. J. Friedrich, N. Vögler, Nucl. Phys. A **373**, 192 (1982)
38. S. Giorgini, L.P. Pitaevski, S. Stringari, Rev. Mod. Phys. **80**, 1216 (2008)
39. N.K. Glendenning, *Compact Stars* (Springer, Heidelberg, 1997)
40. H. Goldstein, *Classical Mechanics* (Addison-Wesley, Cambridge MA, 1950)
41. S. Grebenev, B. Sartakov, J.P. Toennies, A.F. Vilesov, Science **289**, 1532 (2000)
42. W. Greiner, J.A. Maruhn, *Kernmodelle, Theoretische Physik*, vol. 11 (Verl. Harri Deutsch, Thun, Frankfurt am Main, 1995)
43. E.K.U. Gross, E. Runge, O. Heinonen, *Many–Particle Theory* (Adam Hilger, Bristol, Philadelphia, New York, 1991)
44. H. Haberland (ed.), *Clusters of Atoms and Molecules 1- Theory, Experiment, and Clusters of Atoms*, vol. 52 (Springer Series in Chemical Physics, Berlin, 1994)
45. H. Haken, *Introduction to Laser Physics* (Springer, Berlin, 1991)
46. H. Haken, H.C. Wolf, *The Physics of Atoms and Quanta* (Springer, Berlin, 2000)
47. R. Haussmann, W. Rantner, S. Cerrito, W. Zwerger, Phys. Rev. A **75**, 023610 (2007)
48. K.L.G. Heyde, *The Nuclear Shell Model*, 1st edn. (Springer, Berlin, Heidelberg, New York, 1990)
49. S. Hofmann, G. Münzenberg, Rev. Mod. Phys. **72**, 733 (2000)
50. P. Hohenberg, W. Kohn, Phys. Rev. **136**, 864 (1964)
51. G. Jaeger, *Quantum Information: An Overview* (Springer, Berlin, 2006)
52. H.A. Jahn, E. Teller, Proc. Roy. Soc. A **161**, 220 (1937)
53. W.R. Johnson, *Atomic Structure Theory* (Springer, Berlin, 2007)
54. R.O. Jones, O. Gunnarsson, Rev. Mod. Phys. **61**, 689 (1989)
55. C. Kohl, P.G. Reinhard, Z. Phys. D **39**, 225 (1997)
56. W. Kohn, Rev. Mod. Phys. **71**, 1253 (1999)
57. W. Kohn, L.J. Sham, Phys. Rev. **140**, 1133 (1965)
58. M. Koskinen, P.O. Lipas, M. Manninen, Nucl. Phys. A **591**, 421 (1995)
59. U. Kreibig, M. Vollmer, *Optical Properties of Metal Clusters*, vol. 25 (Springer Series in Materials Science, Berlin, 1993)
60. A. Kumar, S.E. Laux, F. Stern, Phys. Rev. B **42**, 5166 (1990)
61. S. Kümmel, L. Kronik, Rev. Mod. Phys. **80**, 3 (2008)
62. L.D. Landau, E.M. Lifshitz, J. Menzies, *Quantum Mechanics: Non-Relativistic Theory* (Butterworth-Heinemann, Oxford, 1991)
63. J.M. Lattimer, M. Prakash, Science **304**, 536 (2004)
64. H.J. Lipkin, N. Meshkov, A.J. Glick, Nucl. Phys. **62**, 188 (1965)
65. B.D. Marco, S.B. Papp, D.S. Jin, Phys. Rev. Lett. **86**, 5409 (2001)
66. E.S.P. Marmier, *Physics of Nuclei and Particles, Vol I* (Academic, New York, 1969)
67. J. Maruhn, W. Greiner, Z. Physik **251**, 431 (1972)
68. J. Maruhn, P.G. Reinhard, P. Stevenson, I. Stone, M. Strayer, Phys. Rev. C **71**, 064328 (2005)
69. E. Merzbacher, *Quantum Mechanics* (Wiley, New York, 1997)
70. A. Messiah, *Quantum Mechanics* (Dover, New York, 2000)
71. G. Mie, Ann. Phys. (Leipzig) **25**, 377 (1908)
72. A. Mourachkine, *High-Temperature Superconductivity in Cuprates* (Springer, Heidelberg, 2007)
73. J. Navarro, P.G. Reinhard, E. Suraud, Euro. Phys. J. A **30**, 333 (2006)
74. J.W. Negele, Rev. Mod. Phys. **54**, 913 (1982)

75. S.G. Nilsson, Mat.-Fys. Medd. Dan. Vid. Selsk. **29**, 16 (1955)
76. D.P. O'Neill, P.M.W. Gill, Phys. Rev. A **68**, 022505 (2003)
77. T. Padmanabhan, *Theoretical Astrophysics Volume II: Stars and Stellar Systems* (Cambridge University Press, Cambridge, 2001)
78. V.R. Pandharipande, I. Sick, P.K.A. deWitt Huberts, Rev. Mod. Phys. **69**, 981 (1997)
79. R.D. Parks. (ed.), *Superconductivity* (Dekker, New York, 1969)
80. R.G. Parr, W. Yang, *Density-Functional Theory of Atoms and Molecules* (Oxford University Press, Oxford, 1989)
81. J.P. Perdew, K. Burke, M. Ernzerhof, Phys. Rev. Lett. **77**, 3865 (1996)
82. J.P. Perdew, Y. Wang, Phys. Rev. B **45**, 13244 (1992)
83. J.P. Perdew, A. Zunger, Phys. Rev. B **23**, 5048 (1981)
84. D. Pettifor, *Bonding and Structure of Molecules and Solids* (Clarendon, Oxford, 1995)
85. D. Pines, P. Nozières, *The Theory of Quantum Liquids* (Benjamin, New York, 1966)
86. S. Raman, C. Nestor, P. Tikkanen, At. Data Nucl. Data Tab. **78**, 1 (2001)
87. S.M. Reimann, M. Manninen, Rev. Mod. Phys. **74** 1283 (2002)
88. P.G. Reinhard, E.W. Otten, Nucl. Phys. A **420**, 173 (1984)
89. P.G. Reinhard, E. Suraud, *Introduction to Cluster Dynamics* (Wiley, New York, 2003)
90. P.G. Reinhard, C. Toepffer, Int. J. Mod. Phys. E **3**, 435 (1994)
91. G. Rickayzen, *Theory of Superconductivity* (Interscience, New York, 1965)
92. P. Ring, P. Schuck, *The Nuclear Many-Body Problem* (Springer-Verlag, New York, Heidelberg, Berlin, 1980)
93. D.J. Rowe, *Nuclear Collective Motion* (Methuen, London, 1970)
94. U. Rössler, *Solid State Theory* (Springer, Berlin, 2004)
95. E.R. Scerri, *The Periodic Table: Its Story and Its Significance* (Oxford University Press, Oxford, 2006)
96. L.I. Schiff, *Quantum Mechanics* (McGraw Hill, New York, 1968)
97. M. Schmidt, H. Haberland, Eur. Phys. J. D **6**, 109 (1999)
98. T.H.R. Skyrme, Nucl. Phys. **9**, 615 (1959)
99. J.C. Slater, Phys Rev. **81**, 385 (1951)
100. A. Szabo, N.S. Ostlund, *Modern Quantum Chemistry* (Dover, New York, 1996)
101. L. Szasz, *Pseudopotential Theory of Atoms and Molecules* (Wiley, New York, 1985)
102. G. Süssmann, *Einführung in die Quantenmechanik I* (Bibliographisches Institut, Mannheim, 1963)
103. B. Talukdar, A. Sarkar, S. Roy, P. Sarkar, Chem. Phys. Lett. **381**, 67 (2003)
104. S. Tarucha, D.G. Austing, T. Honda, R.J. van der Haage, L. Kouwenhoven, Phys. Rev. Lett. **77**, 3613 (1996)
105. L. Thomas, Proc. Camb. Phil. Soc. **23**, 195 (1926)
106. T. Tietz, Ann. Physik **15**, 6, 186 (1955)
107. D. Varsano, R. di Felice, M. Marques, A. Rubio, Science **110**, 7129 (2006)
108. D.A. Varshalovich, A.N. Moskalev, V.K. Khersonskii, *Quantum Theory of Angular Momentum* (World Scientific Pub. Co., Singapore, 1988)
109. D. Vautherin, D.M. Brink, Phys. Rev. C **5**, 626 (1972)
110. G. Vignale, Phys. Rev. Lett. **74**, 3233 (1995)
111. G.E. Volovik, *The Universe in a Helium Droplet* (Clarendon, Oxford, 2003)
112. E. Wahlström, E.K. Vestergaard, R. Schaub, A. Ronnau, M. Vestergaard, E. Laegsgaard, I. Stensgaard, F. Besenbacher, Science **303**, 511 (2004)
113. S. Weinberg, *The Quantum Theory of Fields, Vol. II* (Cambridge University Press, Cambridge, 1996)
114. M. Weissbluth, *Atoms and Molecules* (Academic Press, San Diego, 1978)
115. B. Yoon, J.W. Negele, Phys. Rev. A **16**, 1451 (1977)
116. J. Zinn-Justin, *Quantum Field Theory and Critical Phenomena* (Oxford Science Publishers, Oxford, 2002)
117. W.A. de Heer, Rev. Mod. Phys. **65**, 611 (1993)
118. J. von Neumann, E. Wigner, Z. Phys. **30**, 467 (1929)

Index

Zeitfracht Medien GmbH
Ferdinand-Jühlke-Straße 7
99095 Erfurt, Deutschland
produktsicherheit@kolibri360.de